About the Authors

Dr. Gagan Matta, is a renowned environmental scientist and academician with over 18 years of research experience and 14 years of teaching. He holds a Ph.D. in Environmental Science from Gurukul Kangri University, Haridwar, India, where he is an Associate Professor in the Department of Zoology and Environmental Science. Specializing in water quality indexing, limnology, aquatic ecology, toxicology, and environmental pollution, Dr. Matta's research spans hydrology, green audits, and environmental communication. He has authored six books, edited 18, published over 90 journal articles, and contributed 18 book chapters. His work has earned 14 research grants and seven copyrights. Dr. Matta has been recognized with prestigious awards, including the "Scientist of the Year Award – 2023." He has mentored 8 Ph.D. candidates and 31 M.Sc. dissertations while contributing to developing academic programs and environmental policies.

As an advocate for science communication, Dr. Matta works on initiatives to raise awareness of water hygiene and environmental sustainability. His global presence, invited talks in the UK, Italy, and China, and his roles in various academic bodies solidify his status as a leading environmental scientist committed to research, teaching, and policy development.

Dr. Anjali Nayak, M.Sc., Ph.D. (Environmental Science) from Gurukul Kangri University, Haridwar, Uttrakhand, India. Worked at the Patanjali Resaech Institute, Haridwar, as an assistant scientist. My major areas of research are water quality monitoring, groundwater and surface water quality, air quality, health risk assessment, industrial and heavy metal toxicity, and science communication. Dr. Anjali Nayak has Published 16 research articles and 3 book chapters in reputed Scopus indexing journals and also given oral presentations at national and international conferences.

Dr. D. P. Uniyal has more than 28 years of Research and Extension Experience in the field of Biodiversity: Taxonomy & Ecology of the fish fauna of Northern India, Water Quality Monitoring, Remote Sensing & Geographic Information Studies Application, Innovation Promotion and Science Communication and Science and Society activities. Dr.Uniyal worked in the Zoological Survey of India (Government of India) for more than 10 years in various capacity and completed 14 major research projects. Dr Uniyal is currently serving as Joint Director at the Uttarakhand State Council for Science and Technology (Government of Uttarakhand), looking after Scientific Administration, implementation of Science related policy matter in the state, Science Popularization, Central Sponsored Scheme Division and National Academy of Science, India, Uttarakhand chapter. Dr.Uniyal completed 23 major Research and Extension activities Projects and Presently, handling 3 major research projects. He has published 58 research papers and 20 Popular Articles in International/ National Journals and Magazines, Conferences, Seminars and Symposia. Apart from authoring a book, Dr Uniyal has edited Eight books and has several scientific project reports to his credit. He has guided many master's degree scholars and is currently guiding two Ph.D. research scholars. He has also delivered many invited lectures in various Institutions/Universities across the Country and Globe including Food & Agriculture Organization (FAO), Italy and at the University of Dresden, Germany. Dr. Uniyal is member of many prestigious scientific bodies including the National Academy of Sciences, India (M.N.A.Sc), Zoological Society of India (M.ZSI). He is an Expert Group Member in DST (Govt. of India), Uttarakhand Biodiversity Board, Uttarakhand Pollution Control Board, Uttarakhand Renewable Energy Development Agency, Uttarakhand Biotechnology Board etc. He is also a Board Member / Reviewer in various International and National Journals. Dr. Uniyal is the recipient of Young Scientist Award in State Science Congress, Youth Silver Medal by Minister of Youth Affair (GOI), Governor's and Chief Minister's appreciation for uplifting Science & Technology in Uttarakhand State etc.

Groundwater Pollution in The Himalayan Region

Gagan Matta
Department of Zoology and Environmental Science
Gurukula Kangri (Deemed to be University)
Haridwar, Uttarakhand, India

Anjali Nayak
Gurukul Kangri University (Deemed to be University)
Haridwar, Uttrakhand, India

D.P. Uniyal
Uttarakhand State Council for Science and Technology
Dehradun, Uttarakhand, India

NIPA® GENX ELECTRONIC RESOURCES & SOLUTIONS P. LTD.
New Delhi-110 034

NIPA® GENX ELECTRONIC RESOURCES & SOLUTIONS P. LTD.

101,103, Vikas Surya Plaza, CU Block
L.S.C. Market, Pitam Pura, New Delhi-110 034
Ph : +91-11-43860225, Mob.: +91 9717133558, 9540816132
E-mail: newindiapublishingagency@gmail.com
Website: www.nipaersources.com

Print ISBN: 978-93-58874-93-8
ebook ISBN: 978-93-58871-46-3

Composed and Designed by NIPA®.

Preface

Water is one of the most fundamental resources sustaining life, ecosystems, and human civilization. Among the various forms of water resources, groundwater plays a particularly critical role as a reliable and often invisible source of freshwater that supports agriculture, industry, domestic consumption, and ecological systems. In mountainous regions such as the Himalayas, groundwater systems are closely interconnected with glaciers, rivers, precipitation patterns, and geological formations, making them both valuable and vulnerable. Rapid population growth, unplanned urbanization, agricultural intensification, and industrial activities have significantly increased the pressure on groundwater resources, leading to growing concerns about contamination, depletion, and long-term sustainability. Understanding the dynamics of groundwater, its sources, movement, and the factors influencing its quality has therefore become essential for sustainable environmental management and water security.

This book, Groundwater Pollution in the Himalayan Region, attempts to present a comprehensive overview of groundwater resources and the emerging challenges related to their pollution and management. It explores the scientific foundations of water systems, the hydrological cycle, and the role of groundwater in maintaining ecological balance and supporting human activities. Particular attention is given to the Himalayan region, where fragile ecosystems and increasing anthropogenic pressures make groundwater resources especially vulnerable. Through discussion of global perspectives, regional assessments, and case studies, the book aims to enhance understanding of groundwater contamination, monitoring techniques, and sustainable conservation strategies. It is hoped that this work will serve as a valuable resource for researchers, students, policymakers, and environmental practitioners seeking to protect and manage groundwater resources for present and future generations.

Authors

Contents

1

Water: A Necessity for Survival

Water, a molecule composed of two hydrogen atoms and one oxygen atom (H_2O), is fundamental to life on Earth and exhibits a unique array of chemical and physical properties that make it indispensable for biological processes, geological phenomena, and climate regulation. It is a polar molecule, with a partial negative charge near the oxygen atom due to its higher electronegativity and partial positive charges near the hydrogen atoms. This polarity facilitates the formation of hydrogen bonds between water molecules, leading to high cohesion, adhesion, surface tension, and a relatively high boiling point for a molecule of its size.

Water exists in three states—solid (ice), liquid (water), and gas (vapor)—under Earth's natural conditions. Its solid form is less dense than its liquid form, an anomaly caused by the hydrogen bonding that arranges water molecules in a crystalline structure when frozen, making ice float. This property is crucial for aquatic life, as it insulates the water beneath and maintains a habitable environment even in cold climates.

In the liquid state, water serves as a universal solvent, a property attributable to its polarity. It can dissolve a vast range of substances, from salts and sugars to gases and proteins, facilitating chemical reactions within living organisms and distributing nutrients and waste products. This solvent capability is essential for metabolic processes, cellular respiration, and photosynthesis. Additionally, water's high specific heat capacity, which is the amount of heat required to change its temperature, helps stabilize Earth's climate by absorbing and releasing heat energy slowly. The hydrological cycle, driven by solar energy, governs the distribution and movement of water through the environment. This cycle includes evaporation, transpiration, condensation, precipitation, and runoff, connecting the atmosphere, lithosphere, hydrosphere, and biosphere. Evaporation from oceans and other bodies of water, along with transpiration from plants, contributes to atmospheric moisture. Condensation forms clouds, leading to precipitation as rain, snow, or sleet. This cycle replenishes freshwater resources, which are essential for drinking water, agriculture, and industrial processes.

Figure 1.1.

Water's role in geology includes weathering and erosion, where it breaks down rocks and minerals, shaping landscapes over geological timescales. It also acts as a medium for transporting sediments and nutrients in rivers and streams, which deposit them in various ecological niches. Groundwater, stored in aquifers, is another critical component, supporting ecosystems and human needs. The movement of groundwater through soil and rock layers also contributes to the purification of water, removing contaminants and replenishing surface waters. In biological systems, water is integral to cell structure and function. It composes 60-70% of the human body, participating in biochemical reactions, maintaining cell turgor pressure, and facilitating the transport of molecules across cell membranes. Water's role in temperature regulation is crucial; through processes like sweating and transpiration, it helps maintain homeostasis in organisms. Aquatic ecosystems, ranging from freshwater lakes and rivers to marine environments, depend on water as a habitat. These ecosystems are diverse and productive, supporting myriad species and complex food webs.

Human civilization has always been closely linked to water resources. Early settlements thrived near rivers and lakes, utilizing water for agriculture, drinking, and transportation. Modern society continues to depend on water for various sectors, including agriculture, industry, and energy production. Irrigation systems, hydropower plants, and sanitation infrastructure are vital components of contemporary water management. However, the increasing demand for freshwater, coupled with pollution and climate change, poses significant challenges. Water scarcity affects millions globally, impacting health, food security, and economic development. Climate change exacerbates these issues by altering precipitation patterns, increasing the frequency of extreme weather events, and causing sea level rise. Melting glaciers and polar ice caps contribute to rising sea levels, threatening coastal communities and ecosystems. Ocean acidification, resulting from increased CO_2 absorption, impacts marine life, particularly calcifying organisms like corals and shellfish. These changes necessitate adaptive water management strategies, conservation efforts, and sustainable practices to ensure the availability and quality of water for future generations.

Technological advancements and scientific research continue to enhance our understanding and management of water resources. Desalination technologies provide alternative freshwater sources by removing salt from seawater, though they are energy-intensive and costly. Advances in wastewater treatment and recycling help to extend the usability of existing water supplies. Remote sensing and data analytics improve monitoring and prediction of water-related phenomena, aiding in efficient resource management. Public policies and international cooperation are crucial in addressing global water challenges, promoting equitable access, and protecting water ecosystems.

Distribution of Water

The distribution of water on Earth is a key factor in sustaining life, regulating climate, and supporting ecosystems. Water covers approximately 71% of the Earth's surface, with the majority of this water, around 97.5%, found in the oceans. Despite the abundance of water, only a small fraction—about 2.5%—is freshwater, which is essential for most terrestrial life forms and human activities. However, even within this small fraction, the distribution of freshwater is uneven, and much of it is not readily accessible. To better understand the distribution of water on Earth, it is essential to consider the different sources, their volumes, and how they are cycled through various processes in the hydrologic cycle.

Oceans: The Largest Water Reservoir

Oceans are by far the largest water reservoirs on Earth, holding about 1.332 billion cubic kilometers of water. They contain roughly 97.5% of all water on the planet, making them the dominant feature of the Earth's hydrosphere. Despite their vast size, ocean water is saline, with an average salinity of about 35 grams of dissolved salts per liter, rendering it unsuitable for direct human consumption, agriculture, or most industrial processes. However, oceans play a critical role in regulating the Earth's climate by absorbing and redistributing heat through currents and interacting with the atmosphere. Oceans are also an integral part of the global water cycle, as they are the primary source of evaporation that leads to precipitation. Water vapor from the oceans rises into the atmosphere, condenses into clouds, and eventually falls as rain or snow, replenishing freshwater systems on land.

Figure 1.2. Distribution of water

Freshwater Distribution

While freshwater comprises only 2.5% of the Earth's total water, it is essential for sustaining life on land. This limited freshwater is unevenly distributed across the planet, both geographically and in terms of availability. Most of Earth's freshwater—about 68.7%—is locked away in ice caps, glaciers, and permanent snow cover, primarily in Antarctica and Greenland. These ice reserves hold vast amounts of water but are largely inaccessible to humans and ecosystems due to their remote locations and the solid state of the water. However, these ice sheets and glaciers are critical to the long-term storage of freshwater and, in some regions, they act as a source of freshwater through seasonal melting. Groundwater, which accounts for approximately 30.1% of all freshwater, represents the largest accessible reservoir of freshwater. It is stored in underground aquifers, permeable rock layers that can store and transmit water. Groundwater is crucial for agriculture, drinking water, and industrial use, particularly in regions where surface water is scarce. Aquifers can vary in size, from small, localized systems to vast reserves that span entire continents, such as the Ogallala Aquifer in the central United States. However, groundwater is not always readily accessible, and its extraction must be managed carefully to avoid depletion or contamination, which can lead to severe consequences for water supply and ecosystems. Surface water, including rivers, lakes, reservoirs, and wetlands, represents only about 1.2% of freshwater and a mere 0.03% of the Earth's total water. Despite its small proportion, surface water is vital for human use and the environment. Rivers and lakes serve as sources of drinking water, irrigation, hydroelectric power, and recreation. They also provide habitats for countless species of plants, animals, and microorganisms. Surface water distribution is highly variable, with some regions, such as the Amazon Basin or the Great Lakes of North America, boasting abundant surface water, while other areas, like deserts or arid regions, face chronic water scarcity.

Ice Caps, Glaciers, and Permafrost

Ice caps and glaciers hold the majority of the Earth's freshwater, approximately 68.7%, making them the largest reservoirs of freshwater. These frozen water reserves are concentrated in the polar regions, particularly in Antarctica and Greenland, which together contain more than 99% of the Earth's glacier ice. Antarctica alone contains about 60% of the planet's freshwater in its ice sheet, which has an average thickness of 2 kilometers. Greenland's ice sheet holds about 7% of the Earth's freshwater. Glaciers, found in mountain ranges around the world, also contribute significantly to the global freshwater supply. They act as slow-moving reservoirs, gradually releasing meltwater during warmer months, which feeds rivers and lakes. This seasonal meltwater is crucial for regions that rely on glacier-fed rivers for agriculture, drinking water, and hydropower. For example, the Himalayan glaciers feed major river systems in South Asia, such as the Ganges, Brahmaputra, and Indus, supporting millions of people in India, Nepal, Pakistan, and Bangladesh. However, climate change poses a significant threat to these glaciers, as rising global temperatures are accelerating their melting rates. This not only threatens long-term freshwater availability but also raises sea levels, with profound implications for coastal communities and ecosystems.

Figure 1.3. Ice Caps on earth

Permafrost, which is ground that remains frozen for two or more consecutive years, also stores a significant amount of freshwater. Permafrost is found in polar and high-altitude regions and covers approximately 24% of the Earth's land surface. When permafrost thaws, it releases water into the environment, which can contribute to surface water and groundwater systems. However, the thawing of permafrost also releases methane and carbon dioxide, potent greenhouse gases that can exacerbate global warming.

Figure 1.4.

Groundwater: A Critical Freshwater Resource

Groundwater, accounting for about 30.1% of freshwater, is a crucial resource for many regions, especially those that lack abundant surface water. Stored in underground aquifers, groundwater is replenished by the infiltration of rainwater and snowmelt into the soil. This process, known as recharge, occurs slowly and can take years or even centuries, depending on the geology and climate of the region. Once stored in an aquifer, groundwater can remain there for extended periods, ranging from years to millennia. There are two types of aquifers: confined and unconfined. Confined aquifers are trapped between impermeable layers of rock or clay, while unconfined aquifers are more directly influenced by surface water and precipitation. Groundwater is typically accessed through wells, and in some cases, it can flow naturally to the surface in the form of springs. In arid regions, where surface water is scarce, groundwater is often the primary source of water for drinking, irrigation, and industrial uses. For example, much of the southwestern United States, parts of the Middle East, and North Africa rely heavily on groundwater to meet their water needs. Groundwater extraction, however, must be carefully managed. Over-extraction can lead to aquifer depletion, land subsidence, and the drying up of surface water bodies that are connected to the groundwater system. Additionally, groundwater contamination from pollutants such as agricultural runoff, industrial waste, and saltwater intrusion in coastal areas can render this vital resource unsafe for use. In many parts of the world, groundwater is being depleted faster than it can be replenished, posing significant challenges for future water security.

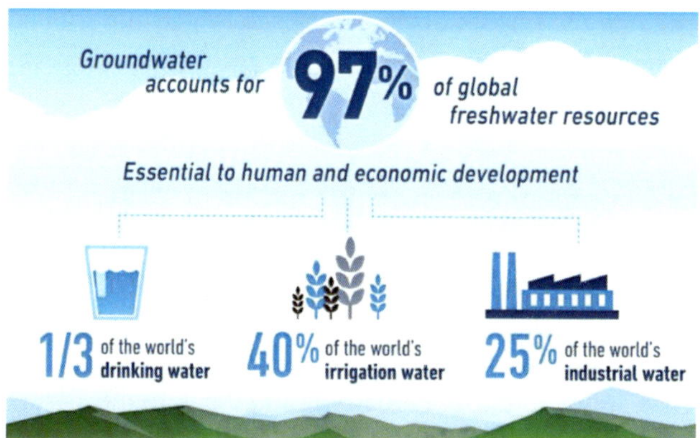

Figure 1.5.

Surface Water: Rivers, Lakes, and Wetlands

Surface water, though comprising only a small fraction of the Earth's total water, plays an outsized role in supporting ecosystems and human activities. Rivers, lakes, reservoirs, and wetlands are the most visible components of the Earth's freshwater system and are integral to agriculture, drinking water supply, energy production, and transportation. Rivers are dynamic systems that transport water, sediment, and nutrients from higher elevations to oceans and lakes. They are fed by precipitation, groundwater discharge, and snowmelt, and their flow can vary seasonally depending on the climate and geography. Major river systems, such as the Amazon, Nile, and Yangtze, are critical for the regions they flow through, providing water for millions of people and supporting diverse ecosystems. However, rivers are also highly vulnerable to human activities such as damming, water diversion, pollution, and habitat destruction. Dams and reservoirs, while providing benefits such as hydroelectric power and irrigation, can disrupt natural river flow, alter ecosystems, and reduce the availability of water downstream. Lakes, both natural and man-made (reservoirs), are important sources of freshwater. Large lakes, such as Lake Baikal in Russia and the Great Lakes in North America, hold substantial volumes of water and are critical for biodiversity, fisheries, and regional water supplies. Reservoirs, created by damming rivers, store water for various uses, including irrigation, flood control, and hydroelectric power generation. However, reservoirs also experience sediment buildup over time, which can reduce their storage capacity and lead to the eventual loss of their functionality. Wetlands, including marshes, swamps, and bogs, are unique ecosystems that serve as natural water filters, removing pollutants and providing critical habitat for a wide range of species. Wetlands also play a crucial role in flood control, as they can absorb excess water during heavy rains and release it slowly, reducing the impact of floods. Despite their importance, wetlands are among the most threatened ecosystems, with large areas being drained or degraded for agriculture, urban development, and infrastructure projects.

The Hydrologic Cycle: Movement and Transformation of Water

The distribution of water on Earth is continually influenced by the hydrologic cycle, the process by which water moves between the atmosphere, oceans, and land. This cycle is powered by solar energy, which drives the evaporation of water from oceans, lakes, and rivers into the atmosphere. As water vapor rises, it cools and condenses into clouds, eventually falling back to the Earth's surface as precipitation in the form of rain, snow, sleet, or hail. Once on land, water follows several pathways. Some of it infiltrates the soil, replenishing groundwater aquifers. Some flows over the surface as runoff, feeding rivers, lakes, and eventually returning to the oceans. Water also evaporates directly from the surface of water bodies and from plant leaves through a process called transpiration. Together, evaporation and transpiration are referred to as evapotranspiration. The water cycle ensures that water is constantly moving and being redistributed across the planet, though the amount of water in each reservoir remains relatively constant over time. The hydrologic cycle also plays a crucial role in regulating the Earth's climate. Water vapor is a greenhouse gas, and its presence in the atmosphere helps trap heat and maintain the Earth's temperature. Clouds, which are formed from condensed water vapor, reflect sunlight and can influence weather patterns and precipitation. The oceans, with their vast heat capacity, absorb and store solar energy, distributing it around the globe through ocean currents. This process helps regulate temperature and weather patterns, influencing everything from monsoons to droughts.

Figure 1.6.

The distribution of water on Earth is becoming increasingly important as human populations grow and climate change alters precipitation patterns, evaporation rates, and the availability of freshwater resources. Many regions are already experiencing water stress, with demand for water exceeding the available supply. Climate change is expected to exacerbate these challenges by intensifying droughts, increasing the frequency of extreme weather events, and accelerating the melting of glaciers and ice caps.

2

Surface Water

Surface waters can be essentially depicted as the water on the Earth's surface. This incorporates the seas, waterways and streams, lakes, and supplies. Surface waters are exceptionally critical. They constitute roughly 80 percent of the water utilized every day. In 1990, the Joined together States alone utilized roughly 327,000 billion gallons of surface water every day. Surface waters include most of the water utilized for open supply and water system. It plays less of a part in mining and animals businesses. Seas, the biggest surface water source, include roughly 97 percent of the Earth's surface water. In any case, since the seas have tall saltiness, the water isn't valuable as drinking water. Endeavors have been made to expel the salt from the water (desalination), but this is often a really expensive. Salt water is used in mining, industry, and control era. The seas moreover play a crucial part within the hydrologic cycle, controlling the worldwide climate, and giving environments for thousands of marine species. Waterways and streams constitute the streaming surface waters. The constrain of gravity actually draws water from a better to a lower elevation. Waterways get their water from two sources: groundwater, and runoff. Streams can get water from the ground in the event that they cut into the water table, where the ground is immersed. This is often known as base stream to the stream. Runoff streams downhill, to begin with as little rivers, at that point steadily combining with other rivulets and streams, expanding in measure until a waterway has shaped. The headwaters are known as little rivulets, or tributaries where the waterway starts. Springs from kept aquifers too can contribute to waterways.

A waterway will inevitably stream into an sea. A river's length can be troublesome to decide, particularly in the event that it has various tributaries. The USGS Web location characterizes a river's length as "the remove to the surge point from the first headwaters where the title characterizes the total length." For water to stream, there must be land upgradient of the river, that's, arrive at a better height than the stream. The arrive that's upgradient of any point on the stream is known as the seepage bowl or watershed. Edges of higher arrive, such as the Mainland Separate, isolated two seepage bowls. Streaming water is amazingly effective and critical in making the scene and human lives. Streaming water is utilized for various reasons counting water system and hydroelectric control generation. Streams disintegrate the landscape and alter the geography of the Soil by carving canyons and transporting soil and dregs to form ripe fields. Streams carry soil and dregs into the stream when rain or snow softens. The quicker the water moves, the bigger the molecule estimate the waterway can carry. The USGS measures how much dregs a stream carries by measuring the streamflow, the sum of water streaming past a given location; and the silt concentration. Silt within the stream can be supportive and hurtful. When stored on the banks and within the surge plain, Silt makes amazing farmlands. In any case, sediment can hurt and even annihilate dams, stores and the life within the stream. Moreover, these silt can be left behind amid floods as sticky, rancid mud in undesirable places.

The streamflow is measured by deciding the stream organize and release. The stream arrange, or datum, is the tallness of the water surface, in feet, over an subjective reference point. The stream release may be a measurement of the sum of water that's streaming at a specific point in time. It is measured in cubic feet per second. A discharge measurement decides the sum of water that's streaming within the stream at any given stream organize. To create this estimation, the width of the stream, and the water's profundity and speed at different focuses must be measured at a few stream stages. A cross-section of the stream is partitioned into interims and the region of each interim is calculated. In the event that the speed was measured at distinctive profundities on the same vertical interim, at that point the speed is found the middle value of. To decide the discharge for the interim, the region is multiplied by the speed. An normal of all the intervals' releases is calculated to discover the complete stream's release. It is critical to require release estimations of the stream at different stream stages, indeed surge organize. A waterway comes to surge arrange when the waterway floods its banks. The surge organize can be decided by measuring the gage tallness, or essentially the stature of the water within the stream measured from the river's foot. The streamflow can increment exponentially as the gage height increases. Thus, a little increment in gage stature may indicate that a waterway has come to its surge organize. Surges are a decently common, however unsafe, common calamity. They ordinarily happen since a storm or quick snow melt has created more runoff than a stream can carry. Dams falling flat, avalanches blocking stream channels, and tall tides are other flooding causes. Climate designs can enormously impact when and where flooding will happen. By examining these designs, geologists can decide a region's vulnerability to surge at certain times of the year. The repeat interim, measured in a long time, portrays the greatness of a surge. Changes within the seepage bowl, such as gathering timber or lodging improvements, can alter the size of a surge. The ordinarily dry arrive secured with water amid a surge is known as the surge plain. Confinements on arrive utilize in surge fields is controlled by flood-plain zoning. Dams and levees have been built to assist decrease harm caused by surges.

When streaming water voyages to an zone of arrive that's totally encompassed by higher arrive, a lake is formed. The water isn't caught in this moo region, the water fair get away at a slower rate than the rate of approaching water. Lakes can shift incredibly in zone, profundity, and water sort. Most lakes are new water, be that as it may a few, such as the Awesome Salt Lake and the Dead Ocean, are salt water. Opposite to common conviction, a supply is different from a lake. A store could be a artificial lake caused by a stream being dammed. The water in a supply is exceptionally moderate moving compared to the stream. In this manner, most of the dregs that the waterway carried settle to the foot of the store. A store will in the long run fill up with dregs and mud and ended up unusable.

The water cycle

The hydrologic cycle or water cycle may be a realistic representation of how water is reused through the environment. Water particles stay consistent, in spite of the fact that they may alter between strong, liquid, and gas shapes. Drops of water within the sea dissipate, which is the method of fluid water getting to be water vapor. Vanishing can happen from water surfaces, arrive surfaces, and snow fields into the discuss as water vapor. Dampness within the discuss can condense, which is the method of water vapor turning into fluid water. Water drops exterior a cold glass of water are condensed water. Condensation is the inverse handle of dissipation. Water vapor condenses on little particles of tidy, smoke, and salt gems to ended up portion of a cloud. After a whereas, the water beads combine with other beads and drop to Soil as precipitation (rain, snow, accost, hail, dew, and ice). Once the precipitation has fallen to Soil, it may go into an aquifer as groundwater or the drop may remain over ground as surface water. The hydrologic cycle is an imperative concept to get it. Water has numerous employments on Soil, such as human and creature utilization, control generation, and mechanical and agrarian needs. Precipitation—in the frame of rain and snow—is too imperative to understand. It is the most way that the water within the skies comes down to Soil, where it fills the lakes and rivers, recharges the underground aquifers, and provides drinks to plants and creatures. Diverse sums of precipitation drop on distinctive regions of the Soil at distinctive rates and times of the year. One problem confronting the cycle of water on Earth is water defilement. Chemicals that go into the water frequently are troublesome, in the event that not inconceivable, to evacuate. One potential source of water defilement is runoff, the overland stream of water. Whereas precipitation causes the runoff, stripping vegetation from arrive can include to the runoff in a specific zone. The silt and soil from these zones, not to specify any pesticides or fertilizers show, are washed into the streams, seas, and lakes. What happens to the rain after it falls depends on numerous variables, such as the escalated and length of precipitation, the geology of the arrive, soil conditions, urbanization, and vegetation thickness. A common misinterpretation approximately rain is that it is tear-shaped, when it is really formed more like a cheeseburger bun. Rain drops are also different due to the initial difference in particle size and the different coalescence rate.

Glaciers and ice caps

Icy masses and icecaps are alluded to as storage facilities for new water. They cover 10 percent of the world's arrive mass. These ice sheets are fundamentally found in Greenland and Antarctica. The icy masses in Greenland nearly cover the complete arrive mass. Icy masses start shaping since of snowfall amassing. Icy masses start to create When snowfall surpasses the melting rate in a certain region. This dissolving happens within the summer. The weight of snow collecting compresses the snow to make ice. Since these ice sheets are so overwhelming, they can gradually move down slopes.

Ice sheets influence the geology of the arrive in a few ranges. Antiquated ice sheets shaped lakes and valleys. The Awesome Lakes are an illustration of this. Ice sheets extend in length from less than the measure of a football field to hundreds of miles long. They too can reach up to 2 miles thick. Ice sheets dissolving can have a huge impact on the ocean level. Concurring to the USGS, on the off chance that all of the ice sheets were to liquefy nowadays, the ocean would rise an assessed 260 feet. Icy masses have had a huge impact on the arrangement of the Earth's surface and are still affecting the geology ordinary.

Surface waters incorporate streams, waterways, lakes, stores, and wetlands. The term stream is utilized here to speak to all streaming surface water, from brooks to expansive streams. Surface waters and their related environments give territory to numerous plant and creature species. Since surface waters are on the arrive surface, they are effectively created for utilize and provide approximately 78 percent of the Joined together States's add up to off-stream water utilize. Stream flow varies in reaction to climatic components and human exercises. A few streams have a little yearly release for the huge measure of their seepage range, such as the Colorado Waterway, and a few have a more noteworthy request for their water than they can supply without supply capacity. Since of their significance as a water source, stream rates for chosen streams are persistently checked by stream gages. Release is the sum of water moving down a stream per unit of time. Release is the item of the normal speed of streaming water and the cross-sectional zone at a chosen location on a stream. Normal speed is decided by measuring streaming water at numerous areas and profundities over the chosen estimation location. The cross-sectional area and the average velocity at each of these measured locations are multiplied to calculate discharge at that point. The discharges for all locations are added to obtain the total discharge of the stream.

Figure 2.1.

Streams are a energetic portion of the environment and are great markers of what is happening in a watershed. Stream stream in a watershed incorporates all water contributed from headwater zones, stream banks, channels, surge fields, patios, associated lakes, lakes, wetlands, and groundwater. Since watersheds are complex frameworks, each reacts in an unexpected way to normal or human exercises.

The physical characteristics of a watershed (arrive utilize, soil sort, geography, vegetation, incline, and perspective) and climate control the amount and quality of water that streams from them. Changes to any of these characteristics can influence water amount and quality. For case, expelling vegetation by characteristic causes such as fire can alter a watershed's water capacity and penetration characteristics. Since burned zones contain less vegetation to moderate runoff and hold soil in put, the rate and amount of water that runs off the surface to streams increments, and so does disintegration. Amid overwhelming downpours, the expanded runoff and disintegration can increment the chance of flooding, mudslides, and impeded water quality. Water looks for the way of slightest resistance. As water streams through a watershed, it picks up and stores dregs, soil and shake particles, making stream hallways. These hallways, comprising of stream channels, banks, and surge fields, are influenced by characteristic and human exercises inside watersheds. The physical forms of silt transport and statement are basic to shaping the stream passage. The transport of dregs inside and from a watershed is one of the major forms that offer assistance shape the surface of the Soil. Silt particles are classified by measure, with the littlest clay and the biggest boulders. Littler particles are as a rule carried in suspension whereas the bigger materials are moved along the channel foot by rolling, sliding, or bouncing.

One of a stream's major exercises is transporting materials inside and out of a watershed. Dregs transport rates of a stream are a work of stream control, which may be a degree of the combined impact of the slant at the streambed (higher inclines produce higher stream speeds) and release (volume of water). Where stream control is reduced, a stream's silt carrying control is additionally diminished, and a parcel of the dregs is stored. For illustration, dregs is stored taking after a flood's top, or most elevated, release. Silt can be stored in channels for brief periods and moved once more or stay stationary as in alluvial fans or expansive stores. Stream channels and their surge fields continually alter to changing water amounts and silt provided by their watersheds. Long-term changes in runoff and silt stack may lead to long-term changes in channel characteristics.

Surface water, the water present on the Earth's surface in various water bodies such as rivers, lakes, ponds, and reservoirs, constitutes a crucial component of the global hydrological cycle and is vital for supporting life, ecosystems, and human activities. It accounts for a mere fraction of the planet's total water resources, roughly 0.3% of all freshwater, yet it plays a disproportionately significant role in meeting the water demands of societies, agriculture, industry, and maintaining biodiversity. Understanding the distribution, availability, and dynamics of surface water is essential for effective water management and ensuring the sustainability of this critical resource in the face of growing pressures such as climate change, pollution, and increasing human consumption. Surface water is primarily derived from precipitation that falls on the Earth's surface, with some flowing over the land as runoff and eventually accumulating in various natural and artificial water bodies. These water bodies can be broadly categorized into rivers and streams, lakes and ponds, wetlands, and reservoirs, each with unique characteristics, ecological roles, and contributions to water availability.

Rivers and Streams: The Lifeblood of the Landscape

Rivers and streams are dynamic systems that fundamentally shape the landscape, support ecosystems, and provide water for human use. They act as natural drainage systems, collecting runoff from precipitation, snowmelt, and groundwater discharge, and transporting it to larger water bodies such as lakes, seas, or oceans. Rivers are classified based on their size, flow characteristics, and the nature of their basins. Major rivers, like the Amazon, Nile, and Yangtze, are vital sources of freshwater and support some of Earth's most densely populated regions. Water availability in rivers and streams is highly variable and depends on several factors, including climate, geography, and seasonality. In temperate and tropical regions, rivers are often fed by regular rainfall, while in arid and semi-arid regions, rivers may rely heavily on seasonal precipitation or snowmelt from mountainous areas. The seasonal variation in river flow can lead to periods of abundance and scarcity, influencing water availability for ecosystems and human activities. In many parts of the world, rivers are subject to significant human intervention, including the construction of dams, diversion for irrigation, and urban development, which can alter their natural flow regimes, impact water quality, and disrupt ecological processes.

Figure 2.2.

Rivers are also vital for sustaining aquatic and riparian ecosystems. They provide habitats for various species, from fish and amphibians to birds and mammals, and support complex food webs. Riparian zones, the areas adjacent to rivers and streams, are particularly important as they act as buffers that protect water quality by filtering pollutants, reducing erosion, and providing critical habitat for wildlife. However, these ecosystems are increasingly threatened by pollution, habitat destruction, and altering natural flow regimes due to human activities.

Lakes and Ponds: Natural and Artificial Reservoirs of Freshwater

Lakes and ponds are inland bodies of standing water that vary greatly in size, depth, and ecological characteristics. While both are similar in water-filled basins, lakes are generally larger and deeper than ponds. Lakes can be classified based on their origin, such as tectonic lakes formed by the movement of Earth's crust, glacial lakes created by glacial activity, volcanic lakes formed in volcanic craters, and man-made lakes, often called reservoirs, created by damming rivers for water storage, hydroelectric power, or flood control. Water availability in lakes and ponds is influenced by their location, climate, and the characteristics of their catchment areas. In some regions, lakes are primarily fed by rivers and streams, while in others, they may rely on direct precipitation, groundwater inflow, or a combination of these sources. The water balance of a lake—its inflows and outflows—determines its size and volume over time. Lakes with no significant outflow may become saline over time, as evaporation exceeds inflow, concentrating salts and minerals.

Figure 23.

Lakes and ponds are critical resources for human socictics and the environment. They provide drinking water, support agriculture through irrigation, offer recreational opportunities, and are sources of fish and other aquatic organisms. Additionally, they play a key role in regulating local climates by moderating temperatures and influencing precipitation patterns. However, lakes and ponds are highly susceptible to environmental changes and human impacts. Pollution from agricultural runoff, industrial discharges, and urban wastewater can lead to eutrophication, a process in which nutrient enrichment causes excessive algae growth, depleting oxygen levels and harming aquatic life. Invasive species, habitat destruction, and climate change further threaten the health of these ecosystems.

Wetlands: The Earth's Natural Water Filters

Wetlands are areas where water covers the soil or is present near the surface for extended periods, creating unique ecosystems among Earth's most productive ecosystems. Wetlands include marshes, swamps, bogs, and fens, each characterized by specific hydrological, soil, and vegetation conditions. They are often found where groundwater emerges at the surface or where surface water collects in depressions, floodplains, and coastal areas. Wetlands play a crucial role in the hydrological cycle and water management. They act as natural water filters, trapping sediments, nutrients, and pollutants, improving water quality. Wetlands also provide flood control by absorbing and slowly releasing stormwater, reducing the risk of flooding downstream. Moreover, wetlands are vital carbon sinks, sequestering carbon dioxide and mitigating the effects of climate change.

Despite their ecological importance, wetlands are among the most threatened ecosystems globally. Human activities such as drainage for agriculture, urban development, and infrastructure projects have led to the loss of a significant portion of the world's wetlands. The degradation of wetlands has severe consequences for biodiversity, water quality, and the ability to mitigate natural disasters. Protecting and restoring wetlands is essential for maintaining the planet's water systems' health and ensuring water resources' sustainability.

Reservoirs: Engineered Water Storage Solutions

Reservoirs are artificial lakes created by damming rivers and streams to store water for various purposes, including water supply, irrigation, hydroelectric power generation, and flood control. Reservoirs are a critical component of modern water management systems, providing a reliable water source in regions where natural water availability is highly variable or insufficient to meet demand. Water availability in reservoirs depends on the flow of water from upstream rivers and streams and the management practices used to regulate water levels and releases. Reservoirs are designed to balance multiple objectives, such as ensuring a stable water supply, generating electricity, protecting against floods, and maintaining environmental flows for downstream ecosystems. However, managing these competing demands can be challenging, particularly in changing climate patterns, which may alter precipitation, evaporation rates, and water availability.

While reservoirs provide significant benefits, they also have environmental and social impacts. The construction of reservoirs often involves the displacement of communities, loss of agricultural land, and alteration of natural habitats. Additionally, reservoirs can disrupt the natural flow of rivers, affecting the migration of fish and other aquatic species, altering sediment transport, and impacting water quality. Over time, reservoirs can also become silted, reducing their storage capacity and effectiveness. To address these challenges, integrated water resource management approaches that consider the needs of both human populations and ecosystems are essential.

Surface Water Availability and Distribution

The distribution and availability of surface water vary widely globally, influenced by climate, topography, and human activity. Regions with abundant rainfall, such as the tropics and temperate zones, typically have extensive river networks, numerous lakes, and large wetland areas. In contrast, arid and semi-arid regions, such as the Middle East, North Africa, and parts of Australia, have limited surface water resources, with rivers that may be seasonal and lakes that can shrink or disappear during dry periods. The availability of surface water is also affected by seasonal variations in precipitation and temperature. In many regions, surface water resources are highly seasonal, with significant differences in water availability between wet and dry seasons. For example, monsoon rains in South Asia lead to swollen rivers and full reservoirs during the wet season, while water scarcity can become a critical issue during the dry season. Similarly, in temperate regions, snowmelt in spring contributes to high river flows, while water levels may drop in late summer and autumn.

Human activities have a profound impact on the distribution and availability of surface water. Urbanization, deforestation, and agricultural practices can alter the natural flow of water, leading to increased runoff, reduced infiltration, and changes in the timing and quantity of water reaching rivers, lakes, and wetlands. The construction of dams, canals, and other infrastructure has transformed many of the world's river systems, providing benefits such as water supply and energy generation but also causing ecological disruptions and affecting downstream water availability.

Surface water is also increasingly under pressure from pollution, which can degrade water quality and reduce its availability for human and ecological use. Pollutants from agricultural runoff, industrial discharges, and urban wastewater can introduce harmful chemicals, nutrients, and pathogens into surface water bodies, leading to problems such as eutrophication, loss of biodiversity, and health risks for people and wildlife.

Surface Water Management and Conservation

The sustainable management of surface water resources is critical for ensuring water security, protecting ecosystems, and supporting economic development. Integrated Water Resources Management (IWRM) is a widely recognized approach that seeks to balance the needs of different water users, protect water quality, and maintain the health of aquatic ecosystems. IWRM involves coordinating the development and management of water, land, and related resources across different sectors and scales, from local to national and transboundary levels.

Conservation of surface water involves a range of strategies to reduce water use, improve water efficiency, and protect water bodies from pollution and degradation. In agriculture, drip irrigation, crop rotation, and soil conservation can help reduce water consumption and prevent runoff. In urban areas, strategies such as rainwater harvesting, green infrastructure, and wastewater recycling can reduce the demand for surface water and protect water quality. Protecting riparian zones, restoring wetlands, and maintaining environmental flows are also essential for preserving the ecological functions of surface water bodies. Climate change poses significant challenges to the management and availability of surface water. Changes in precipitation patterns, increased frequency of extreme weather events, and rising temperatures are expected to alter the hydrological cycle, affecting surface water resources' timing, quantity, and quality. Adapting to these changes will require resilient water management practices that can accommodate uncertainty and variability in water availability. This may include diversifying water sources, enhancing water storage capacity, improving water use efficiency, and protecting natural water systems.

Public awareness and participation are crucial for successfully managing and conserving surface water. Engaging communities in water management decisions, raising awareness about the importance of water conservation, and promoting sustainable water use practices can help build support for water management policies and initiatives. Education and outreach programs can also empower individuals and organizations to take action to protect and conserve surface water resources.

3

Groundwater

Groundwater, often called the Earth's natural water reserve and water bank, is one of the most vital yet often underappreciated components of the planet's hydrological cycle. Representing about 30% of the world's freshwater resources, groundwater is critical in supporting ecosystems, human consumption, agriculture, and industry. It is stored in aquifers, which are permeable rock formations or sediments capable of holding and transmitting water. However, the global distribution and availability of groundwater vary significantly, influenced by geology, climate, and human activity. Understanding groundwater dynamics, managing its use sustainably, and protecting it from contamination are crucial for ensuring long-term water security.

Groundwater originates from precipitation that infiltrates the Earth's surface, percolating through soil and porous rock layers until it reaches the water table—the upper surface of the saturated zone where all the pores in the soil and rock are filled with water. The movement of groundwater is typically slow, influenced by gravity and pressure differences, and it can take years, decades, or even centuries for water to move from the surface to deeper aquifers. This slow movement is one of the reasons groundwater is often considered a renewable but limited resource; the rate at which it is replenished through natural processes is typically much slower than the rate at which it can be extracted.

Aquifers, the geological formations that store groundwater, are classified into two main types: confined and unconfined. Unconfined aquifers are directly recharged by surface water infiltrating the water table. They are more susceptible to contamination from surface activities because there is no confining layer above them to protect them from pollutants. On the other hand, confined aquifers are overlain by impermeable rock or clay layers, which protect them from surface contamination and make them more difficult to recharge. Water in confined aquifers is often under pressure, which can cause it to rise above the aquifer's level when tapped by a well, a phenomenon known as artesian pressure.

The availability of groundwater is not uniform across the globe. Regions with abundant rainfall and favorable geological conditions often have substantial groundwater reserves. In contrast, arid and semi-arid regions may have limited groundwater resources, despite extensive aquifers, because the recharge and evaporation rates are low. Some of the world's most significant groundwater reserves are found in large sedimentary basins, such as the Great Artesian Basin in Australia, the Ogallala Aquifer in the United States, and the Guarani Aquifer in South America. These aquifers store vast amounts of water, providing a critical resource for these regions' agriculture, drinking water, and industry.

Groundwater is a crucial source of drinking water for billions of people worldwide. In many rural areas, where surface water sources are either unavailable or unreliable, groundwater serves as the primary or sole source of potable water. Groundwater is typically of higher quality than surface water because it undergoes natural filtration as it percolates through soil and rock, removing many impurities and pathogens. However, this natural filtration process is not foolproof, and various human activities, including agricultural runoff, industrial discharges, and improper waste disposal can still contaminate groundwater.

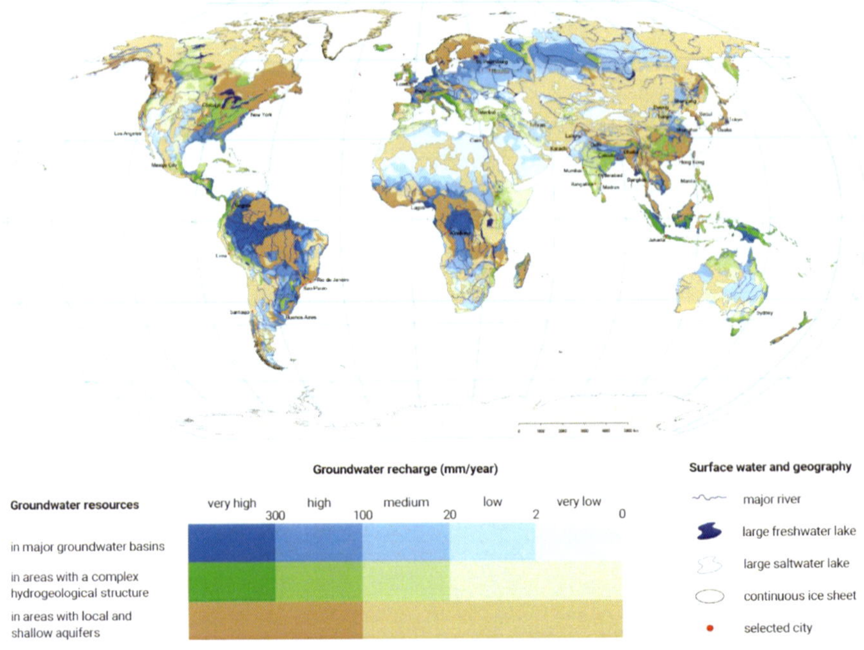

Figure 3.1.

https://www.unesco.org/reports/wwdr/2022/sites/default/files/2022-03/wwdr2022_Figure_6.png

Agriculture is the largest consumer of groundwater, accounting for approximately 70% of global groundwater withdrawals. Groundwater is essential for irrigation, especially in regions where surface water is scarce or seasonal. The use of groundwater for irrigation has enabled the expansion of agriculture into areas that would otherwise be unsuitable for farming, contributing to food security and economic development. However, intensive groundwater extraction for irrigation has also led to significant challenges, including the depletion of aquifers, land subsidence, and the deterioration of water quality.

Groundwater depletion occurs when water is extracted from an aquifer faster than it is replenished through natural recharge processes. This is a growing concern in many parts of the world, particularly in regions with intensive agriculture or rapidly growing populations. In some cases, groundwater levels have dropped so much that wells have run dry, leading to water shortages and increased costs for water extraction. Land subsidence, the gradual sinking of the ground, can also result from excessive groundwater extraction, as the removal of water causes the collapse of the pore spaces in the aquifer. This can damage infrastructure, reduce aquifer storage capacity, and increase flood risk.

The quality of groundwater is another critical concern. While groundwater is generally of high quality, it can be contaminated by natural and anthropogenic sources. Naturally occurring contaminants include dissolved minerals such as arsenic, fluoride, and radon, which can pose significant health risks in high concentrations. Anthropogenic sources of contamination include agricultural runoff (containing fertilizers and pesticides), industrial discharges, leaking underground storage tanks, and improperly disposed of hazardous wastes. Once groundwater is contaminated, cleaning up can be extremely difficult and expensive, making protecting groundwater quality a high priority.

Groundwater management is a complex and multifaceted challenge that requires understanding the hydrological, geological, and socio-economic factors that influence groundwater availability and use. Sustainable groundwater management involves balancing the needs of various users, including agriculture, industry, and domestic use, while ensuring that groundwater extraction does not exceed natural recharge rates. This can be achieved through a combination of regulatory measures, such as groundwater abstraction limits and well-permitting systems, as well as the promotion of water-saving technologies and practices. In some regions, managed aquifer recharge (MAR) enhances groundwater reserves. MAR involves intentionally recharging aquifers with surface water or treated wastewater during periods of surplus, to be stored and used during drought or high demand. This approach can help to stabilize groundwater levels, reduce the risk of land subsidence, and improve water security. However, MAR requires careful management to avoid the introduction of contaminants and to ensure that the recharged water is compatible with the native groundwater.

Protecting groundwater resources also requires addressing the broader challenges of water governance, including coordinating groundwater management across political and administrative boundaries. Many of the world's major aquifers are transboundary, meaning two or more countries share them. The management of transboundary aquifers requires international cooperation and the development of agreements that ensure the equitable and sustainable use of shared groundwater resources. The United Nations Educational, Scientific and Cultural Organization (UNESCO) and the International Groundwater Resources Assessment Centre (IGRAC) have been involved in efforts to promote the sustainable management of transboundary aquifers through initiatives such as the International Shared Aquifer Resources Management (ISARM) program.

Climate change is expected to significantly impact groundwater resources, with water availability, quality, and management implications. Changes in precipitation patterns, increased frequency and intensity of droughts, and rising temperatures will likely alter groundwater recharge rates and demand. In some regions, reduced recharge could lead to declining groundwater levels, exacerbating water scarcity and increasing competition for water resources. In other areas, increased flooding could result in higher recharge rates and a greater risk of contamination from surface pollutants. Adapting to these changes will require a comprehensive approach to groundwater management that considers the potential impacts of climate change and promotes resilience in water systems.

Groundwater is a vital resource for human use and plays a critical role in maintaining ecological health. Groundwater contributes to the flow of rivers and streams, particularly during dry periods, and supports wetlands, which are important habitats for various plant and animal species. The interaction between groundwater and surface water is complex and varies depending on a given area's geological and hydrological conditions. In many cases, the sustainable management of surface water and groundwater must be integrated to protect the ecological functions of these interconnected systems.

Public awareness and education are essential components of effective groundwater management. Many people are unaware of the importance of groundwater or the challenges associated with its management. Raising awareness about the value of groundwater, the risks of over-extraction and contamination, and the importance of conservation can help to build support for sustainable groundwater management practices. Education initiatives can also empower communities to actively protect and manage their local groundwater resources.

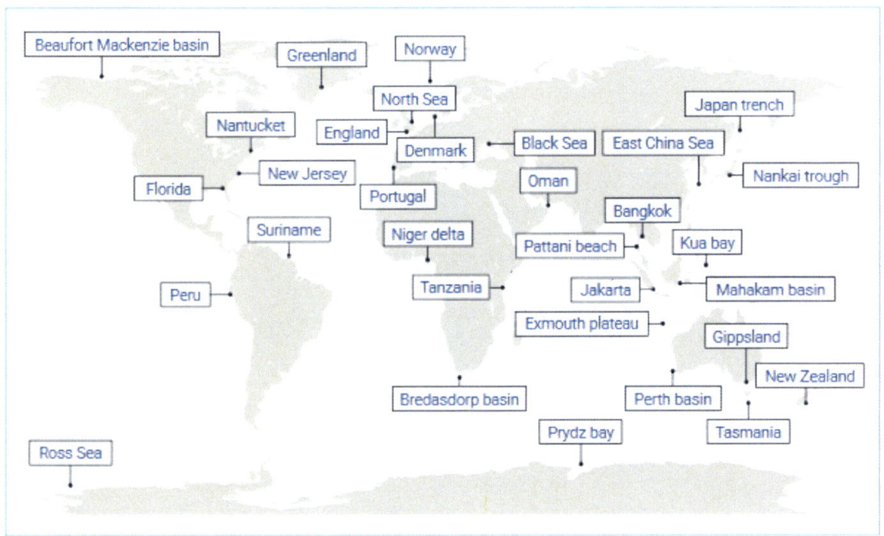

Figure 3.2. World map of known occurrences of fresh and brackish offshore groundwater
Source: Based on Post et al. (2013, fig. 1, p. 72).
https://unesdoc.unesco.org/ark:/48223/pf0000380721/PDF/380721eng.pdf.multi

4

Groundwater: Water Reserve for Future

Groundwater, often termed the Earth's hidden water reserve, is an essential lifeline for the future of global water security. It is the water found beneath the Earth's surface, stored in the pores and fractures of rock formations and sediments, known as aquifers. This vast underground reservoir represents about 30% of the world's freshwater resources, making it a critical component of the planet's hydrological system. Groundwater's role extends beyond merely being a water source; it acts as a buffer against droughts, sustains rivers and lakes during dry periods, supports ecosystems, and provides a reliable supply for agriculture, industry, and domestic use. Given its importance, understanding groundwater's availability, dynamics, and management is crucial for ensuring its sustainable use, especially in the face of growing demands and climate change.

Groundwater Formation and Aquifers

Groundwater originates primarily from precipitation that infiltrates the ground, percolating through the soil and rock layers until it reaches a zone of saturation, where all the voids in the rock or sediment are filled with water. This zone is known as the saturated zone, and its upper boundary is called the water table. The process of water moving from the surface to the aquifer is known as recharge. Depending on the geological characteristics of the area, groundwater can be stored in various types of aquifers.

Aquifers are broadly categorized into two types: unconfined and confined. Unconfined aquifers have no overlying impermeable layer, allowing direct interaction with surface water bodies and easier recharge from precipitation. These aquifers are more susceptible to contamination from surface activities due to their direct connection with the surface. On the other hand, confined aquifers are trapped between impermeable rock or clay layers, which protect them from surface contaminants but also make recharge more challenging. Water in confined aquifers is often under pressure, leading to the natural phenomenon of artesian wells, where water flows to the surface without pumping.

The quality and quantity of groundwater in an aquifer depend on several factors, including the type of rock or sediment, the rate of recharge, the depth of the aquifer, and the presence of any confining layers. Porous and permeable materials such as sandstone and gravel make excellent aquifers, as they can store and transmit significant amounts of water. In contrast, materials like clay and shale have low permeability, limiting their ability to store and transmit water. The depth of the aquifer also plays a crucial role, as deeper aquifers are typically more isolated from surface contaminants but may be more challenging and costly to access.

Global Distribution and Availability of Groundwater

Groundwater is unevenly distributed globally, with its availability varying significantly from region to region based on geological, climatic, and hydrological factors. Some of the world's largest aquifers are found in sedimentary basins, where extensive deposits of sand, gravel, and other permeable materials create vast underground reservoirs. These include the High Plains Aquifer (Ogallala Aquifer) in the United States, the Guarani Aquifer in South America, the Great Artesian Basin in Australia, and the North China Plain Aquifer.

Table 1. Groundwater withdrawals in 2017, aggregated by the world's major regions

Continent and region	Groundwater withdrawal (km^3/year)	% change since 2010	% of total freshwater withdrawal
North America	156	0	26
Northern North America	113	-1	24
Central America	37.1	+12	38
Caribbean	6.5	-37	27
South America	27	+6	13
Northern and Eastern South America	7.9	-32	9
Andean countries	4.7	-22	11
Southern South America	14.7	+83	19
Europe	65	-6	23
Northern Europe	4.7	-3	20
Western Europe	15.2	+1	22
Eastern Europe	15.2	-24	18
Southern Europe	29.7	+1	31
Africa	45	+10	20
Northern Africa	26.2	+24	21
Western Africa	8.0	+9	28

Central Africa	2.1	-21	72
Eastern Africa	6.3	-6	13
Southern Africa	2.8	-16	14
Asia	657	-4	26
Northern Asia	3.1	-10	15
Central Asia	2.7	-85	2
Western Asia	63.7	-3	39
South Asia	401	-5	39
East Asia	132	-6	18
Southeast Asia	54.2	+54	11
Australia & Oceania	8	+21	31
Australasia	7.5	+30	29
Micro-, Mela-and Polynesia	0.6?	-36?	79?
World	959	-2	25

Source: Estimates based on Aquastat (n.d.), Eurostat (n.d.), and Margat and Van der Gun (2013). https://unesdoc.unesco.org/ark:/48223/pf0000380721/PDF/380721eng.pdf.multi

In regions with abundant rainfall, such as the Amazon Basin or the Congo Basin, groundwater reserves are regularly replenished, leading to a steady water supply. In contrast, arid and semi-arid regions, such as the Arabian Peninsula, North Africa, and parts of India, rely heavily on groundwater for water supply, often tapping into ancient reserves known as fossil water, which may have been stored in the aquifers thousands of years ago and are not actively recharged. In these areas, groundwater is a finite resource, and its extraction must be carefully managed to avoid overexploitation and depletion.

Groundwater is critical in sustaining agriculture, especially in regions where surface water is scarce or seasonal. In many parts of the world, such as the Indo-Gangetic Plain in South Asia, the North China Plain, and the Central Valley in California, groundwater provides a reliable source of irrigation water, enabling the production of food and other crops. However, the intensive use of groundwater for agriculture has led to significant challenges, including the depletion of aquifers, land subsidence, and the deterioration of water quality due to salinization and contamination.

Table 2. Breakdown of 2017 groundwater withdrawal by water use sector and by continent

	Groundwater withdrawal		Volumes by water use sector			Percentages by water use sector		
	(km³/ year)	% of world	Agri-culture	Do-mestic	Indus-try	Agricul-ture	Domes-tic	Industry
North America	156	16	97	48	12	62	30	7
South America	27	3	13	9	5	49	32	20
Europe	65	7	24	29	12	36	45	19
Africa	45	5	29	14	2	65	32	4
Asia	657	68	496	107	53	76	16	8
Australia & Oceania	8	1	4	4	0	48	48	3
World	959	100	664	211	84	69	22	9

Source: Estimates based on https://unesdoc.unesco.org/ark:/48223/pf0000380721/PDF/380721eng.pdf.multi

Groundwater Depletion and Overexploitation

Groundwater depletion occurs when water is extracted from an aquifer faster than it is naturally recharged. This is a growing concern in many parts of the world, particularly in regions with intensive agriculture or rapidly expanding populations. The consequences of groundwater depletion are severe and far-reaching. As water levels decline, wells may run dry, leading to water shortages and increased costs for water extraction. In some cases, the overextraction of groundwater has led to the drying up of rivers, lakes, and wetlands, devastatingly impacting ecosystems and biodiversity. Land subsidence is one of the most visible and concerning effects of groundwater depletion, which occurs when water removal from an aquifer causes the ground to sink. This can result in the loss of valuable agricultural land, damage to infrastructure, and increased flood risk. In some urban areas, such as Mexico City, Jakarta, and parts of California, land subsidence has caused significant damage to buildings, roads, and other infrastructure, leading to costly repairs and the need for ongoing monitoring and management. In addition to physical depletion, groundwater is also at risk of chemical degradation. The overuse of fertilizers and pesticides in agriculture, industrial pollution, and the improper disposal of hazardous wastes can lead to groundwater contamination with harmful substances. Common contaminants include nitrates, heavy metals, volatile organic compounds (VOCs), and pathogens. Once groundwater is contaminated, it can be extremely difficult and expensive to clean up, and in some cases, the water may become unusable for drinking or irrigation.

Groundwater as a Buffer Against Climate Variability

One of the key benefits of groundwater is its ability to act as a buffer against climate variability and change. Unlike surface water, which is subject to rapid fluctuations in availability due to seasonal changes in precipitation, groundwater provides a more stable and reliable water source. This makes it particularly valuable in regions prone to droughts, where surface water sources may dry up or become insufficient to meet demand.

During drought, groundwater can sustain rivers, lakes, and wetlands by providing baseflow, the portion of river flow that comes from groundwater discharge. This is critical for maintaining the health of aquatic ecosystems and ensuring water availability for human use during dry periods. Groundwater also supports vegetation and agriculture in drylands, where surface water is often scarce. By tapping into groundwater reserves, farmers can continue to irrigate crops and maintain food production despite prolonged drought. However, relying on groundwater as a buffer against climate variability also has limits. In regions where groundwater is heavily exploited, the ability to rely on it during droughts may be compromised. Overextraction during drought can lead to the rapid depletion of groundwater reserves, reducing the resource's long-term sustainability. Moreover, climate change is expected to impact groundwater recharge in many regions, with shifts in precipitation patterns and increased evaporation rates potentially leading to reduced recharge rates and lower groundwater levels.

Groundwater and Ecosystems

Groundwater is vital for human consumption and agriculture and plays a crucial role in sustaining ecosystems. Many rivers, lakes, and wetlands rely on groundwater discharge to maintain flow, particularly during dry periods when surface water inputs are limited. This interaction between groundwater and surface water is critical for maintaining the health and functioning of aquatic ecosystems. Wetlands, in particular, are highly dependent on groundwater. These ecosystems, which include marshes, swamps, and peatlands, are among the most productive on Earth and provide essential services such as water filtration, carbon storage, and habitat for various species. Groundwater helps maintain the water levels in wetlands, supporting vegetation growth and aquatic organisms' survival.

In some cases, the overextraction of groundwater can lead to the drying out of wetlands, resulting in the loss of biodiversity and the degradation of ecosystem services. Groundwater also supports terrestrial ecosystems, particularly in arid and semi-arid regions where surface water is scarce. Many plants in these regions rely on deep root systems that tap into groundwater to survive. The depletion of groundwater can lead to the loss of vegetation, which can increase soil erosion, reduce carbon sequestration, and contribute to desertification.

Sustainable Groundwater Management

Given the importance of groundwater and the challenges associated with its depletion and contamination, sustainable groundwater management is essential for ensuring the long-term availability of this critical resource. Sustainable management involves balancing groundwater extraction with the natural recharge rates, protecting groundwater quality, and considering the needs of both human populations and ecosystems. One of the key principles of sustainable groundwater management is the concept of the "sustainable yield," which refers to the amount of groundwater that can be extracted from an aquifer without causing long-term depletion or degradation. Determining the sustainable yield requires understanding the aquifer's recharge rate, the demands placed on it by various users, and the potential impacts of extraction on surface water bodies and ecosystems. In many regions, however, groundwater is extracted at rates that exceed the sustainable yield, leading to resource depletion.

To address the issue of overexploitation, many countries have implemented regulatory measures such as groundwater abstraction permits, well registration, and groundwater monitoring programs. These measures aim to control the amount of water extracted from aquifers, prevent illegal drilling, and ensure that groundwater use is sustainable. In some cases, governments have also introduced pricing mechanisms, such as groundwater extraction fees, to encourage more efficient resource use. Another important aspect of sustainable groundwater management is the protection of groundwater quality. This involves implementing land-use practices that minimize the risk of contamination, such as regulating the use of fertilizers and pesticides, controlling industrial discharges, and managing waste disposal. In areas where groundwater contamination has already occurred, remediation efforts may be necessary to clean up polluted aquifers and restore water quality. However, prevention is generally more cost-effective and sustainable than remediation. Groundwater recharge enhancement is another strategy that can contribute to the sustainability of groundwater resources. This can be achieved through managed aquifer recharge (MAR) techniques, where surface water is deliberately directed into aquifers to replenish groundwater levels. MAR can involve methods such as the construction of infiltration basins, the use of recharge wells, or the diversion of stormwater and treated wastewater into aquifers. These techniques can help augment groundwater supplies, particularly in areas with low natural recharge rates.

Groundwater and Climate Change Adaptation

Climate change poses significant challenges to groundwater management, as it is expected to alter the patterns of precipitation and temperature that influence groundwater recharge. In some regions, climate change may lead to reduced rainfall, increased evaporation, and longer drought periods, reducing groundwater recharge and increasing the demand for groundwater. Conversely, in other regions, increased rainfall and extreme weather events may lead to more frequent flooding, which could enhance recharge and pose risks to groundwater quality. Adapting groundwater management to the impacts of climate change will require a flexible and forward-looking approach. This may involve diversifying water sources to reduce reliance on groundwater, enhancing groundwater recharge through MAR and other techniques, and improving water use efficiency across all sectors. It will also be important to monitor groundwater levels and quality closely, using data to inform management decisions and respond to changes in the availability and demand for groundwater.

Climate change adaptation will often require integrating groundwater management with broader water resource management strategies. This includes considering the interactions between surface water and groundwater, managing water use in agriculture to reduce the demand for groundwater, and protecting ecosystems that depend on groundwater. By taking an integrated approach, it will be possible to manage groundwater resources in a way that is resilient to climate change's impacts and ensures this critical resource's long-term sustainability.

5

Global Groundwater Crisis

The global groundwater crisis is an escalating issue that significantly threatens water security, public health, ecosystems, and economic stability worldwide. Groundwater, which constitutes about 30% of the world's freshwater resources and nearly 99% of the available freshwater supply (excluding glaciers and ice caps), is critical for drinking water, agriculture, and industry. However, the overextraction, pollution, and mismanagement of this vital resource have led to a decline in groundwater reserves and the degradation of water quality, raising serious concerns about the sustainability of groundwater for future generations. This crisis is exacerbated by population growth, urbanization, industrialization, and climate change, all placing increasing pressure on groundwater resources.

Groundwater Over Extraction: Depleting the Earth's Water Reserve

One of the most pressing aspects of the global groundwater crisis is the overextraction of groundwater, which is occurring at an alarming rate in many parts of the world. Groundwater is extracted from aquifers for various purposes, including drinking water, irrigation, industrial processes, and domestic use. In many regions, groundwater is the primary water source, particularly in arid and semi-arid areas where surface water is scarce or seasonal. Overextraction occurs when groundwater is withdrawn from an aquifer faster than it is naturally recharged through precipitation, infiltration, and surface water interaction. This unsustainable practice leads to declining groundwater levels, causing various negative consequences. In many agricultural regions, such as the Indo-Gangetic Plain in South Asia, the North China Plain, and the Central Valley in California, intensive groundwater pumping for irrigation has significantly dropped groundwater levels. For example, in parts of India, groundwater levels have been declining by more than 1 meter per year, leading to a situation where farmers are forced to drill deeper and deeper wells, often at great financial and environmental cost.

The depletion of groundwater reserves has several far-reaching impacts. First, as groundwater levels drop, wells may run dry, leading to water shortages and increased costs for water extraction. This is particularly concerning in regions where groundwater is the primary or sole source of drinking water. In some areas, such as in parts of the Middle East and North Africa, overextraction has led to the exhaustion of ancient fossil water reserves, which were accumulated over millennia and are not replenished under current climatic conditions. Once these reserves are depleted, they cannot be easily or quickly replaced, leading to a permanent water availability loss. Second, groundwater depletion can cause land subsidence, a process in which the ground sinks due to the compaction of sediments that occurs when water is removed from the aquifer. Land subsidence can lead to agricultural land loss, infrastructure damage, increased flood risk, and changes in the hydrological landscape. In cities such as Mexico City, Jakarta, and Bangkok, land subsidence has caused significant damage to buildings, roads, and other infrastructure, leading to costly repairs and the need for ongoing management.

Third, groundwater depletion can disrupt the natural balance between groundwater and surface water. In many cases, groundwater contributes to the flow of rivers, lakes, and wetlands through a process known as baseflow. When groundwater levels decline, the baseflow to surface water bodies may decrease or even cease, leading to the drying up of rivers and wetlands, with severe consequences for ecosystems and biodiversity.

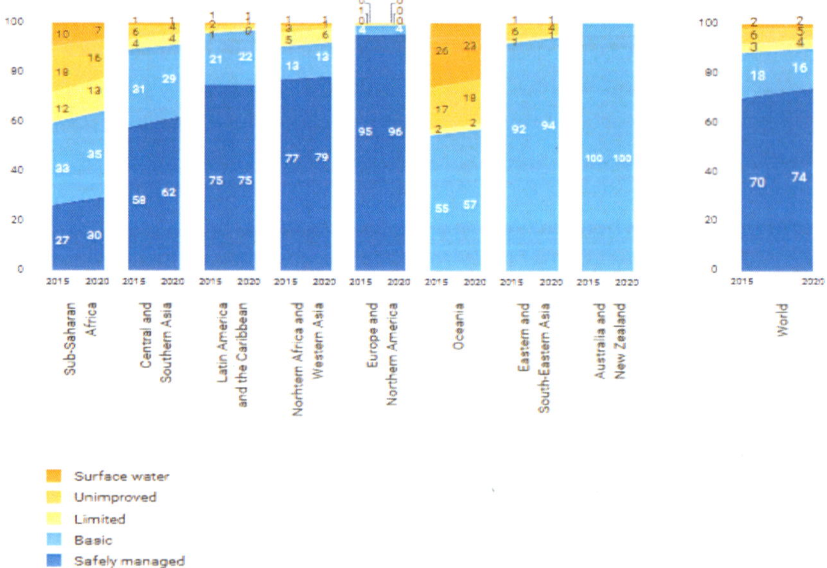

Figure 5.1. Regional and global drinking water coverage, 2015-2020 (%)
Note: Five SDG region had estimates for safety managed drinking water aervicea in 2020.
Source: Adapted from WHO/UNICEF (2021, fig. 2 and 3. p. 8)

Groundwater Pollution: A Growing Threat to Water Quality

In addition to the problem of overextraction, groundwater pollution is another critical aspect of the global groundwater crisis. Groundwater pollution occurs when harmful substances from human activities infiltrate the ground and contaminate aquifers. Once groundwater is contaminated, it can be extremely difficult, expensive, and sometimes impossible to clean up, leading to long-term impacts on water quality and availability. One of the most common sources of groundwater pollution is agricultural activities, particularly chemical fertilizers and pesticides. These substances can leach into the soil and percolate to the groundwater, leading to elevated levels of nitrates, phosphates, and other harmful chemicals. Nitrate contamination is a widespread problem in agricultural regions around the world. High nitrate levels in drinking water are a serious health concern, particularly for infants, as they can cause methemoglobinemia, also known as "blue baby syndrome," a condition that reduces the ability of blood to carry oxygen. Industrial activities are another major source of groundwater pollution. Industries such as manufacturing, mining, and chemical processing generate a variety of pollutants, including heavy metals (e.g., lead, mercury, arsenic), volatile organic compounds (VOCs), and other hazardous substances. These pollutants can enter groundwater through spills, leaks, improper waste disposal, and inadequate industrial wastewater treatment. Once in the groundwater, these contaminants can persist for long periods, posing significant risks to human health and the environment. For example, industrial pollution in some parts of China has led to groundwater contamination with toxic substances such as benzene, cadmium, and chromium, resulting in severe health impacts for local communities.

Urbanization and population growth also contribute to groundwater pollution. In rapidly growing cities, the expansion of urban areas often leads to increased runoff of pollutants from roads, parking lots, and other impervious surfaces. This runoff can carry oil, heavy metals, and other contaminants into the groundwater. In addition, inadequate sanitation and wastewater management in urban areas can contaminate groundwater with pathogens, pharmaceuticals, and other pollutants. In many developing countries, the lack of proper sewage systems and the widespread use of pit latrines and septic tanks contribute to the contamination of shallow aquifers with fecal coliform bacteria and other pathogens, posing a significant public health risk.

Groundwater Toxicity: Health Risks and Ecosystem Impacts

The toxicity of contaminated groundwater poses severe risks to human health and ecosystems. Depending on the nature and concentration of the contaminants, exposure to polluted groundwater can lead to various health problems, including acute poisoning, chronic diseases, and developmental issues. One of the most well-known cases of groundwater toxicity is arsenic contamination, affecting millions worldwide. Arsenic is a naturally occurring element found in high groundwater concentrations in certain regions, particularly in parts of South Asia, Southeast Asia, and South America. Long-term exposure to arsenic in drinking water is associated with a range of health issues, including skin lesions, cancer (especially skin, bladder, and lung cancer), cardiovascular diseases, and developmental effects. The situation is particularly severe in Bangladesh and West Bengal, India, where millions of people are exposed to groundwater with arsenic levels far exceeding the World Health Organization's (WHO) guidelines. Efforts to address arsenic contamination include providing alternative water sources, such as surface water or treated water, and the development of low-cost arsenic removal technologies.

Another significant concern is groundwater contamination with fluoride, which is also a naturally occurring element. While fluoride is beneficial in small amounts for preventing dental caries, excessive fluoride exposure can lead to fluorosis, which affects teeth and bones. Dental fluorosis results in the discoloration and pitting of teeth, while skeletal fluorosis can cause joint pain, stiffness, and, in severe cases, deformities. High fluoride levels in groundwater are found in parts of India, China, Africa, and the Middle East, affecting millions of people. In addition to the direct health risks, groundwater toxicity can significantly impact ecosystems. Many plants, animals, and microorganisms depend on clean groundwater for survival. When groundwater becomes contaminated, it can disrupt the balance of ecosystems, leading to biodiversity loss and ecosystem services' degradation. For example, wetlands, among the most productive ecosystems on Earth, often rely on groundwater inputs to maintain their water levels and support their diverse flora and fauna. Contamination of groundwater that feeds into wetlands can lead to the decline of plant and animal species, reduce the ability of wetlands to filter pollutants, and impair their role in carbon sequestration.

Industrial Discharge and Its Impact on Groundwater

Industrial discharge is a significant contributor to the global groundwater crisis, as it introduces a wide range of pollutants into the environment, many of which can infiltrate and contaminate groundwater resources. Industries such as mining, chemical manufacturing, textile production, and metal plating are notorious for potentially releasing hazardous substances into groundwater. For example, mining activities often involve using large quantities of water to process ores and minerals. This water, often laden with heavy metals, acids, and other toxic substances, can seep into the ground and contaminate local aquifers. Acid mine drainage (AMD) is a specific and severe problem associated with mining. It occurs when sulfide minerals exposed to air and water during mining operations generate sulfuric acid, which then leaches heavy metals from surrounding rocks into the water. This acidic and metal-rich water can infiltrate groundwater, leading to long-term contamination that is challenging and costly to remediate. AMD is a significant issue in mining regions worldwide, including the Appalachian region in the United States, parts of Canada, and South Africa.

Chemical manufacturing industries are another major source of groundwater contamination. The production of chemicals often involves the use of solvents, heavy metals, and other toxic substances, which can enter groundwater through leaks, spills, and improper waste disposal practices. These pollutants persist in the environment in many cases, meaning they do not break down easily and can remain in groundwater for decades. For instance, trichloroethylene (TCE) and perchloroethylene (PCE), chlorinated solvents used in industrial processes, are common groundwater contaminants. These chemicals are classified as carcinogens and have been linked to various health problems, including cancer and damage to the liver and kidneys.

The textile industry also significantly impacts groundwater, particularly in regions where textile manufacturing is a major economic activity. The dyeing and finishing processes in textile production require large amounts of water and involve various chemicals, including dyes, detergents, and bleaches. Wastewater from textile factories often contains high concentrations of organic pollutants, heavy metals, and dyes, which can infiltrate groundwater if not properly treated. In countries such as Bangladesh and India, where the textile industry is a major source of income, the discharge of untreated or inadequately treated wastewater has led to severe groundwater pollution, affecting both human health and the environment.

The Role of Policy and Management in Addressing the Groundwater Crisis

Addressing the global groundwater crisis requires a multifaceted approach that includes policy interventions, improved groundwater management practices, and the implementation of sustainable technologies. Effective groundwater management is essential to ensure groundwater resources' long-term availability and quality, particularly in regions where groundwater is a critical water source. One key aspect of groundwater management is the establishment of legal and regulatory frameworks that govern groundwater extraction and use. In many parts of the world, groundwater resources are poorly regulated, leading to uncontrolled extraction and the overexploitation of aquifers. Implementing groundwater laws that set limits on extraction, require permits for drilling wells, and enforce sustainable groundwater use is essential to prevent overextraction and protect groundwater resources. For example, some countries, such as Denmark and Australia, have implemented groundwater management policies that include water rights systems, which allocate specific amounts of groundwater to users based on sustainable yield estimates. These policies have helped to reduce overextraction and promote the sustainable use of groundwater. In addition to regulatory measures, adopting advanced technologies and practices can play a crucial role in mitigating the impacts of the groundwater crisis. Managed aquifer recharge (MAR) is an approach that involves the intentional recharge of groundwater aquifers using surface water, stormwater, or treated wastewater. MAR can help to replenish depleted aquifers, improve groundwater quality, and provide a buffer against droughts. Techniques such as the construction of infiltration basins, recharge wells, and the use of permeable pavements can facilitate the recharge of groundwater and enhance water security.

Monitoring and data collection are also critical components of effective groundwater management. Reliable data on groundwater levels, quality, and extraction rates are necessary to inform decision-making and assess management interventions' effectiveness. Using remote sensing technology, geographic information systems (GIS), and groundwater modeling tools can enhance the ability to monitor groundwater resources and predict future trends.

Public awareness and community engagement are essential for successfully implementing groundwater management strategies. Educating the public about the importance of groundwater, the risks associated with overextraction and pollution, and the need for sustainable water use can foster a sense of responsibility and encourage water conservation practices. Community-based groundwater management approaches, where local communities participate in the monitoring and management of groundwater resources, have been successful in some regions in promoting sustainable groundwater use.

6

Groundwater Contamination

Groundwater contamination and pollution have emerged as critical challenges in global water security, posing severe risks to human health, ecosystems, and the availability of this vital resource for future generations. Groundwater, which accounts for nearly 30% of the world's freshwater reserves, is a crucial source of drinking water, irrigation, and industrial processes. However, the widespread contamination and pollution of groundwater due to various human activities, including agriculture, industrial discharge, and urbanization, have led to significant degradation in groundwater quality, compromising its utility and availability. The situation is exacerbated by the overextraction of groundwater, which depletes aquifers and increases the concentration of pollutants in the remaining water, leading to a cycle of deteriorating water quality. This detailed exploration will discuss the sources, mechanisms, and impacts of groundwater contamination and pollution, focusing on their implications for water reserves, public health, and environmental sustainability.

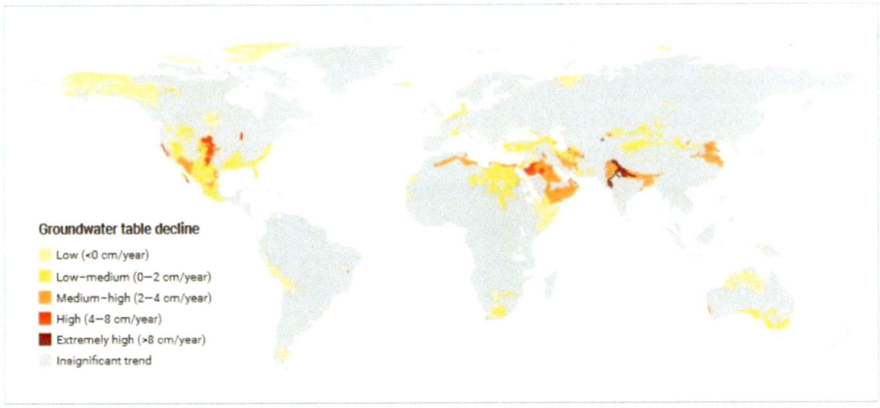

Figure 6.1. Groundwater table decline in a selection of the world's major aquifers

Note Groundwater table decline indication the average yearly decline of the groundwater table over

Sources and Mechanisms of Groundwater Contamination

Groundwater contamination occurs when pollutants from various sources infiltrate the soil and percolate down to the water table, entering aquifers and degrading the water's quality. These contaminants can originate from a wide range of human activities, each contributing to the global groundwater crisis in distinct ways.

1. **Agricultural Activities**: Agriculture is one of the most significant contributors to groundwater contamination, primarily due to chemical fertilizers, pesticides, and herbicides. These substances are applied to crops to enhance yield and control pests but often leach into the soil and reach groundwater. Nitrate, a common component of fertilizers, is particularly problematic. When it enters groundwater, nitrate can persist for long periods, leading to elevated concentrations that pose serious health risks. Nitrate contamination is widespread in agricultural regions worldwide, affecting drinking water supplies and causing conditions such as methemoglobinemia, or "blue baby syndrome," in infants. Additionally, pesticides used in agriculture, including organophosphates and carbamates, can leach into groundwater, contaminating drinking water sources and posing risks to human health and wildlife.
2. **Industrial Discharge**: Industrial activities are another major source of groundwater contamination. Industries such as manufacturing, mining, petrochemicals, and textiles generate a variety of pollutants, including heavy metals (e.g., lead, mercury, arsenic), volatile organic compounds (VOCs), and other hazardous chemicals. These pollutants can enter groundwater through spills, leaks, improper waste disposal, and inadequate treatment of industrial effluents. For example, in regions with significant mining activities, acid mine drainage (AMD) can release sulfuric acid and heavy metals into groundwater, causing long-term contamination. Similarly, the improper disposal of industrial waste can lead to the infiltration of toxic substances into aquifers, making the groundwater unsuitable for drinking or agricultural use. The persistence of these pollutants in groundwater is a significant concern, as they can remain in the environment for decades, posing ongoing risks to human health and ecosystems.
3. **Urbanization and Population Growth**: Rapid urbanization and population growth contribute to groundwater contamination through increased waste generation, impervious surfaces' expansion, and the proliferation of poorly managed sanitation systems. In many urban areas, stormwater runoff from roads, parking lots, and other impervious

surfaces carries pollutants such as oil, heavy metals, and chemicals into the ground, where they can infiltrate and contaminate groundwater. In addition, inadequate sewage treatment and the widespread use of septic tanks in areas without centralized sanitation systems can contaminate groundwater with pathogens, pharmaceuticals, and other pollutants. The contamination of shallow aquifers with fecal coliform bacteria and other pathogens is a concern in many developing countries, where the lack of proper sanitation infrastructure poses significant public health risks.

4. **Landfills and Waste Disposal**: Landfills and waste disposal sites are also significant sources of groundwater contamination. As waste decomposes in landfills, leachate—a liquid that forms as water percolates through the waste—can infiltrate the ground and enter aquifers. This leachate often contains a complex mixture of organic and inorganic pollutants, including heavy metals, VOCs, and pathogens. The improper management of landfills, particularly in regions with high water tables, can lead to severe groundwater contamination, making the water unsuitable for consumption and requiring expensive remediation efforts. In addition, illegal dumping of hazardous waste, particularly in developing countries, exacerbates the problem of groundwater pollution, as these sites are often unregulated and lack the necessary safeguards to prevent contamination.
5. **Septic Systems and On-Site Wastewater Treatment**: In rural and peri-urban areas, the use of septic systems and on-site wastewater treatment systems is common. However, when these systems are poorly designed, improperly maintained, or located in areas with high water tables, they can become sources of groundwater contamination. Septic systems that are not adequately maintained can fail, releasing untreated or partially treated wastewater into the ground. This wastewater can contain pathogens, nutrients (such as nitrates and phosphates), and chemicals that can contaminate groundwater, posing risks to public health and the environment. The widespread use of septic systems in areas with high population densities or poor soil conditions exacerbates the risk of groundwater contamination.

Impact of Groundwater Pollution on Water Reserves

Groundwater pollution profoundly impacts the availability and quality of groundwater reserves, which are critical for meeting the water needs of a growing global population. As groundwater becomes increasingly polluted, its utility as a source of drinking water, irrigation, and industrial processes diminishes, leading to various negative consequences.

Reduced Availability of Safe Drinking Water: Groundwater is a primary source of drinking water for billions of people worldwide. However, as groundwater becomes contaminated with pollutants such as nitrates, heavy metals, and pathogens, the availability of safe drinking water decreases. In many regions, groundwater contamination has reached levels that exceed safe drinking water standards, forcing communities to rely on alternative water sources or invest in expensive water treatment technologies. For example, in parts of South Asia, the widespread contamination of groundwater with arsenic has led to a public health crisis, with millions of people exposed to unsafe drinking water. The cost of providing alternative water sources or treating contaminated groundwater can be prohibitively high, particularly in low-income regions, leading to a situation where large populations are left without access to safe drinking water.

Impact on Agricultural Productivity: Groundwater is essential for agriculture, particularly in arid and semi-arid regions where surface water is scarce. However, the contamination of groundwater with pollutants such as nitrates, pesticides, and heavy metals can have a detrimental impact on agricultural productivity. When contaminated groundwater is used for irrigation, it can accumulate harmful substances in the soil, affecting crop yields and the quality of agricultural produce. In some cases, using contaminated groundwater for irrigation can lead to the uptake of pollutants by crops, resulting in food safety issues and potential health risks for consumers. Additionally, soil quality degradation due to contaminated water use can lead to long-term declines in agricultural productivity, affecting food security and the livelihoods of farming communities.

Depletion of Groundwater Reserves: The overextraction of groundwater, combined with pollution, leads to the depletion of groundwater reserves, reducing the availability of this critical resource for future generations. As groundwater levels decline due to overextraction, the concentration of pollutants in the remaining water often increases, leading to further water quality degradation. In regions where groundwater is heavily relied upon for drinking water and irrigation, the depletion and contamination of aquifers can lead to severe water shortages, forcing communities to rely on unsustainable alternatives such as desalination, water imports, or the exploitation of deeper, non-renewable aquifers. This cycle of depletion and contamination undermines the long-term sustainability of groundwater resources, posing a significant challenge for water security in many parts of the world.

Long-Term Environmental Impacts: Groundwater contamination has significant long-term environmental impacts, particularly on ecosystems that survive clean groundwater. Wetlands, rivers, and lakes that rely on groundwater inputs can be severely affected by the contamination of the underlying aquifers. As contaminated groundwater discharges into surface water bodies, it can lead to the degradation of aquatic ecosystems, affecting biodiversity and the health of plant and animal species. In some cases, the contamination of groundwater with nutrients such as nitrates and phosphates can lead to eutrophication, a process that results in the excessive growth of algae and other aquatic plants, depleting oxygen levels in the water and leading to the death of fish and other aquatic organisms. The long-term degradation of ecosystems due to groundwater contamination has far-reaching implications for biodiversity, ecosystem services, and the environment's overall health.

Toxicity of Contaminated Groundwater: Health Risks & Environmental Impacts

The toxicity of contaminated groundwater is a major concern, as it poses significant risks to human health and the environment. The nature and severity of these risks depend on the types and concentrations of contaminants in the groundwater.

- **Health Risks Associated with Contaminated Groundwater**: Contaminated groundwater can expose populations to various health risks, depending on the nature of the pollutants. Some of the most common health risks associated with contaminated groundwater include:
  - **Nitrate Contamination**: High levels of nitrates in drinking water are a serious health concern, particularly for infants and pregnant women. Nitrate contamination is associated with methemoglobinemia, or "blue baby syndrome," a condition that reduces the ability of blood to carry oxygen. Long-term exposure to high nitrate levels has also been linked to an increased risk of certain cancers and thyroid disorders.
  - **Arsenic Contamination**: Arsenic is a naturally occurring element found in high groundwater concentrations in certain regions. Long-term exposure to arsenic in drinking water is associated with an increased risk of skin, lung, bladder, and kidney cancers. Arsenic exposure can also cause skin lesions, cardiovascular disease, and neurological effects.
  - **Heavy Metal Contamination**: Groundwater contaminated with heavy metals such as lead, mercury, and cadmium poses significant health risks. Lead exposure, even at low levels, can cause cognitive impairment and developmental delays in children, while mercury exposure can damage the nervous system and kidneys. Cadmium

exposure is associated with kidney damage and an increased risk of bone fractures.

- **Pathogen Contamination**: Groundwater contaminated with pathogens, such as bacteria, viruses, and parasites, can cause various waterborne diseases, including diarrhea, cholera, and hepatitis. In many developing countries, the lack of proper sanitation infrastructure leads to contaminating shallow aquifers with pathogens, posing significant public health risks.

- **Environmental Impacts of Groundwater Contamination**: The environmental impacts of groundwater contamination are far-reaching and can affect entire ecosystems. Some of the key environmental impacts include:
  - **Eutrophication**: The groundwater contamination with nitrates and phosphates can lead to eutrophication when these nutrients are discharged into surface water bodies. Eutrophication results in the excessive growth of algae and other aquatic plants, leading to oxygen depletion and the death of fish and other aquatic organisms.
  - **Biodiversity Loss**: The contamination of groundwater with toxic substances such as heavy metals and pesticides can lead to the degradation of ecosystems and the loss of biodiversity. Species that rely on clean groundwater for survival, such as certain fish, amphibians, and plants, are particularly vulnerable to the effects of groundwater contamination.
  - **Degradation of Wetlands**: Wetlands that rely on groundwater inputs can be severely affected by the contamination of the underlying aquifers. The degradation of wetlands due to groundwater contamination can lead to the loss of important ecosystem services, such as water filtration, flood control, and habitat provision for wildlife.
  - **Soil Degradation**: Using contaminated groundwater for irrigation can lead to the accumulation of toxic substances in the soil, affecting soil quality and fertility. This can result in reduced agricultural productivity and the long-term degradation of land resources.

Industrial Discharge and Groundwater Pollution

Industrial discharge is a major contributor to groundwater pollution, particularly in regions with significant industrial activity. Industries such as manufacturing, mining, petrochemicals, and textiles generate a wide range of pollutants that can contaminate groundwater if not properly managed.

- **Sources of Industrial Pollutants**: Industrial pollutants can enter groundwater through various pathways, including spills, leaks, improper waste disposal, and inadequate treatment of industrial effluents. Some of the most common industrial pollutants that contaminate groundwater include:
    - **Heavy Metals**: Industries such as mining, electroplating, and metal processing generate heavy metals such as lead, mercury, arsenic, and cadmium, contaminating groundwater. These metals are toxic to humans and wildlife and can persist in the environment for long periods.
    - **Volatile Organic Compounds (VOCs)**: VOCs are a group of organic chemicals that are commonly used in industrial processes, including the manufacture of paints, solvents, and plastics. VOCs can contaminate groundwater through leaks, spills, and improper disposal of industrial waste. Some VOCs, such as benzene and trichloroethylene, are known carcinogens.
    - **Petroleum Hydrocarbons**: The leakage of petroleum products from underground storage tanks, pipelines, and refineries can lead to groundwater contamination with hydrocarbons such as benzene, toluene, and xylene. These substances are toxic and can cause long-term contamination of groundwater.
    - **Acid Mine Drainage (AMD)**: In regions with significant mining activity, the extraction of minerals can lead to acid mine drainage (AMD) formation. AMD occurs when sulfide minerals are exposed to air and water, producing sulfuric acid, which can leach heavy metals from surrounding rocks. The acidic water and heavy metals can contaminate groundwater, causing long-term environmental damage.
- **Impact of Industrial Discharge on Groundwater Quality**: The discharge of industrial pollutants into groundwater can severely impact water quality, making the water unsafe for drinking, irrigation, and other uses. In many cases, groundwater contamination with industrial pollutants is difficult to remediate, as these substances can persist in the environment for decades or centuries. Heavy metals, VOCs, and other toxic substances in groundwater can lead to long-term health risks for populations that rely on contaminated water for drinking and domestic use.
- **Regulatory Measures and Best Practices**: Addressing the issue of industrial discharge and groundwater pollution requires stringent regulatory measures and the adoption of best practices for waste

management and pollution control. Some of the key strategies for mitigating the impact of industrial discharge on groundwater quality include:

- **Regulation and Enforcement**: Governments need to implement and enforce regulations that limit the discharge of pollutants into the environment. This includes setting standards for the treatment of industrial effluents, requiring permits for waste disposal, and monitoring industrial activities for compliance with environmental regulations.
- **Pollution Prevention**: Industries should adopt pollution prevention practices, such as reducing hazardous substances, improving process efficiency, and implementing waste minimization strategies. Pollution prevention can help to reduce the generation of pollutants and minimize the risk of groundwater contamination.
- **Waste Treatment and Disposal:** Proper treatment and disposal of industrial waste are essential to prevent groundwater contamination. This includes using advanced treatment technologies to remove pollutants from industrial effluents and the safe disposal of hazardous waste in regulated landfills.
- **Remediation and Cleanup**: In cases where industrial pollutants have already contaminated groundwater, remediation and cleanup efforts are necessary to restore water quality. This can involve techniques such as pump-and-treat, in-situ bioremediation, and using permeable reactive barriers to remove or neutralize contaminants in groundwater.

Addressing Groundwater Pollution: A Path Forward

Addressing the global challenge of groundwater pollution requires a multifaceted approach that involves policy interventions, improved groundwater management practices, and the implementation of sustainable technologies.

- **Policy Interventions and Regulatory Frameworks**: Governments play a crucial role in addressing groundwater pollution by implementing and enforcing regulatory frameworks that govern groundwater use and protection. This includes setting standards for groundwater quality, regulating the discharge of pollutants, and monitoring groundwater resources. In addition, governments should promote the sustainable use of groundwater by implementing water rights systems, encouraging water conservation, and supporting the adoption of advanced technologies for water management.

- **Groundwater Management and Monitoring:** Effective groundwater management is essential to protect and sustain groundwater resources. This includes using integrated water management approaches that consider the interactions between surface water and groundwater and implementing managed aquifer recharge (MAR) techniques to replenish depleted aquifers. Monitoring and data collection are also critical components of groundwater management, as reliable data on groundwater levels, quality, and extraction rates are necessary to inform decision-making and assess the effectiveness of management interventions.
- **Sustainable Technologies and Practices:** Adopting sustainable technologies and practices can help mitigate the impact of groundwater pollution. This includes using advanced water treatment technologies to remove pollutants from industrial effluents and wastewater, as well as implementing pollution prevention practices in industries and agriculture. In addition, using natural treatment systems, such as constructed wetlands and biofiltration, can help improve groundwater quality by removing contaminants from water before it infiltrates into aquifers.
- **Public Awareness and Community Engagement:** Public awareness and community engagement are essential for successfully implementing groundwater management strategies. Educating the public about the importance of groundwater, the risks associated with pollution, and the need for sustainable water use can foster a sense of responsibility and encourage water conservation practices. Community-based groundwater management approaches, where local communities participate in the monitoring and management of groundwater resources, have been successful in some regions in promoting sustainable groundwater use.

7

Himalayan Region

The Himalayan region, a colossal mountain range stretching across five countries—Bhutan, China, India, Nepal, and Pakistan—forms the backbone of the Asian continent. Renowned for its dramatic landscapes, diverse ecosystems, and significant natural resources, the Himalayas play a crucial role in shaping South Asia's climate, hydrology, and biodiversity. The region, often referred to as the "Third Pole" due to its vast glaciers and snowfields, is a vital source of freshwater for millions of people and a key player in global environmental dynamics. This detailed examination explores the natural resources of the Himalayan region, with a particular focus on its water resources, including rivers, glaciers, and groundwater, and the challenges associated with their availability and sustainability.

Geographical and Geological Overview

The Himalayas span approximately 1,500 kilometers (930 miles) from west to east, forming a natural barrier between the Tibetan Plateau to the north and the Indian subcontinent to the south. The range is characterized by its towering peaks, including Mount Everest (8,848 meters), K2 (8,611 meters), and Kangchenjunga (8,586 meters), which are among the highest in the world. The geology of the Himalayas results from the ongoing collision between the Indian Plate and the Eurasian Plate, a process that began around 50 million years ago and continues to shape the region's topography.

The Himalayas are divided into sub-ranges, including the Greater Himalayas (Himalayan Main Range), the Lesser Himalayas, and the Shivalik Hills. The region's geological complexity includes diverse rock types, from ancient metamorphic rocks to younger sedimentary deposits, influencing its hydrology and natural resource distribution.

Water Resources in the Himalayan Region

Water resources in the Himalayan region are paramount, serving as a crucial source of freshwater for over 1.5 billion people across South Asia. The region's water resources are primarily from glaciers, snowfields, rivers, and groundwater. These resources are vital for drinking water, agriculture, hydropower generation, and maintaining ecological balance.

- **Glaciers and Snowfields**: The Himalayas host some of the largest and most important glaciers outside the polar regions. These glaciers, including the Gangotri, Yamunotri, and Siachen, act as "water towers" that store vast amounts of freshwater in ice and snow. The seasonal melting of these glaciers contributes significantly to the flow of major rivers in the region. Glacial meltwater is particularly crucial during the dry seasons when river flows from insufficient rainfall.

The Himalayan glaciers are facing significant challenges due to climate change. Rising temperatures have led to accelerated glacier retreat and reduced ice mass, which impacts the seasonal water flow of the rivers fed by these glaciers. Studies indicate that many Himalayan glaciers are retreating at an alarming rate, with potential consequences for water availability, agriculture, and hydropower production in the downstream regions.

- **Rivers**: The rivers originating from the Himalayas are among the largest and most significant in Asia, including the Ganges, Brahmaputra, Indus, and Yangtze rivers. These rivers are essential for providing freshwater, supporting agriculture, and generating hydropower. The Himalayan rivers are fed by both glacial meltwater and rainfall, with their flows varying seasonally.
    - **The Ganges (Ganga):** Originating from the Gangotri Glacier, the Ganges is a major river in India and Bangladcsh. It supports one of the most densely populated regions in the world and is crucial for agriculture, industry, and domestic use. The river faces challenges related to pollution, overextraction, and the impacts of climate change on its glacial sources.
    - **The Brahmaputra:** This river originates in Tibet, flows through India and Bangladesh, and is known for its high discharge and sediment load. The Brahmaputra is vital for agriculture and flood control in the region. Its flow is influenced by glacial melt and seasonal monsoon rains.
    - **The Indus:** Originating in Tibet and flowing through northern India and Pakistan, the Indus is critical for agriculture and water supply in Pakistan. The river system includes several major tributaries fed by Himalayan glaciers and snowfields.
    - **The Yangtze:** Although its source is in the Tibetan Plateau, the Yangtze River has significant implications for the Himalayan region's hydrology. It is the longest river in Asia and plays a crucial role in China's water resources and economy.

- **Groundwater:** Groundwater in the Himalayan region is an essential resource, particularly in areas where surface water is scarce or unreliable. Aquifers in the foothills and lower valleys store groundwater that is accessed through wells and springs. Groundwater availability is influenced by the region's geology, topography, and precipitation patterns.

In some areas, groundwater recharge is directly linked to river and glacier systems. However, overexploitation of groundwater, often due to increased agricultural and industrial demands, poses a risk to sustainable water use. Additionally, groundwater quality can be affected by natural processes and human activities, including contamination from agricultural runoff and industrial discharge.

Natural Resources Beyond Water

Besides its vital water resources, the Himalayan region is rich in other natural resources, including minerals, forests, and biodiversity. These resources contribute significantly to the economies of the countries in the region but also pose challenges related to sustainable management and conservation.

- Mineral Resources: The Himalayas are known for their rich mineral deposits, including precious metals, semi-precious stones, and industrial minerals. Notable resources include:
    - **Precious Metals:** The region contains deposits of gold and silver, primarily found in areas such as the Kulu Valley in India and parts of Nepal. Mining activities have significant economic value but also pose environmental and ecological risks.
    - **Semi-Precious Stones:** The Himalayas are a source of semi-precious stones like jade, sapphire, and garnet. These stones are valued in the gem and jewelry industries.
    - **Industrial Minerals:** The region also contains deposits of industrial minerals such as limestone, gypsum, and coal. These resources are important for construction, agriculture, and energy production.

The extraction of minerals from the Himalayas presents challenges related to environmental degradation, including habitat destruction, soil erosion, and water pollution. Sustainable mining practices and effective regulation are necessary to mitigate these impacts.

- **Forests and Biodiversity:** The Himalayan region is home to diverse forest ecosystems, ranging from subtropical forests in the lower elevations to alpine meadows and coniferous forests at higher altitudes.

These forests provide important ecological services, including carbon sequestration, soil conservation, and habitat for wildlife.

- **Flora:** The region boasts a rich diversity of plant species, including medicinal herbs, timber trees, and high-altitude flora. Some notable plant species include the rhododendron, oak, and fir. Medicinal plants, such as those used in traditional Himalayan medicine, have significant cultural and economic value.
- **Fauna:** The Himalayas support a variety of wildlife, including iconic species such as the Bengal tiger, snow leopard, red panda, and Himalayan bear. The region's biodiversity is crucial for maintaining ecological balance and supporting local livelihoods through ecotourism and sustainable use of natural resources.

Deforestation, habitat fragmentation, and poaching pose significant threats to Himalayan biodiversity. Conservation efforts and sustainable forest management are essential to protect these ecosystems and ensure forest resource availability.

Challenges and Sustainability Issues

The Himalayan region faces various challenges related to managing and sustaining its natural resources, particularly water resources. These challenges include:

- **Climate Change**: Climate change impacts the Himalayan region through rising temperatures, changing precipitation patterns, and accelerated glacier melting. The retreat of glaciers and reduced snow cover affect rivers' flow, leading to water availability variability. Changes in the monsoon patterns and increased frequency of extreme weather events further complicate water management and resource planning.
- **Water Pollution**: Pollution from agricultural runoff, industrial discharge, and urban wastewater affects the quality of Himalayan rivers and groundwater. Contaminants such as nitrates, heavy metals, and pathogens pose risks to human health and aquatic ecosystems. Efforts to control pollution and improve wastewater treatment are critical for protecting water resources.
- **Overexploitation of Resources**: The overexploitation of natural resources, including water, minerals, and forest products, threatens the sustainability of the Himalayan environment. Unsustainable agricultural practices, mining activities, and deforestation contribute to resource depletion and environmental degradation.

- **Transboundary Water Management**: The rivers originating in the Himalayas often flow across international borders, creating challenges for transboundary water management. Cooperative agreements and effective management frameworks are necessary to address countries' shared interests and ensure equitable water resource distribution.

The Himalayan region is a vital source of natural resources, with its glaciers, rivers, groundwater, minerals, and forests playing crucial roles in sustaining the environment and supporting the livelihoods of millions of people. However, the region faces significant challenges related to climate change, pollution, overexploitation, and transboundary water management. Addressing these challenges requires a comprehensive approach that includes sustainable resource management, effective conservation strategies, and international cooperation. By taking proactive measures to protect and manage its natural resources, the Himalayan region can continue to provide essential services and support the well-being of its diverse populations for generations to come.

8

Groundwater in Himalayan Region

The Himalayan region, an extraordinary and dynamic mountain range extending across several countries in South Asia, encompasses a variety of environmental and geological features that influence its groundwater resources. Groundwater in this region is a crucial component of the water cycle, supporting millions of people and playing a significant role in agriculture, industry, and ecosystem sustainability. This detailed analysis explores groundwater's characteristics, availability, and challenges in the Himalayan region, examining its sources, hydrological processes, and the impact of environmental and human factors on its sustainability.

Hydrogeological Characteristics of the Himalayan Region

The hydrogeology of the Himalayan region is complex due to its varied topography, geology, and climatic conditions. The region's hydrogeological framework can be broadly categorized into three main zones: the high-altitude mountain areas, the foothill regions, and the river plains.

High-Altitude Mountain Areas: In the high-altitude regions of the Himalayas, groundwater occurs mainly in the form of ice and snow stored in glaciers and snowfields. The groundwater system here is primarily influenced by seasonal snowmelt and glacier runoff. These areas feature limited and localized aquifers, typically formed within fractured rocks and morainal deposits. The recharge rates are relatively low, and groundwater flow is often slow and confined to narrow channels.

- **Glacial Meltwater Contribution**: The glaciers and snowfields in the Himalayas act as significant sources of groundwater recharge. As glaciers melt, they contribute to the formation of seasonal streams and springs, which feed into aquifers at lower elevations. The melting of glaciers has become a crucial water source during the dry season, particularly in regions where surface water availability decreases.
- **Permafrost Influence**: Permafrost can influence groundwater flow and recharge in some high-altitude areas. The presence of permafrost can limit the infiltration of surface water into the ground, affecting the

replenishment of aquifers and potentially leading to a slower rate of groundwater recharge.

Foothill Regions: The foothill regions of the Himalayas, where the mountains transition to the plains, are characterized by more extensive and accessible groundwater resources. In these areas, groundwater occurs in both alluvial and fractured rock aquifers.

- **Alluvial Aquifers**: The foothills feature extensive alluvial deposits formed by river sediments. These alluvial aquifers are typically characterized by high permeability and provide a significant groundwater source for local communities. The recharge of these aquifers is influenced by surface water from rivers and rainfall infiltration.
- **Fractured Rock Aquifers**: Besides alluvial aquifers, the foothill region contains fractured rock aquifers. These aquifers are formed in crystalline and sedimentary rocks, with groundwater stored in fractures and joints. The hydraulic conductivity of these aquifers varies, and groundwater flow is often irregular and influenced by the geological structure.

River Plains: The rivers, such as the Indo-Gangetic Plain, are characterized by extensive alluvial aquifers with high groundwater availability. These aquifers are crucial for agriculture and urban water supply.

- **High Recharge Rates**: The river plains receive substantial recharge from river infiltration, rainfall, and irrigation. The high permeability of the alluvial deposits allows for significant groundwater recharge, making these areas important for agricultural and domestic water use.
- **Aquifer Depletion**: In some river plains, over-extraction of groundwater for irrigation and other uses has led to a decline in groundwater levels. Reduced recharge rates often exacerbate this depletion due to changes in land use and water management practices.

Groundwater Availability and Usage

- Groundwater availability in the Himalayan region varies widely depending on location, geology, and climate. The interplay of recharge rates, aquifer properties, and human activities influences groundwater availability.

Recharge Sources and Rates: Groundwater recharge in the Himalayan region primarily comes from precipitation and surface water infiltration.

- **Precipitation**: Rainfall is a primary source of groundwater recharge in many parts of the Himalayas. In high-altitude areas, precipitation falls as snow, which melts and contributes to groundwater recharge during the spring and summer months. In lower elevations, rainfall infiltrates directly into the ground, replenishing aquifers.

  - **Surface Water Infiltration**: Rivers and streams originating from the Himalayas contribute to groundwater recharge through infiltration. This process is particularly important in the foothill regions and river plains, where surface water interacts with alluvial deposits to replenish groundwater supplies.

- **Groundwater Extraction and Usage**: Groundwater is extensively used in the Himalayan region for various purposes, including agriculture, drinking water supply, and industrial activities.
  - **Agriculture**: In many parts of the Himalayan region, particularly in the foothills and river plains, groundwater is a crucial resource for irrigation. Farmers rely on groundwater to supplement surface water for crop cultivation, especially during dry periods. Using groundwater for irrigation has led to developing wells and boreholes, which provide access to this vital resource.
  - **Drinking Water Supply**: Groundwater is a primary source of drinking water for rural and urban populations in the Himalayan region. Springs, wells, and boreholes are common potable water sources, especially in areas where surface water is less accessible or polluted.
  - **Industrial Use**: In some parts of the Himalayan region, groundwater is used for industrial purposes, including mining, manufacturing, and energy production. The extraction of groundwater for industrial use can impact local aquifers and water availability.

Challenges and Issues

Despite its significance, groundwater in the Himalayan region faces several challenges that affect its availability and sustainability.

- **Climate Change**: Climate change is altering the hydrological cycle in the Himalayan region, affecting groundwater resources in several ways.
  - **Glacier Retreat**: The retreat of glaciers due to rising temperatures impacts the availability of glacial meltwater, a crucial source of groundwater recharge. Reduced glacier mass and melting rates can lead to decreased flow in rivers and streams that feed groundwater aquifers.
  - **Changing Precipitation Patterns**: Altered precipitation patterns, including rainfall intensity and distribution changes, can affect groundwater recharge rates. Increased frequency of extreme weather

events, such as heavy rains and droughts, can lead to variability in groundwater availability.

- **Groundwater Depletion**: Over-extraction of groundwater for agricultural and domestic use is a significant concern in the Himalayan region.
  - **Over-Irrigation**: Excessive groundwater use for irrigation can lead to depletion of aquifer and reduced groundwater levels. This depletion can impact water availability for other uses and affect the sustainability of agricultural practices.
  - **Unregulated Extraction**: In some areas, groundwater extraction is unregulated, leading to overuse and potential depletion of aquifers. The lack of effective management and regulation can exacerbate the depletion of groundwater resources.
- **Pollution and Contamination:** Groundwater pollution is a growing concern in the Himalayan region, affecting water quality and availability.
  - **Agricultural Runoff**: Using fertilizers and pesticides in agriculture can lead to groundwater contamination with nitrates, phosphates, and other chemicals. This pollution can affect water quality and pose risks to human health.
  - **Industrial Discharge**: Industrial activities, including mining and manufacturing, can lead to the discharge of pollutants into groundwater. Contaminants such as heavy metals and chemicals can compromise groundwater quality and require remediation.
  - **Waste Disposal**: Improper disposal of solid and liquid waste can contaminate groundwater. Leachate from landfills and sewage systems can infiltrate aquifers, affecting water quality.
- **Infrastructure and Access**: The mountainous terrain and remote location of many areas in the Himalayas pose challenges for groundwater infrastructure and access.
  - **Well Construction and Maintenance**: The construction and maintenance of wells and boreholes in mountainous and high-altitude regions can be challenging due to geological conditions and limited accessibility. Proper infrastructure is necessary to ensure reliable access to groundwater.
  - **Water Management**: Effective water management practices are essential for the sustainable use of groundwater resources. In some areas, inadequate water management and planning can lead to inefficiencies and conflicts over water use.

Sustainable Management and Conservation

Addressing the challenges associated with groundwater in the Himalayan region requires a comprehensive approach to sustainable management and conservation.

- **Integrated Water Resource Management**: Implementing integrated water resource management (IWRM) strategies can help balance the needs of different water users and ensure the sustainable use of groundwater resources.
    - Data Collection and Monitoring: Accurate data on groundwater levels, quality, and usage is essential for informed decision-making and effective management. Regular monitoring and assessment can help identify trends and potential issues.
    - Stakeholder Involvement: Engaging local communities, government agencies, and other stakeholders in groundwater management can promote cooperation and ensure that the needs and concerns of all users are addressed.
- **Pollution Prevention and Remediation**: Preventing and mitigating groundwater pollution is crucial for protecting water quality and availability.
    - **Regulation and Enforcement**: Implementing and enforcing regulations related to agricultural practices, industrial discharges, and waste disposal can help reduce groundwater pollution.
    - **Remediation Technologies**: Technologies such as bioremediation, pump-and-treat systems, and natural attenuation can be used to address contaminated groundwater and restore water quality.
- **Climate Adaptation Strategies**: Adapting to the impacts of climate change on groundwater resources involves implementing strategies to manage variability and ensure resilience.
    - **Climate-Resilient Infrastructure**: Designing and implementing infrastructure that can withstand the impacts of climate change, such as extreme weather events and changing precipitation patterns, can enhance groundwater sustainability.
    - **Water Conservation Practices**: Promoting water conservation practices, such as efficient irrigation techniques and water-saving technologies, can help reduce groundwater demand and improve resource management.

- **Community Engagement and Education**: Educating communities about the importance of groundwater conservation and involving them in management efforts can promote sustainable practices and stewardship.
    - **Awareness Campaigns**: Conducting awareness campaigns and educational programs can help raise awareness about groundwater issues and encourage responsible water use.
    - **Capacity Building**: Providing training and resources to local communities and water managers can enhance their ability to manage groundwater resources effectively.

9

Current Status of Freshwater Monitoring and Assessment

Freshwater is vital for human survival, economic development, and ecosystem health. It comprises only about 2.5% of the Earth's total water resources, and of this, less than 1% is readily accessible for human use in rivers, lakes, and groundwater. As we progress into 2024, freshwater resources face unprecedented stress due to population growth, industrialization, agricultural expansion, pollution, and the impacts of climate change. The sustainable management of freshwater is critical, and it is central to achieving the United Nations Sustainable Development Goals (SDGs), particularly SDG 6, which calls for clean water and sanitation for all. This detailed scientific discussion will examine the current status of freshwater in 2024, encompassing availability, distribution, quality, climate change impacts, monitoring and assessment techniques, and strategies for sustainable conservation.

Availability and Distribution of Freshwater

The distribution of freshwater resources is highly uneven globally, with significant disparities between regions. In 2024, approximately 2.2 billion people continue to live in regions that experience high water stress. This stress primarily results from population density, climate conditions, and economic activities that increase water demand.

Sub-Saharan Africa, the Middle East, and parts of South Asia continue to experience acute freshwater shortages. For example, countries in the Middle East and North Africa (MENA) region rely heavily on non-traditional freshwater sources, such as desalination and wastewater reuse, due to limited natural water availability. In contrast, areas such as the Amazon Basin and North America have abundant freshwater resources. However, even in water-rich regions, localized water scarcity may occur due to over-extraction and pollution.

Water scarcity is classified as either physical or economic. Physical water scarcity refers to insufficient water resources to meet demand, while economic water scarcity refers to the inability to access water due to infrastructure or financial constraints. Sub-Saharan Africa, in particular, continues to suffer from economic water scarcity, where inadequate infrastructure and governance limit access to freshwater supplies. Only 58% of the population in this region can access safely managed drinking water.

Monitoring and Assessment of Freshwater Resources

Freshwater is critical for human health, ecosystems, and economic development. With increasing pressures from population growth, industrialization, agricultural intensification, and climate change, the need for effective monitoring and assessment of freshwater resources is more pressing than ever. Monitoring freshwater systems allows governments, scientists, and water managers to track quantity, quality, and distribution changes, enabling timely interventions to prevent water scarcity, pollution, and ecosystem degradation. In 2024, technological advancements, data integration, and monitoring practices offer new opportunities to improve freshwater resource management. However, challenges remain, particularly in regions facing resource constraints or transboundary water conflicts. This comprehensive review discusses the status of freshwater monitoring and assessment, highlighting the scientific techniques, challenges, and solutions that support sustainable water management in 2024.

Importance of Freshwater Monitoring and Assessment

Freshwater systems are highly dynamic and vulnerable to various natural and anthropogenic stressors, making consistent monitoring and assessment essential for effective management. Monitoring is necessary to detect changes in water quantity, assess pollution levels, track biodiversity loss, and manage shared water resources. It also plays a central role in the United Nations Sustainable Development Goals (SDGs), particularly SDG 6, which aims to ensure the availability and sustainable management of water and sanitation for all by 2030.

The key goals of freshwater monitoring and assessment include:

- **Assessing water quantity**: Monitoring River flows, lake levels, and groundwater to track changes in water availability.
- **Measuring water quality**: Identifying pollutants such as nutrients, chemicals, and pathogens that degrade water systems.
- **Protecting biodiversity**: Monitoring ecosystems to assess the health of aquatic life and detect changes due to pollution, habitat loss, and invasive species.
- **Supporting transboundary management**: Tracking shared water resources to prevent conflicts and ensure equitable distribution among riparian nations.

Monitoring freshwater resources requires a combination of scientific methodologies, ranging from traditional in-situ measurements to advanced satellite technologies and biological assessments.

Freshwater Quantity Monitoring

River Flow and Surface Water Monitoring

Rivers and lakes represent the most visible components of freshwater systems, and monitoring their flow and levels is essential for managing water availability. In 2024, river gauging stations remain the primary tools for monitoring river discharge. These stations measure water levels and flow velocities at various points along river systems, providing real-time data on water availability for drinking, agriculture, and industrial use. However, many countries, especially developing regions, face gaps in river gauging networks due to a lack of resources and infrastructure.

Satellite-based remote sensing technologies are increasingly used to supplement ground-based measurements to monitor surface water. For example, NASA's Surface Water and Ocean Topography (SWOT) satellite, launched in late 2022, is a game-changer for freshwater monitoring in 2024. SWOT provides high-resolution data on the spatial distribution of surface water, allowing scientists to track changes in lake levels, river discharge, and wetland areas globally. This technology is especially valuable in remote areas where ground-based monitoring is limited.

Groundwater Monitoring

Groundwater represents a critical source of freshwater for drinking, irrigation, and industrial purposes, especially in arid and semi-arid regions. However, monitoring groundwater is more challenging than surface water due to the complexity of underground aquifers and the lack of direct visibility. Groundwater over-extraction and depletion remain pressing in regions such as the Middle East, North Africa, South Asia, and parts of the United States.

Traditional groundwater monitoring relies on observation wells, which measure changes in water table levels. These measurements provide localized data but may not capture the broader trends across large aquifer systems. In response to this challenge, scientists in 2024 increasingly rely on satellite-based gravity measurements to monitor groundwater depletion. The Gravity Recovery and Climate Experiment Follow-On (GRACE-FO) mission, launched by NASA and the German Aerospace Center, has proven instrumental in assessing changes in groundwater storage on a global scale. GRACE-FO measures variations in the Earth's gravity field caused by changes in water mass, enabling researchers to detect groundwater depletion over large areas, even in regions with minimal monitoring infrastructure.

Glacier and Snowpack Monitoring

Glaciers and snowpacks act as natural freshwater reservoirs, storing water during winter months and releasing it during the summer. In 2024, many regions, such as the Himalayas, Andes, and western North America, rely on seasonal snowmelt for water supplies. However, rising global temperatures are accelerating the melting of glaciers and reducing snowpack levels, threatening the long-term availability of freshwater.

Scientists use ground-based measurements, satellite imagery, and climate models to monitor these changes. Remote sensing technologies, including radar and optical satellite imagery, allow researchers to monitor the extent and volume of glaciers and snowpacks. The European Space Agency's Sentinel-1 and Sentinel-2 satellites provide detailed imagery that helps track glacier retreat and snowpack distribution. Climate models are also used to predict future changes in water availability based on projected temperature increases and precipitation patterns.

Freshwater Quality Monitoring

Water quality monitoring is critical for protecting human health, supporting agriculture, and maintaining aquatic ecosystems. Due to agricultural runoff, industrial discharges, urbanization, and climate change, freshwater pollution is a growing concern in many regions. By 2024, several advanced technologies and methodologies are used to monitor and assess water quality across different environments.

Traditional In-Situ Water Quality Monitoring

Conventional water quality monitoring relies on collecting samples from rivers, lakes, and groundwater and analyzing them in laboratories. Parameters typically measured include dissolved oxygen, pH, conductivity, temperature, nutrient levels (such as nitrogen and phosphorus), heavy metals, pesticides, and microbial contaminants. In-situ sensors have become more sophisticated, enabling continuous real-time monitoring of water quality parameters.

Automated water quality monitoring stations, equipped with sensors for real-time data collection, are widely used in 2024. These stations are deployed in critical water bodies and provide continuous data on parameters like turbidity, oxygen levels, and temperature. This real-time monitoring helps authorities respond quickly to pollution events like chemical spills or algal blooms.

Nutrient and Pesticide Monitoring

Nutrient pollution, particularly nitrogen and phosphorus from agricultural runoff, is one of the leading causes of water quality degradation. Excessive nutrient levels lead to eutrophication, which causes harmful algal blooms, oxygen depletion, and the loss of aquatic biodiversity. Monitoring nutrient levels in freshwater systems is essential to assess the impacts of agricultural practices and guide policies for reducing pollution.

In 2024, advances in spectroscopy and chemical sensors allow for more accurate and timely detection of nutrient levels in water bodies. Spectroscopic sensors, which use light to measure chemical concentrations, are increasingly deployed in monitoring stations to continuously track nutrient loads in rivers and lakes. Additionally, remote sensing technologies, such as NASA's Landsat and the European Space Agency's Sentinel-2, monitor algal blooms and assess water quality over large spatial scales.

Emerging Contaminants: Pharmaceuticals, Microplastics, and eDNA

Emerging contaminants, including pharmaceuticals, personal care products, and microplastics, are becoming a major concern for freshwater quality. These pollutants, often not removed by conventional wastewater treatment processes, can have long-term impacts on aquatic ecosystems and human health.

Pharmaceutical residues, such as antibiotics and hormones, are increasingly detected in rivers and lakes, where they can disrupt aquatic life and contribute to the development of antibiotic-resistant bacteria. In 2024, monitoring techniques for these contaminants have improved, with mass spectrometry and high-performance liquid chromatography being used to detect trace amounts of pharmaceuticals in water samples.

Microplastics are another emerging contaminant that poses a significant threat to freshwater ecosystems. These tiny plastic particles, less than 5 millimeters in size, are found in virtually all freshwater systems worldwide. Their impacts on aquatic organisms, including ingestion and toxicity, are areas of active research. Techniques for monitoring microplastics in 2024 involve a combination of field sampling and laboratory analysis using microscopy and spectroscopy.

Environmental DNA (eDNA) is a groundbreaking tool used to monitor biodiversity and detect the presence of specific organisms, including invasive species, in freshwater environments. By analyzing DNA fragments found in water samples, researchers can identify species without physically capturing them. In 2024, eDNA is widely used to monitor the health of freshwater ecosystems and track the spread of invasive species, such as zebra mussels and Asian carp.

Advances in Freshwater Monitoring Technologies

The field of freshwater monitoring has benefited from significant technological advancements over the past decade. In 2024, these technologies allow for more comprehensive, real-time, and cost-effective monitoring of freshwater resources.

Remote Sensing and Satellite Monitoring

Remote sensing technologies have transformed freshwater monitoring by providing large-scale, high-resolution data on surface water, snowpack, and glacier extent. In 2024, satellites with radar, optical, and thermal sensors provide continuous global coverage of freshwater resources. One of the most important remote sensing missions for freshwater monitoring is the SWOT satellite, which provides detailed data on the elevation and area of surface water bodies. This data is critical for tracking changes in river discharge, lake levels, and wetland extent, especially in regions with limited ground-based monitoring.

Internet of Things (IoT) and Smart Sensors

The Internet of Things (IoT) has revolutionized real-time water quality monitoring by enabling the deployment of smart sensors in freshwater systems. These sensors, connected via wireless networks, continuously measure water quality parameters such as temperature, dissolved oxygen, turbidity, and conductivity. The data collected by these sensors is transmitted to cloud-based platforms, where it is analyzed and visualized in real-time.

In 2024, IoT-based monitoring networks are widely used in developed and developing regions, providing real-time data for water managers and policymakers. These networks enable early detection of pollution events, such as chemical spills or sewage discharges, allowing for rapid response and mitigation.

Artificial Intelligence and Machine Learning

Artificial intelligence (AI) and machine learning (ML) are increasingly used in freshwater monitoring to analyze large datasets and predict future water quality and quantity trends. These technologies allow scientists to identify patterns in historical data and develop predictive models that can forecast changes in freshwater systems based on climate change, land use, and population growth.

In 2024, AI and ML are being used to predict the occurrence of harmful algal blooms, assess the impacts of nutrient pollution, and model the impacts of climate change on water availability. These predictive models help water managers make informed decisions and plan for future water needs.

Challenges in Freshwater Monitoring and Assessment

Despite the advances in freshwater monitoring technologies, significant challenges remain in 2024, particularly in developing regions.

Data Gaps and Inconsistent Monitoring Networks

Many regions, especially in developing countries, lack the infrastructure and resources to maintain comprehensive freshwater monitoring networks. Data gaps in river flow measurements, groundwater levels, and water quality parameters limit the ability to assess the status of freshwater resources and make informed management decisions. Expanding monitoring networks and ensuring data consistency across regions are key challenges for global water management.

Transboundary Water Monitoring

Many of the world's freshwater resources are shared between countries, making transboundary water monitoring a complex and politically sensitive issue. Rivers such as the Nile, Mekong, and Danube cross multiple national borders, and monitoring these transboundary waters requires cooperation between riparian nations. In 2024, many transboundary monitoring agreements are in place, but tensions remain in regions where water scarcity and political conflicts intersect.

Climate Change and Extreme Events

Climate change adds complexity to freshwater monitoring by increasing the frequency and intensity of extreme weather events, such as floods and droughts. Monitoring these events in real-time and predicting their impacts on freshwater systems are critical challenges for water managers in 2024. Climate models and remote sensing technologies are being used to predict and monitor the impacts of extreme events, but the uncertainty surrounding future climate scenarios makes long-term planning difficult.

The monitoring and assessment of freshwater resources have become more sophisticated and comprehensive in 2024, thanks to advances in remote sensing, IoT, AI, and environmental DNA technologies. These innovations provide valuable data for managing water resources, protecting ecosystems, and ensuring sustainable development. However, challenges remain, including data gaps in developing regions, the complexity of transboundary water management, and the impacts of climate change. Addressing these challenges requires continued investment in monitoring infrastructure, international cooperation, and the integration of cutting-edge technologies into water management practices. By improving freshwater monitoring and assessment, the global community can work toward achieving the SDGs and ensuring the availability of clean water for future generations.

10

Sustainable Conservation

Sustainable conservation is a multifaceted approach aimed at preserving natural resources and ecosystems while meeting the needs of present and future generations. In 2024, sustainable conservation encompasses a broad spectrum of strategies, technologies, and policies to balance human development with ecological preservation. This detailed examination delves into the principles of sustainable conservation, exploring its core components, challenges, advancements, and the role it plays in achieving the United Nations Sustainable Development Goals (SDGs), particularly those related to biodiversity, climate action, and responsible consumption.

Principles of Sustainable Conservation

Sustainable conservation is grounded in several core principles that guide its implementation and effectiveness:

- **Ecological Integrity:** At the heart of sustainable conservation is maintaining ecological integrity. This involves protecting ecosystems' ability to function effectively and sustain biodiversity. Ecosystems provide essential services such as nutrient cycling, water purification, and climate regulation. To uphold ecological integrity, conservation efforts focus on preserving natural habitats, restoring degraded areas, and maintaining ecological processes. This principle ensures that conservation strategies protect individual species and maintain the health and resilience of entire ecosystems.
- **Biodiversity Preservation:** Biodiversity, the variety of life forms on Earth, is crucial for ecosystem stability and resilience. Sustainable conservation aims to protect the full spectrum of genetic, species, and ecosystem diversity. This involves safeguarding endangered species, preserving genetic diversity within species, and maintaining diverse habitats. Biodiversity preservation is not only about protecting charismatic megafauna but also involves conserving less visible organisms that play critical roles in ecosystem functioning, such as pollinators, soil microbes, and aquatic invertebrates.

- **Sustainable Resource Use:** Sustainable conservation requires that natural resources be used in ways that do not deplete or harm the environment. This principle applies to various resources, including water, land, forests, and minerals. Sustainable resource use involves implementing practices that minimize environmental impacts, such as sustainable forestry, responsible fishing, and sustainable agriculture. It also promotes circular economy principles, which aim to reduce waste and increase resource efficiency by reusing and recycling materials.
- **Socioeconomic Considerations:** Integrating socioeconomic considerations into conservation efforts is crucial for achieving sustainability. This principle recognizes that human well-being and environmental health are interconnected. Sustainable conservation seeks to balance ecological goals with the needs and aspirations of local communities. This includes promoting equitable access to resources, supporting livelihoods through sustainable practices, and ensuring conservation initiatives do not disproportionately impact marginalized groups. By involving local communities in decision-making and providing incentives for conservation, sustainable conservation can achieve better outcomes for both people and nature.

Strategies for Sustainable Conservation

Sustainable conservation employs various strategies to achieve its goals, from protected areas and restoration projects to sustainable management practices and policy frameworks.

- **Protected Areas:** Protected areas are a cornerstone of conservation strategies. These areas are designated to safeguard ecosystems, species, and natural processes from human exploitation and degradation. By 2024, protected areas cover approximately 15% of the Earth's terrestrial surface and 7% of the marine environment. These areas include national parks, nature reserves, and marine protected areas (MPAs).

  Protected areas play a vital role in preserving biodiversity and providing ecosystem services. They serve as refuges for endangered species, protect critical habitats, and maintain ecological processes. However, the effectiveness of protected areas depends on their design, management, and enforcement. Ensuring adequate funding, effective governance, and addressing threats such as poaching and habitat fragmentation are essential for maintaining the integrity of protected areas.
- **Ecosystem Restoration:** Ecosystem restoration involves actively restoring degraded ecosystems to their natural state or improving their functionality. Restoration projects aim to rehabilitate ecosystems that

have damaged by human activities, such as deforestation, mining, and pollution. By 2024, ecosystem restoration has become a key strategy for enhancing biodiversity, improving ecosystem services, and mitigating climate change.

Restoration efforts can include reforestation, wetland restoration, and riparian zone rehabilitation. For example, reforestation projects in tropical regions aim to restore lost forest cover, enhance carbon sequestration, and provide habitat for wildlife. Similarly, wetland restoration projects focus on reestablishing the ecological functions of wetlands, such as flood regulation and water purification.

- **Sustainable Management Practices:** Sustainable management practices are designed to balance human resource use with conservation objectives. These practices are applied to various sectors, including agriculture, forestry, fisheries, and water management. By 2024, sustainable management practices are increasingly being adopted to reduce environmental impacts and promote resource efficiency.
- **Sustainable Agriculture:** Sustainable agriculture aims to produce food in ways that minimize environmental harm and promote soil health, water conservation, and biodiversity. Techniques such as agroecology, conservation tillage, and organic farming are employed to enhance the sustainability of agricultural systems. Agroecology, for example, integrates ecological principles into farming practices to improve resilience and reduce reliance on synthetic inputs.
- **Sustainable Forestry:** Sustainable forestry practices focus on managing forests to meet current needs while ensuring that future generations can benefit from forest resources. These practices include selective, reduced-impact, and the certification of sustainably managed forests. Forest certification programs, such as those offered by the Forest Stewardship Council (FSC), assure that forest management practices meet rigorous environmental and social standards.
- **Sustainable Fisheries:** Sustainable fisheries management seeks to prevent overfishing, protect marine ecosystems, and ensure the long-term viability of fish stocks. Techniques such as quota systems, marine protected areas, and bycatch reduction measures are employed to achieve fishery sustainability. Adopting ecosystem-based fisheries management (EBFM) approaches, considering the interactions between species and their environments, is increasingly recognized as a best practice.
- **Water Management:** Sustainable water management involves optimizing the use of water resources to meet human needs while

preserving ecological health. This includes implementing practices such as water-efficient irrigation, reducing water waste, and protecting watershed areas. Integrated water resources management (IWRM) approaches, which promote the coordinated management of water, land, and related resources, are increasingly used to achieve sustainable outcomes.

- **Policy Frameworks and International Agreements:** Effective conservation requires robust policy frameworks and international cooperation. Policies and agreements provide the necessary governance structures, regulations, and incentives to support conservation efforts and ensure compliance.
- **National Policies:** Many countries have developed national policies and strategies to guide conservation efforts. These policies may include legislation related to protected areas, wildlife conservation, and environmental impact assessments. National biodiversity action plans (NBAPs) outline strategies for preserving biodiversity and are often aligned with international commitments.
- **International Agreements:** International agreements are crucial in addressing global conservation challenges and fostering cooperation between countries. Key agreements include:
    - **The Convention on Biological Diversity (CBD)**: The CBD aims to conserve biodiversity, promote sustainable use of natural resources, and ensure fair and equitable sharing of benefits from genetic resources. The post-2020 global biodiversity framework, adopted in December 2022, sets out targets for biodiversity conservation and sustainable use up to 2030.
    - **The Paris Agreement**: The Paris Agreement focuses on climate change mitigation and adaptation, with implications for conservation efforts. By addressing climate change, the agreement helps to protect ecosystems and species from the impacts of global warming.
    - **The Ramsar Convention on Wetlands**: The Ramsar Convention promotes the conservation and sustainable use of wetlands, recognizing their importance for biodiversity, water regulation, and climate mitigation.
    - **The Convention on International Trade in Endangered Species of Wild Fauna and Flora (CITES)**: CITES regulates international trade in endangered species to prevent exploitation and ensure survival.

Advancements in Conservation Science and Technology

Advancements in science and technology have significantly enhanced conservation efforts by providing new monitoring, analysis, and management tools.

- **Remote Sensing and Geographic Information Systems (GIS):** Remote sensing and GIS technologies have revolutionized conservation by providing detailed spatial data on land use, habitat loss, and environmental changes. Remote sensing satellites, such as those from NASA and the European Space Agency, offer high-resolution imagery for monitoring deforestation, land degradation, and changes in biodiversity.

  GIS enables the analysis and visualization of spatial data, supporting conservation planning and decision-making. For example, GIS can be used to map critical habitats, assess the connectivity of protected areas, and evaluate the impacts of human activities on ecosystems.
- **Environmental DNA (eDNA):** Environmental DNA (eDNA) technology allows scientists to detect and monitor species by analyzing DNA fragments found in environmental samples, such as water, soil, and air. eDNA is a non-invasive and cost-effective method for monitoring biodiversity, detecting rare or elusive species, and tracking the spread of invasive species. In 2024, eDNA is increasingly used in freshwater and marine ecosystems to assess species distributions, monitor ecosystem health, and support conservation efforts. This technology provides valuable data for informing management decisions and prioritizing conservation actions.
- **Conservation Genomics:** Conservation genomics involves using genetic tools and techniques to study and protect biodiversity. By analyzing species' genetic diversity, scientists can assess their vulnerability to extinction, identify genetically distinct populations, and develop strategies for conservation. Conservation genomics is used to monitor the genetic health of endangered species, track population dynamics, and inform breeding programs. For example, genomic studies of the critically endangered Amur tiger have provided insights into its genetic diversity and guided conservation efforts.

Challenges in Sustainable Conservation

Despite significant progress, sustainable conservation faces several challenges that require ongoing attention and innovation.

- **Climate Change Impacts:** Climate change poses a major threat to ecosystems and biodiversity, with impacts such as habitat loss, species range shifts, and altered ecological processes. Conserving ecosystems and species in the face of climate change requires adaptive management strategies that account for changing conditions. This includes identifying climate refugia, protecting migration corridors, and implementing climate-smart conservation practices.
- **Human-Wildlife Conflicts:** Human-wildlife conflicts arise when wildlife causes damage to crops, livestock, or property, leading to negative interactions between humans and animals. Managing these conflicts is essential for achieving conservation goals and ensuring the coexistence of people and wildlife. Strategies include community-based approaches, compensation schemes, and non-lethal methods of deterring wildlife.
- **Funding and Resource Constraints:** Sustainable conservation efforts often face financial and resource constraints, limiting the scope and effectiveness of initiatives. Securing adequate funding for conservation projects, particularly in developing regions, remains a challenge. Innovative financing mechanisms, such as conservation trust funds, public-private partnerships, and eco-tourism revenues, are being explored to address funding gaps and ensure the sustainability of conservation efforts.
- **Policy and Governance Issues:** Effective conservation requires robust policies and governance structures. Weak governance, inadequate enforcement, and conflicting interests can undermine conservation efforts. Strengthening institutional capacity, promoting transparency, and fostering stakeholder participation are crucial for improving governance and achieving conservation goals.

Sustainable conservation is a dynamic and evolving field that encompasses a range of strategies, technologies, and policies aimed at preserving natural resources and ecosystems while meeting human needs. In 2024, advancements in remote sensing, environmental DNA, and conservation genomics provide valuable tools for monitoring and managing biodiversity. However, challenges such as climate change, human-wildlife conflicts, and funding constraints continue to impact conservation efforts. Addressing these challenges requires a holistic approach that integrates ecological, socioeconomic, and policy considerations. By advancing sustainable conservation practices and fostering international cooperation, the global community can work towards achieving the SDGs and ensuring a healthy and resilient planet for future generations.

Freshwater Conservation and SDG Goals: An In-depth Scientific Analysis

Freshwater conservation is pivotal to achieving several Sustainable Development Goals (SDGs), particularly SDG 6 (Clean Water and Sanitation), but its impact resonates across a broader spectrum of global sustainability objectives. As of 2024, the urgent need for freshwater conservation is underscored by the fact that over 2 billion people worldwide still lack access to safely managed drinking water services, a stark reminder of the disparities in water access and the critical importance of preserving and managing this finite resource. Freshwater conservation involves maintaining the quality and availability of freshwater resources, including rivers, lakes, wetlands, and aquifers, essential for supporting human populations and ecosystems.

The hydrological cycle forms the foundation of our understanding of freshwater resources. This cycle, driven by solar energy, encompasses the processes of precipitation, evaporation, infiltration, and runoff. Precipitation replenishes surface water bodies and groundwater aquifers, while evaporation and transpiration return water to the atmosphere. Human activities, such as urbanization, deforestation, and industrialization, disrupt these natural processes by altering land surfaces, increasing runoff, and decreasing groundwater recharge. These disruptions can lead to significant water availability and quality alterations, highlighting the need for comprehensive water management strategies to mitigate these impacts.

Freshwater ecosystems, which include rivers, lakes, wetlands, and groundwater systems, are among the most biodiverse habitats on the planet. They provide critical services such as water purification, nutrient cycling, and flood regulation. However, these ecosystems are increasingly threatened by pollution, habitat destruction, and climate change. For instance, wetlands, which play a crucial role in filtering pollutants and regulating water flow, are being drained for agriculture and development, resulting in biodiversity loss and decreased water quality. Protecting these ecosystems is essential for maintaining biodiversity and preserving the ecosystem services they provide, which directly support human well-being and economic activities.

Pollution is a major threat to freshwater resources, with contaminants originating from various sources including agricultural runoff, industrial discharges, and untreated sewage. Nutrient pollution, driven by excess nitrogen and phosphorus from fertilizers, leads to eutrophication, which causes harmful algal blooms and oxygen depletion in water bodies, negatively impacting aquatic life and water quality. Emerging contaminants, such as pharmaceuticals and microplastics, pose additional challenges due to their persistence and potential to disrupt ecosystems and human health. Addressing water pollution requires a multifaceted approach, including improved wastewater treatment technologies, stricter regulations on industrial discharges, and sustainable agricultural practices to reduce nutrient runoff.

Climate change exacerbates freshwater challenges by altering precipitation patterns, increasing the frequency and intensity of extreme weather events, and affecting snowmelt and glacier runoff. Changes in precipitation can lead to more frequent and severe floods and droughts, which impact water availability and quality. For example, the melting of glaciers, critical sources of freshwater for many regions, affects the timing and volume of river flows, impacting water supply and agriculture. Adaptation strategies, such as enhancing water storage infrastructure, improving water use efficiency, and integrating climate projections into water management plans, are essential to building resilience against the impacts of climate change.

Effective water governance and management are crucial for ensuring sustainable freshwater conservation. Integrated Water Resources Management (IWRM) is an approach that promotes the coordinated development and management of water, land, and related resources. It emphasizes the need for stakeholder participation, equitable access to water resources, and the consideration of environmental, social, and economic factors in decision-making. Policies and regulations supporting water conservation, promoting water reuse, and incentivizing sustainable practices are vital to effective water governance. Moreover, recognizing the role of local communities and incorporating indigenous knowledge into water management practices can enhance the effectiveness of conservation efforts and promote social equity.

Technological innovations play a significant role in advancing freshwater conservation. Remote sensing technologies, including satellite imagery and geographic information systems (GIS), enable the monitoring and managing water resources with greater precision. These technologies help assess water availability, track pollution sources, and manage water distribution. Advances in water purification technologies, such as membrane filtration and advanced oxidation processes, improve the safety and availability of clean water. Smart water management systems that use real-time data and automation to optimize water distribution and reduce wastage also contribute to more efficient water use. Investing in research and developing new technologies is critical for addressing emerging challenges in freshwater conservation.

The socioeconomic implications of freshwater conservation are profound. Access to clean water is directly linked to health outcomes, economic development, and social equity. Inadequate access to safe drinking water and sanitation disproportionately affects marginalized communities, exacerbating existing inequalities. Ensuring equitable access to water resources requires addressing supply and demand issues, improving infrastructure, and implementing policies promoting fairness and inclusivity. Community-based approaches that involve local stakeholders in water management and decision-making processes can enhance the effectiveness of conservation initiatives and foster greater social cohesion.

Looking to the future, research in freshwater conservation must address several key areas. Understanding the impacts of emerging contaminants and climate change on water resources requires continued investigation and adaptation. The development of innovative conservation strategies, such as nature-based solutions that leverage natural processes for water management, holds promise for enhancing sustainability. Interdisciplinary research integrating hydrology, ecology, social sciences, and technology is essential for addressing the complex challenges of freshwater conservation and achieving the SDGs.

Freshwater Conservation and Its Interconnection with the Sustainable Development Goals

Freshwater conservation is fundamental in achieving several of the United Nations Sustainable Development Goals (SDGs), integral to sustaining both human and ecological systems. The need for effective freshwater management and conservation has never been more critical, as evidenced by current global water scarcity, pollution, and ecosystem degradation. The following detailed analysis explores how freshwater conservation intersects with each SDG, highlighting the multifaceted role of water in global sustainability.

SDG 1: No Poverty

Freshwater Conservation and Poverty Reduction

Access to clean water is a fundamental human right and a crucial factor in poverty alleviation. In many developing regions, inadequate access to safe drinking water and sanitation facilities perpetuates cycles of poverty. Water scarcity can undermine economic opportunities by limiting agricultural productivity and affecting health, exacerbating poverty. Freshwater conservation is pivotal in ensuring reliable water supplies that support livelihoods, agriculture, and economic development. Efficient water management and infrastructure improvements can enhance water availability and quality, vital for economic growth and poverty reduction. Addressing water-related issues through conservation efforts improves living conditions and fosters economic opportunities for impoverished communities.

SDG 2: Zero Hunger

Water for Agriculture and Food Security

Agriculture heavily depends on freshwater resources, and sustainable water management is crucial for ensuring food security. Freshwater conservation impacts agricultural productivity by maintaining water availability for irrigation and supporting healthy ecosystems contributing to crop resilience. Water scarcity can lead to reduced crop yields, increased food prices, and diminished food security. Implementing water-efficient agricultural practices, such as drip irrigation and rainwater harvesting, alongside conservation efforts, is essential for optimizing water use in agriculture and ensuring food security. Additionally, preserving freshwater ecosystems, such as wetlands, can enhance soil fertility and support sustainable agricultural practices, thereby contributing to the goal of zero hunger.

SDG 3: Good Health and Well-Being

Water Quality and Public Health

The quality of freshwater directly affects human health. Contaminated water sources can spread waterborne diseases such as cholera, dysentery, and hepatitis. Freshwater conservation efforts focusing on pollution control, wastewater treatment, and maintaining healthy aquatic ecosystems are critical for safeguarding public health. Ensuring access to clean and safe water reduces the incidence of water-related diseases and improves overall well-being. Furthermore, protecting freshwater ecosystems supports natural water purification processes, essential for maintaining high water quality and promoting good health.

SDG 4: Quality Education

Educational Opportunities Through Water Access

Access to clean water is indirectly linked to educational outcomes. In many areas, the burden of collecting water falls on children, particularly girls, which can impede their ability to attend school. Improved freshwater infrastructure and conservation efforts can reduce the time and effort required to obtain clean water, allowing children to focus on their education. Additionally, integrating water-related topics into educational curricula can raise awareness about the importance of water conservation and foster a sense of responsibility among students. By addressing water scarcity and improving access to clean water, conservation efforts can create more equitable educational opportunities and support quality education.

SDG 5: Gender Equality

Empowerment Through Water Access

Water scarcity and poor sanitation disproportionately affect women and girls, who are often responsible for managing household water needs. This can limit their opportunities for education and economic participation. Freshwater conservation that improves water access and reduces the burden of water collection can empower women and girls by freeing up time for educational and economic activities. Furthermore, involving women in water management decision-making processes can enhance the effectiveness of conservation efforts and promote gender equality. Addressing water-related challenges through conservation initiatives supports gender equity by improving access to resources and opportunities for women and girls.

SDG 6: Clean Water and Sanitation

Direct Relationship with Freshwater Conservation

SDG 6 focuses explicitly on ensuring availability and sustainable management of water and sanitation for all. Freshwater conservation is intrinsically linked to this goal, as it involves preserving water quality, managing water resources efficiently, and providing safe sanitation facilities. Effective conservation practices, such as reducing pollution, protecting watersheds, and promoting water reuse, are essential for achieving clean water and sanitation. Additionally, integrating sustainable water management practices into policies and infrastructure development is crucial for meeting the targets of SDG 6. By addressing water scarcity, improving sanitation, and protecting aquatic ecosystems, freshwater conservation directly contributes to realizing this goal.

SDG 7: Affordable and Clean Energy

Water-Energy Nexus

The relationship between water and energy is known as the water-energy nexus. Freshwater is used in various energy production processes, including hydropower, cooling in thermal power plants, and bioenergy production. Conversely, energy is required for water extraction, treatment, and distribution. Freshwater conservation efforts focusing on sustainable water management can reduce the strain on water resources used in energy production, while energy-efficient technologies can minimize water use. Ensuring the sustainability of both water and energy systems is critical for achieving affordable and clean energy. By promoting efficient water use and exploring renewable energy sources that minimize water impact, conservation efforts contribute to the objectives of SDG 7.

SDG 8: Decent Work and Economic Growth

Economic Implications of Water Management

Freshwater resources are essential for various economic activities, including agriculture, industry, and tourism. Sustainable water management and conservation can support economic growth by ensuring the reliability of water supplies and reducing the costs associated with water scarcity and pollution. Moreover, water-related infrastructure projects can create job opportunities and stimulate economic development. Effective freshwater conservation can lead to more stable and resilient economies by protecting critical water resources for productive sectors and mitigating the economic risks associated with water-related challenges.

SDG 9: Industry, Innovation, and Infrastructure

Water Management in Industrial Processes

Industries often require significant amounts of water for their operations, and efficient water use is crucial for sustainable industrial development. Freshwater conservation efforts that focus on improving water use efficiency and reducing wastewater can support the development of sustainable industrial practices. Innovations in water treatment technologies and adopting water-saving practices can enhance industrial processes and infrastructure. By integrating sustainable water management into industrial operations and infrastructure development, conservation efforts contribute to the goals of SDG 9 by promoting innovation and ensuring the sustainability of industrial activities.

SDG 10: Reduced Inequalities

Equitable Access to Water Resources

Inequities in access to clean water and sanitation disproportionately affect marginalized and disadvantaged communities. Freshwater conservation initiatives prioritizing equitable access to water resources can help reduce these inequalities by addressing water availability and quality disparities. Ensuring all communities, particularly those in underserved areas, have access to safe and reliable water services is essential for promoting social inclusion and reducing inequalities. Conservation efforts that focus on improving water access and addressing systemic disparities contribute to achieving the objectives of SDG 10.

SDG 11: Sustainable Cities and Communities

Urban Water Management and Sustainability

Rapid urbanization places significant pressure on freshwater resources and infrastructure. Sustainable water management practices are essential for developing resilient and sustainable cities and communities. Freshwater conservation efforts that focus on managing urban water supplies, reducing pollution, and enhancing water infrastructure contribute to the sustainability of urban areas. Implementing green infrastructure solutions, such as rain gardens and permeable pavements, can mitigate the impacts of urbanization on water resources and improve the resilience of cities to water-related challenges.

SDG 12: Responsible Consumption and Production

Water Efficiency and Resource Management

Responsible consumption and production practices are closely linked to water use. Efficient water management and conservation are essential for minimizing the environmental impact of production processes and ensuring the sustainability of water resources. Reducing water waste, promoting water reuse, and adopting sustainable production practices contribute to the responsible use of water resources. By integrating water efficiency into consumption and production patterns, conservation efforts support the objectives of SDG 12 and contribute to more sustainable resource management.

SDG 13: Climate Action

Climate Change and Water Resources

Climate change has significant implications for freshwater resources, including altered precipitation patterns, increased frequency of extreme weather events, and impacts on snow and glacier melt. Freshwater conservation efforts that address climate change impacts, such as enhancing water storage and implementing adaptive management strategies, are critical for building resilience to climate-related challenges. Integrating climate considerations into water management practices and conservation efforts supports the goals of SDG 13 by mitigating the impacts of climate change and promoting adaptive strategies.

SDG 14: Life Below Water

Protection of Aquatic Ecosystems

Freshwater ecosystems, including rivers, lakes, and wetlands, support aquatic biodiversity and provide essential ecosystem services. Conservation efforts that protect these ecosystems and address pollution are crucial for maintaining healthy aquatic environments. By preserving freshwater habitats and reducing the impacts of human activities on aquatic systems, freshwater conservation contributes to the objectives of SDG 14 by supporting marine and freshwater biodiversity and ensuring the sustainability of aquatic resources.

SDG 15: Life on Land

Ecosystem Conservation and Water Resources

The conservation of terrestrial ecosystems is closely linked to freshwater resources. Healthy forests, wetlands, and other terrestrial ecosystems are vital in regulating water cycles, preventing erosion, and maintaining water quality. Freshwater conservation efforts that focus on protecting and restoring terrestrial ecosystems contribute to the objectives of SDG 15 by promoting ecosystem health and resilience. Integrating water conservation into land management practices supports the sustainability of both terrestrial and freshwater ecosystems.

SDG 16: Peace, Justice, and Strong Institutions

Water Governance and Conflict Resolution

Effective governance and management of freshwater resources are essential for promoting peace and justice. Water-related conflicts can arise due to competition for scarce resources, particularly in transboundary river basins. Strong institutions and effective governance frameworks are crucial for managing water resources equitably and resolving conflicts. Freshwater conservation efforts that promote transparent and inclusive water governance contribute to the objectives of SDG 16 by supporting peace and justice in water resource management.

SDG 17: Partnerships for the Goals

Collaboration in Water Management

Sustainable freshwater conservation requires collaboration among governments, organizations, communities, and the private sector. Partnerships and cooperative approaches are essential for addressing water-related challenges and implementing effective conservation strategies. Sharing knowledge, resources, and expertise through partnerships supports the achievement of SDG 17 by fostering collaborative efforts and enhancing the effectiveness of water conservation initiatives.

11

Case Study – Introduction

"प्राणावैजगतानापोभूतानिभुवनानिच। बहुनात्राकिमुक्तेनचराचरमिदंजगतः।।" लि.पु. 134/9
"जल जगत के प्राण है। जिसमें सब भूत और भुवन है। सम्पूर्ण चर और अचर जगत जल के आधर पर स्थित है।"

"Water is the soul of the world where all creatures and Bhuvan exist. The entire variable and constant is based on water."

"When the well's dry, we know the worth of Water"– Benjamin Franklin

Water is acknowledged as among the earth's precious, necessary, abundant, and free assets. The environment, society, and economy are all substantially dependent on freshwater. Ecosystems like wetlands, rivers, aquifers, and lakes are necessary for life to exist on our planet. These ecosystems are also crucial for directly ensuring various benefits and services, including drinking water, food and industry water, energy, aquatic life habitats, and natural water purification solutions. They are also essential for buffering floods and bridging periods of drought. Freshwater ecosystems have a role in reducing risks, fostering measures that promote stability and trust, and ensuring their resilience in the face of conflict, catastrophes, and climate change when they are adequately managed to balance conflicting needs. However, the issue of industrialization, urbanization, and over-exploitation of the available natural and man-made resources has been exacerbated by the world's constantly growing population, which has also accelerated the issue of different sorts of pollution. Water pollution is one of these pollutants that are seen as being especially hazardous.Water resources are now being overused due to using fresh water for drinking, household, commercial, and agricultural purposes.

India recorded the highest annual freshwater withdrawals in 2019, at over 647 billion cubic meters. The water used varies greatly worldwide (World Bank, 2022). The graph (Figure 1.1) displays the average annual water extraction rate per person.The withdrawal of water is the amount of freshwater abolished for household, commercial, or agricultural uses from lakes, rivers, or other surface water sources. It may include desalinated water, agricultural irrigation water, the direct use of (treated) effluent, and water derived from primary renewable and secondary freshwater resources. It may also include excessive consumption of reusable groundwater. It excludes in-stream applications, such as amusement transportation, hydropower, inland commercial fishing, etc., with relatively low net consumption rates (Ritchie and Roser, 2017).

Annual freshwater withdrawals, 2019

Annual freshwater withdrawals refer to total water withdrawalas, not counting evaporation losses from storage basins, measured in cubic metres (m3) per year. Total water withdrawals are the sum of withdrawals for agriculture, industry and municipal (Domestic uses). Withdrawls also include water from desalination plants in countries where they are a signification source.

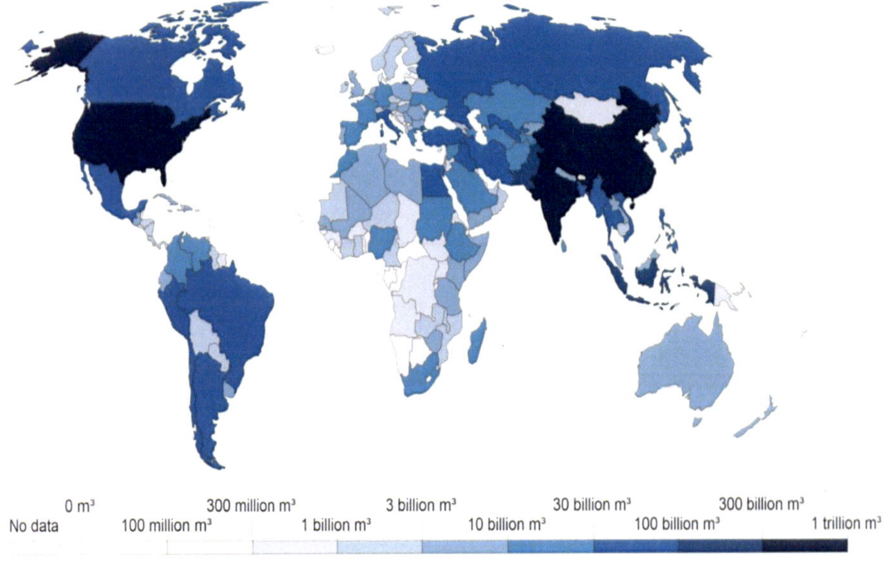

Figure 11.1. Water withdrawals per capita (Source: World Bank, 2022)
Source: Food Agriculture Organization of the United Nations (via World Bank) OurWouldData.org/water-use-stress.CC BY

Freshwater is a finite and precious resource necessary for all species on earth to survive. The primary source of flow for many surface streams is groundwater. On the other hand, the primary source of groundwater recharge is water from surface watercourses. As a result, the two heads of supply depend on one another, and using one source may reduce the water supply from the other. Increasing drinking water demand, insufficient management of the nation's water resources, and unpredictable water supply brought on by anthropogenic and climatic causes contribute to India's ongoing water crisis (Natarajan *et al.,* 2016; Thatte, 2018). With 18% of the worldwide population and 4% of its water reserves, India is a water-stressed country, and its annual national water consumption of 1588 m^3 per person is rapidly decreasing (Chakraborti *et al.*, 2019; World Bank, 2022). In 2020, a total of 5.8 billion people, or 74% of the global population, consumed on-site drinking water that was easily accessible and free of contaminants (WHO, 2022).

Status of Groundwater quality

The International Association of Hydrogeologists (2020) estimates that approximately one-third of people worldwide utilise water for drinking. Groundwater provides most of the world's population with drinking water. Many nations struggle with a lack of groundwater. To calculate the relative risk of groundwater shortages, the study examined data from 189 countries in 2019. From these data, an index was formed (Figure 1.2). The index calculates the proportion of total water withdrawals to the renewable surface and groundwater resources; withdrawals cover household, industrial, agricultural, and animal uses (Armstrong, 2022).

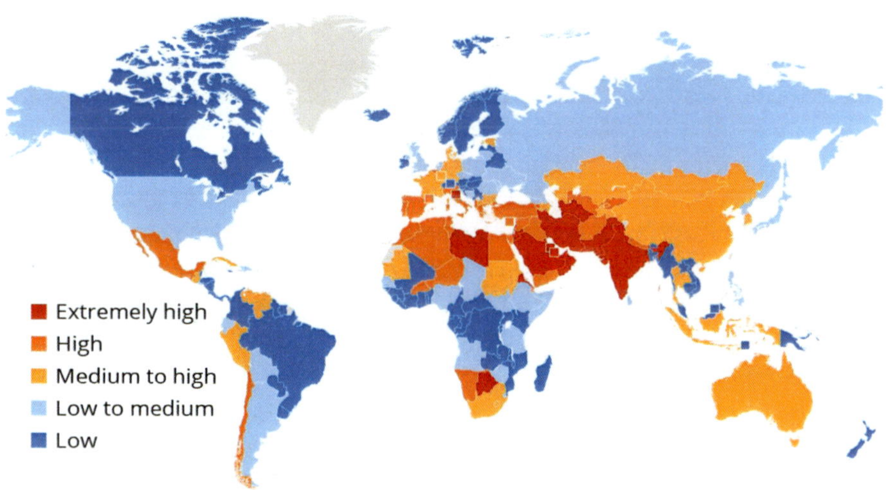

Figure 11.2. The risk of groundwater shortages in 2019. (Source: Armstrong, 2022)

Worldwide water scarcity results from the unequal distribution of water as a resource and the heavy use of industrializedand rising nations. Poorer communities are severely impacted and are having a harder time dealing with droughts and dry spells as a result of climate change's effects. For this reason, several sources forecast an increase in wars and disputes over water in the future (Armstrong, 2022). People have shifted to groundwater sources for drinking due to the rise in surface water pollution. Groundwater sources are considered more secure than surface water sources because they lack bacteria, though various metals and dissolved minerals can contaminate them (Berhe, 2020; Elumalai *et al.,* 2020). India has become one of the world's most populated and agriculturally dependent countries. As a result, many of its uses require a sizable volume of Water. Due to the ease with which groundwater may be used year-round, its requirements are sharply rising (Taloor *et al.,* 2020; 2021). As a result, groundwater emergencies can occur in various parts of India, varying in severity depending on the season (Jasrotia *et al.,* 2019). Recent population expansion has raised water demand to such a degree that it has significantly impacted the quality and quantity of Water (Taloor *et al.,* 2020; Verma, 2021). Other causes, such as industrialization, agriculture, and urbanization, have exacerbated this demand.

As a consequence of anthropogenic activities, including fast population growth, a developing economy, and other activities, the demand for groundwater increased, as did contamination and degradation of quality (Chowdhury *et al.,* 2021; Wang and Li, 2021; Zhao *et al.,* 2021). An increase in groundwater contamination leads to serious health problems in the human body and for other living organisms (Panneerselvam *et al.,* 2023; Mao *et al.,* 2023). Groundwater quality has been rapidly declining in many regions of the nation, particularly in the states of Punjab, Haryana, Rajasthan, Gujarat, Madhya Pradesh, Odisha, West Bengal, Karnataka, Andhra Pradesh, as well as Telangana, making it inappropriate for a variety of uses (Singh *et al.,* 2017; Mukherjee *et al.,* 2019; Haque *et al.,* 2020; Khan *et al.,* 2020). In India, in-depth groundwater research has been conducted in several of the nation's areas (Adimalla *et al.,* 2022; Patil *et al.,* 2020; Karunanidhi *et al.,* 2021). The Indian government periodically created committees to estimate the country's groundwater supply. But due to a lack of scientific data and a complete understanding of the recharge and discharge processes, all prior estimates were speculative or approximate (Adimalla *et al.,* 2022; Karunanidhi *et al.,* 2021).

Groundwater, the Key to Sustainability

One of the most crucial resources for the continuation of life, as we know it on the planet, is groundwater. Groundwater levels are declining alarmingly due to the recent groundwater exploration spurred by manufacturing and rapid economic expansion. Worldwide, groundwater is divided into three major categories: drinking water (65%), agricultural and animal husbandry (20%), and industrial and mining (15%) (Saeid *et al.*, 2018; Adimalla and Qian, 2020; Nayak *et al.*, 2023). Water demand has grown due to its usage for various purposes, including urbanization, agriculture, industry, and rising living standards. Due to several factors, such as climate change, population growth and insufficient precipitation, groundwater consumption has increased (IPCC Report, 2021).

As reported in the Sustainable Development Report 2022, freshwater withdrawals have risen substantially globally between 2015 and 2019 due to water stress. Northern Africa, Central Asia, and Southern Asia have been reported to experience the most significant levels of water stress (United Nations, 2018). Increasing water usage efficiency also contributes to decreasing water stress, which grew 12% worldwide between 2015 and 2019. To achieve the stated SDG 6, decisive steps are required to increase water usage efficiency globally to preserve natural freshwater supplies, including groundwater (United Nations, 2022). The Jal Jeevan Mission is national water supply surveillance and monitoring initiative the Government of India launched to fulfill SDG 6 (JJM, 2019). The State governments are mandated to assess the chemical quality of the water sources supplying rural regions once a year under this continuing effort to monitor water quality. The initiative encourages community participation at the grassroots level to conduct water quality surveillance by implementing field testing kits, which only offer qualitative information about the water source (JJM, 2019). A thorough quantitative examination of the water's quality concerning significant physicochemical characteristics is often neglected. Even after executing the necessary treatment, the deterioration of the water supply infrastructure can impact the water quality during the supply process (Pundir *et al.*, 2021).

Industrial wastewater

Although freshwater supplies are rapidly running out and demand is rising, wastewater reclamation or reuse is one of the most crucial requirements of the present situation. Water with dissolved and suspended elements discharged from different industrial processes, such as the water produced during manufacturing, cleaning, and other commercial activities, is called industrial wastewater. Various poisons can be found in industrial effluent, depending on the kind of facility and industry. Businesses that create wastewater include pharmaceuticals, mineral extraction, steel/iron production, commercial laundry, power, natural gas and oil, metal finishing, and food/beverage industries. Among the numerous pollutants typically found in industrial water outflows are chemicals, heavy metals, lubricants, pesticides, silt, medications, and other industrial byproducts (United Nations, 2022).

In all but the most highly industrialized countries, the vast majority of wastewater is released into the environment untreated, with detrimental impacts on the environment, human wellness, revenue generation, and ambient water supplies. 70% of the wastewater produced by municipalities and industries in high-income countries is routinely treated. The ratio is 28% for lower-middle-income countries and 38% for upper-middle-income nations. In low-income countries, just 8% of individuals receive treatment. These numbers support the often-cited statement that more than 80% of all wastewater is discharged untreated globally (UN-Water, 2018). SDG Target 6.3 states that by 2030, "improved the quality of water by reduced contamination, preventing disposal, reduction of discharge of hazardous substances and chemicals, reducing the amount of untreated wastewater, and substantially boosting recovery and secure usage worldwide". Due to the deficient level of wastewater treatment, it is clear that technological advancements and safe water reuse options are urgently needed to support the accomplishment of Target 6.3, which is essential to fulfilling the entire Agenda (Pundir *et al.*, 2021).

Low- and lower-middle-income nations will be at a financial disadvantage relative to high- and higher income nations because of the increased costs associated with the activities necessary to attain this Target. 99% of wastewater is made up of water, and 1% is made up of suspended, colloidal, and dissolved particles. Three categories can be used to group the effects of discharging wastewater that has not been properly or sufficiently treated: 1) unfavourable consequences on human health, 2) detrimental effects on the environment, and 3) harmful effects on economic activity (UNEP, 2015). Actual amounts of wastewater may not have the same substantial effects on water resources, human health, or the environment as industrial pollutant's toxicity, mobility, and loading. At the point of origin, from idea to design to operations and maintenance, the first task is to minimize pollutant quantities and toxicity to a minimum. This includes replacing existing raw materials and process chemicals with more environmentally friendly counterparts and staff education and training to address pollution-related issues. The second phase maximizes water recycling within a plant to reduce outflow. Small and medium-sized enterprises (SMEs) and unregulated companies regularly release their wastewater into municipal networks or the environment. Because industries that release into drainage systems or surface water must abide by discharge restrictions to avoid fines, end-of-pipe treatment is typically required at the site before release. In other instances, however, industries may determine that paying fines is more cost-effective than spending money on treatment to comply with regulations. Cooperation between facilities through industrial symbiosis is a popular alternative for utilizing and recycling industrial wastewater. This is best illustrated in eco-industrial parks, grouped by organisations near one another to optimise the use of reclaimed water, diverse wastewater flows, and product reuse.

Impact of Industrial wastewater on drinking water quality, human health, and environment

Rapid industrialization has dramatically increased the amount of contaminants in the environment. All types of biological forms have been negatively impacted, either directly or indirectly, by the improper management of some hazardous industrial wastes dumped into water bodies. Untreated wastewater discharge into the environment is still frequent, particularly in developing nations.

Health

- Increased disease impact due to hazardous food (infected fish, vegetables, and other produce irrigated with sewage)
- Increased risk of disease when working or playing in wastewater-irrigated areas
- Increased burden of disease due to decreased drinking, bathing, and other water quality

Environment

- Reduced recreational opportunities; diminished biodiversity; degraded aquatic ecosystems (such as dead zones and eutrophication); foul odours; increased greenhouse gas emissions; raised water temperature; and bioaccumulation of toxins

Economy

- Reduced opportunities for water-based recreational activities; decreased industrial and agricultural production; decreased market value of produced commodities
- Higher expenses of water treatment (for human supply and other uses)
- Lower prices of properties near contaminated water bodies
- Higher hurdles to international trade (exports)

Figure 11.3. Negative impacts of untreated wastewater on human health, the environment, and productive activities (Source: UNEP (2015))

This process has an immediate negative impact on human health (specifically the health of women), the environment, and economic efficiency (Figure 1.3).

When wastewater that has either not been treated or has only been partially treated is discharged into the environment, surface water, soil, and groundwater are all affected. Wastewater dumped into bodies of water may affect the quality (and consequently the availability) of freshwater sources because it is either diluted, moved downstream, or seeps into aquifers. The ocean is frequently the final destination of effluent that is released into rivers and lakes.Three groups can be used to classify the effects of discharging untreated or insufficiently treated wastewater: adverse impact on human health linked to decreased water quality; negative effects on the environment related to the degradation of water bodies and ecosystems; and potential effects on economic activity (UNEP, 2015).

Human health effects

According to estimates, only 26% of sanitation and wastewater services in urban areas and 34% in rural areas effectively limit human contact with excrement along the entire sanitation chain and may therefore be regarded as securely handled (Hutton and Varughese, 2016). Diseases including cholera, diarrhea, dysentery, hepatitis A, typhoid, and polio can spread due to contaminated water and poor sanitation. People are exposed to avoidable health hazards when water and sanitation facilities are unavailable, inadequate, or improperly managed. The drinking water of hundreds of millions of people is critically poisoned or chemically polluted due to improper management of urban, industrial, and agricultural wastewater. Arsenic and fluoride are two chemicals that naturally occur and may have health effects, while lead levels in drinking water may increase due to leaching from water supply components that come into contact with the water, according to estimates from the World Health Organisation (WHO, 2021), poor sanitation, hand hygiene, and contaminated drinking water cause 82, 9000 deaths annually from diarrhoea. The percentage of industrial and total wastewater flows that are safely treated is monitored by SDG Indicator 6.3.1. However, wastewater statistics are still in their infancy in many nations, and there is a dearth of reporting, particularly from industrial sources. A 58% estimate of household wastewater generated in 2022 was treated correctly, according to data from 140 countries and territories (Pirlea *et al.*, 2023). In Figure (1.4) 6% of the global population had no access to a better supply of water in 2020 (WHO-UNICEF (JMP), 2020).

Environmental effects

The amount of water resources accessible for direct use is impacted by the release of untreated wastewater into the environment, affecting water quality. Due to increased wastewater production brought on by population growth, increased economic activity, and expanding agriculture, as well as the discharge of sewage with no (or only minimal levels of) treatment, water pollution has been rising in most rivers in Africa, Asia, and Latin America since 1990 (UNEP, 2016). Insufficient wastewater management directly impacts ecosystems and their services (Corcoran *et al.*, 2010).

Share of the population without access to an improved water source, 2020

Improved drinking water sources are those that can deliver safe water. They include piped water, boreholes or tube wells, protected dug wells, protected springs, rainwater, and packaged or delivered water.

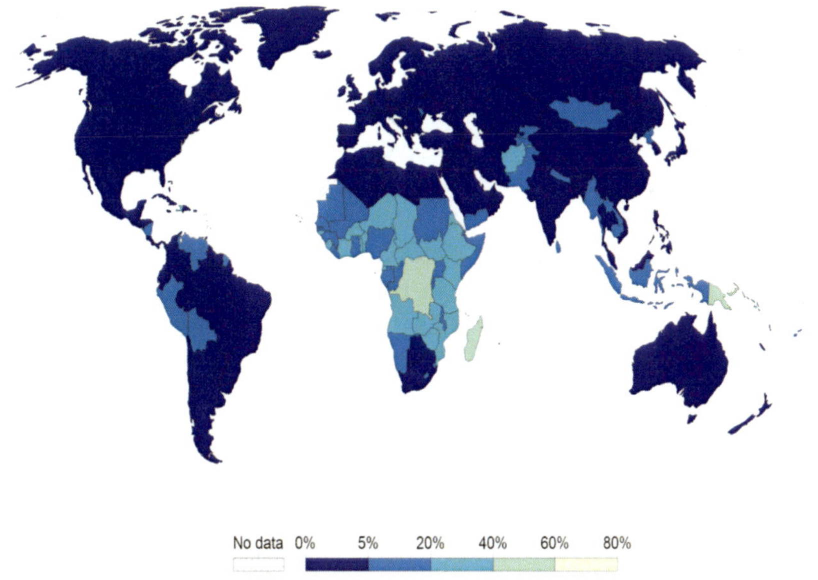

Source: WHO/UNICEF Joint Monitoring Programme (JMP) for Water Supply and Sanitation OurWorldInData.org/water-access • CC BY

Figure 11.4. Share of the population without access to an improved water source (Source:WHO-UNICEF (JMP) 2020)

Excess nitrogen and phosphorus cause eutrophication, harming algal blooms and reducing biodiversity. De-oxygenated dead zones are quickly expanding, partly due to the discharge of untreated sewage into seas and oceans. According to Corcoran *et al.*, (2010), this influences food chains, fisheries, livelihoods, and an estimated 245,000 km^2 of marine ecosystems.

Economic effects

A further barrier to economic growth is poor water quality, as freshwater availability is essential to maintaining the financialwell-being of every human population. In peri-urban and rural areas, poor water quality reduces agricultural output. Due to restrictions (and even outright bans) on contaminated products, contaminated water can directly impact economic activities that use water, such as industrial production, fisheries, aquaculture, and tourism (UNEP, 2015). It can also have an indirect impact on the exportation of specific goods. The potential advantages of utilizing wastewater are lost when the disposal of wastewater results in environmental harm, which generates external costs (externalities). To minimise possible harm and maximise potential advantages, an economic argument must be made for improved wastewater management. If wastewater is acknowledged as good for the economy, properly treated wastewater may have benefits for both those who produce it and those who use it (UNEP, 2015).

Water quality indices

The influence of advancement on the accessibility of resources for clean freshwater has been demonstrated in several studies and this impact has an impact on human health. Long-term water quality forecasting hasn't received much attention, despite being essential for organizing efforts to limit water pollution and manage water resources. Most source apportionment studies conducted recently occurred in urban or large-scale metropolitan regions. The measurements of a few chosen factors are combined to create a single dimensionless number known as a water quality index (WQI), which describes water quality. The advantage of these WQIs is that less linked characteristics must be interpreted to ascertain the water quality state (Abbasi and Abbasi, 2012; Hossain and Patra, 2020; Tripathi and Singal, 2019). As a result, they offer a straight forward yet inclusive method of interpreting the variety of water quality parameter values available (Abbasi and Abbasi, 2012; Asadollahfardi, 2015). These indices may and have been used to compare the conditions of various water bodies through time and space, which helps determine where to focus management efforts and financial resources. They could also help assess and communicate the overall results of ongoing, upcoming, or proposed management decisions and activities related to water quality. Compared to traditional WQIs, objective index formulations provide greater application flexibility by including several determinant sets to reflect location-specific conditions and changing water quality challenges (Banda and Kumaraswamy, 2020).

WQIs can be combined with other metrics, including biological integrity indices, to give a whole picture of the condition of a water resource. Certain sub-indices or other potentially important factors might be thoroughly explored to gain insight on present and potential hazards. Continuous use of the indices provides long-term data that helps make decisions about water quality (Gracia *et al.*, 2018). As with many computing methods; there are some general problems with the WQIs. Choosing the appropriate index, eclipsing, ambiguity and rigidity, expert bias, and relation to water quality standards are frequent issues because of their aggregated nature, as shown in Figure 1.5. WQIs are potentially significant tools for managing and making decisions on water quality (Gitau *et al.*, 2016). After reviewing the previous research, the following six basic uses for WQI were identified: resource assignment, standard enforcement, trend analysis, location ranking, public information, and scientific study. Indexes are thus descriptive and objective tools for identifying trends in water quality. Because of data and volume collision, parameters are not clearly illustrated in comparison to one another or against time, and they do not explicitly display patterns. In scientific study, indices may be used to simplify enormous amounts of data into a process that focuses on inquiry and enables analysis of many of these environmental initiatives.Through indexes, the general public might be made aware of environmental issues.Trend analysis: By collecting environmental data from several periods, indices may be utilized to examine changes in environmental quality over time.

Figure 11.5. Exploring potentials for use of Water Quality Indices for decision-making and management (*Source*: Gitau *et al.*, 2016 Water Quality Indices as Tools for Decision Making and Management)

Indexes can be used to evaluate the degree to which statutory requirements and benchmarks are fulfilled or exceeded in particular industries. According to environmental characteristics in various places or geographical areas, resources are assigned according to those locations'rankings (Kachroud, 2019). Accordingly, WQI often compares a number to an artificial or scientific norm or a pre-determined criterion. Consequently, the WQI tracked and reported environmental changes and status statistically. The WQI approachs prevents using redundant data to describe water quality and instead provides useful information on the degree of both the purity and contamination of water. Additionally, the WQI tool gives users access to a comprehensive quality control system. The National Academy of Sciences Planning Committee on Environmental Indices claims that environmental indicators are crucial for aiding in program design, determining the specifics of policy, and helping people who are impacted.

Human health risk assessment and Questioner Survey: An effective technique

A health risk assessment, also known as a health risk appraisal, is a technique for gathering health-related data and is usually used in conjunction with a process that incorporates verification of identity to assess a person's health, hazards, and behavior. According to Barzegar *et al.,* (2015), heavy metals are very hazardous metals and metalloids with atomic densities of more than 4000 kg/m^3, or five times that of water. This group of elements includes metals and non-metals, with a high risk of contamination and toxicity. When consumed in large quantities (via food or water), trace elements can offer substantial health hazards due to their toxicity, bioaccumulation, and intrinsic persistence. In trace amounts, trace elements are critical micronutrients for people (Fito *et al.,* 2019). Through natural processes such as water and soil interaction, rock deterioration, precipitation, seismic emission, etc., trace elements can contaminate groundwater. Human sources include agriculture, domestic wastewater, mining, industrial activities, dumping sites, etc. The three main routes via which humans are exposed to trace metals are by ingestion, inhalation, and dermal absorption through the skin (Duggal and Rani, 2018). Heavy metals can accumulate in the bodies and tissues of humans and animals, which can result in issues with the nervous, immune, and circulatory systems, as well as intellectual and developmental disabilities like hearing loss, trouble concentrating, problems with reading and learning, behavioral issues, and disruptions of perceptual-motor function.

Potentially toxic elements (PTEs) have been a source of significant worry over the past few years due to their ability to accumulate in the human body for practically the entirety of a person's lifespan and lead to a variety of disorders (Li *et al.*, 2017; Papazotos *et al.,* 2021). It is essential to constantly check the concentrations of heavy metals in the environment and the human body. As a result, several researchers have examined the potential risks to public health caused by heavy metals in waterways (Tabassum *et al.*, 2019; Bodrud-Doza *et al.*, 2020; Sharma *et al.*, 2020). A series of questions or other prompts designed to gather information from a respondent constitute a questionnaire, a sort of research tool. In research surveys, both open-ended and closed-ended questions are common. Long-form, open-ended inquiries give the respondent space to go into further detail about their views. In 1838, the London Statistical Society introduced research questionnaires. Depending on the data collection method used the information acquired may be qualitative or quantitative (Figure 1.6).

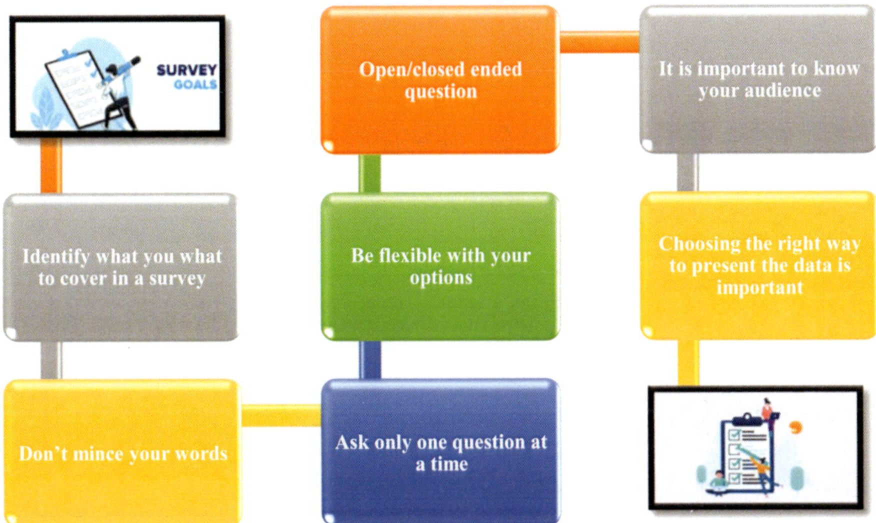

Figure 11.6. Types of Questionnaires based on distribution (*Source*:https://www.questionpro.com/blog/what-is-a-questionnaire/)

A survey must always contain a questionnaire, whether presented in a study or not. According to the benefits of a well-crafted questionnaire, one may get many data with a survey questionnaire quickly. There is less chance of prejudice leaking if users use consistent questions with your target audience. Although one can apply logic to answer questions based on respondents'answers, the questionnaire will remain the same for a group of respondents in the same segment. Online (Google forms, Snap Survey, Scattport Survey) and offline survey methods may be used to conduct surveys rapidly and economically. It has a range of features that allow it to generate, sabotage, and analyse response data. It could be changed to match their organization's voice. Thus, it might be applied to enhance your brand's reputation. The replies may be compared to earlier data to see how the respondents' preferences and experiences have changed over time. Public opinion research is a technique for gauging the opinions of a large population by looking at a representative sample of that population via surveys, in-depth interviews, and focus groups. Public opinion research is crucial because it identifies the information, studies public opinion on a particular issue, and represents the broader public worries, attitudes, and values. It helps identify areas where investments are needed, among other things and offers a proper understanding of the problems and essential fixes.

Research gap in this region and scope of this study

Water quality is one of the primary challenges in managing water resources, and it may be categorized in three ways: physically, chemically, and biologically. There are specific traits that belong to each group (Uddin *et al.*, 2022). Evaluation of these three categories through field monitoring offers key data for trend detection, giving water quality information to water authorities, and advising future behaviour. In North Indian areas like Uttarakhand, the Central Ground Water Board (CGWB, 2019) of India observed a severe decline in groundwater levels of between 70 and 80 percent. The state of Uttrakhand'scapital and one of its most developed cities is Dehradun. Most people rely on surface and ground waters for their everyday needs. Regarding population growth and the use of natural resources, the valley is one of the sub-Himalayan region's expanding cities (Sathyaseelan *et al.*, 2017; Pundir *et al.*, 2022).

Although drinking water from natural sources is thought to be pure and suitable for human consumption, there is always a chance of various forms of contamination since natural water sources are always in contact with anthropogenic and geological pollution causes. In addition, differences in space and time can impact some physical characteristics, including color, turbidity, alkalinity, hardness, pH, and others. Domestic waste, fecalmatter, and run-off with various contaminants from various sources can easily contaminate the natural water sources due to insufficient water purification techniques, sanitation facilities, and sewage treatment plants in hilly and rural parts of the district, which poses a serious health risk (Maurya *et al.*, 2021; Kothari *et al.*, 2021). Water may thus provideseveral health problems if delivered without first being subjected to mostphysico-chemical analyses. Conducting scientific evaluations of the water quality of the sources of drinking water is required to identify the cause and origin of contamination in drinking natural and supplied water. It is essential to characterise the physico-chemical water quality indicators and determine the origins and causes of pollution to pinpoint the specific geogenic or anthropogenic sources polluting the drinking water sources in the Dehradun region. This will make taking the required precautions to prevent pollutants from entering water sources easier. The responsible agency will be able to supply the general population with high-quality, clean, and safe drinking water for household applications with the help of such scientific input and information. Putting suitable safeguards in place to protect drinking water quality is impossible without such comprehensive information. The current study has been conducted to close the gaps mentioned above.

In Uttarakhand state, there has been an observed rise in the establishment of industrial infrastructure. The quick economic growth of the region was one of the primary goals for the foundation of the independent state of Uttarakhand. Integrated Industrial Estate (IIE) Dehradun is one of four industrial parks the State Industrial Infrastructural Development Corporation of Uttaranchal (SIDCUL) built in the State. Most industries need water for processing, which results in wastewater. This contaminated wastewater is either released on the surface or drained into a neighboringbody of water. It always makes it to the subsurface water table, contaminating the groundwater and rendering it unsafe for ingestion. Few information is available on these effluents'current and potential future effects on groundwater. The current study was carried out to record the water quality features of groundwater and industrial effluents in the Dehradun region. Its originality is characterised by the fact that it is one of the very few studies to be conducted in the area. The investigation aids in determining the elements that threaten the quality of the groundwater. The research will offer baseline information on the wastewater produced by industrial units, making it easier for regulators to monitor current regulations and enforce them more strictly. Additionally, the information gathered on the groundwater quality characteristics of the industrial region near Dehradun would assist policymakers in developingsolutions to reduce pollution. The results of this study will open the door for more thorough research into the complex interaction between industrial wastewater discharge and its effects on groundwater quality.

Objective of the study

The following objectives were established to accomplish the research's purpose:

- To assess the impact of industrial wastewater on groundwater quality in terms of physico-chemical parameters and heavy metals.
- Assessment of health risk using survey method.
- To understand the level of pollution and contamination with the applicability of Water quality indexing on selected parameters using Standard methods.

12

Literature and Research Gap

Groundwater pollution-An Overview

The study of related literature gives he researcher an idea on the quantum of work made in the field and makes to analyze the methodology used directly to work along with usefulness. Thus, this literature study will serve as a background to the researcher to arrive at a design to segregate the exact area to cover. Several research scientists from India and abroad have discussed and conducted studies related to the proposed research.

Groundwater constitutes the majority (95%) of the freshwater sources on Earth and is extensively used for industrial, domestic, and agricultural uses (Li *et al.,* 2021; Zhang *et al.,* 2021). It is the primary source of drinking water for around one-third of the world's population (International Association of Hydrogeologists, 2020). Earth is currently experiencing an severe environmental catastrophe. Concerns about problems like pollution, global warming, ozone layer thinning, overpopulation, climate change, loss of biodiversity, depletion of natural resources, waste management, etc., are already clearly apparent. However, the deposition of dissolved or suspended materials into groundwater, streams, rivers, and oceans continues to be one of the significant worldwide problems. In addition to other factors, the exponential rise in commercialization has severely degraded the environment and devoured vast tracts of agricultural land. The overabundance of trace metals, weighty metals, in our waterways results from rapid industry and urbanization (Jasrotia *et al.,* 2019).

India is one of the most populous and reliant on agricultural nations worldwide. It thus requires a substantial amount of water for many uses. Because groundwater is readily available throughout the year, its demands and needs are increasing rapidly (Taloor *et al.,* 2020, 2021). As a result, groundwater crises in various areas of India fluctuate in size and severity according to the seasons (Singh *et al.,* 2017; Jasrotia *et al.,* 2019; Taloor *et al.,* 2020; Kumar *et al.,* 2021). In recent years, the growing population has increased the demand for water to such a degree that it has negatively impacted the quantity and quality of water. This need has been exacerbated by other activities, such as agriculture, industrialization, and urbanization (Taloor *et al.*, 2020; Verma *et al.*, 2021). The demand for groundwater rose as a result of factors such as rapid population expansion, an expanding economy, and numerous human activities, which also contaminated and decreased the quality of the groundwater (Chowdhury *et al.*, 2021; Zhao *et al.*, 2021; Asmamaw *et al.*, 2023). The most significant human factors influencing the quantity and quality of these resources are rising levels of hygiene, continued growth in industrial production and the expansion of their variety, and the dumping of different types of waste, both solid and liquid, from industrial, urban, and agricultural processes (Singh *et al.*, 2023; Sharma *et al.*, 2023). Urban wastewater, according to Einsiedl *et al.*, (2010), is a significant source of harmful microbial contamination.

Industrial operations (either individual or in a zone) are recognized as one of the biggest contributors of both environmental pollutants and reseources in the ground and surface waters due to the range of waste from diverse industries and factories. The country's groundwater quality has been rapidly declining, especially in the states of Punjab, Haryana, Gujarat, Rajasthan, Madhya Pradesh, West Bengal, Odisha, Andhra Pradesh, Karnataka, Telangana, and Uttrakhand (Singh *et al.*, 2017; Mukherjee *et al.*, 2019; Khan *et al.*, 2020; Rahman *et al.*, 2023). As a result, the degraded groundwater contributes to a variety of chronic ailments throughout the nation, including cancer, bronchitis, fluorosis, arsenicosis, hemochromatosis, and many more (Mukherjee *et al.*, 2019; Adimalla and Taloor, 2020). Due to this, scientific groups and environmentalists have always been interested in identifying and analyzing how industrial activity affects the quality of water supplies (Ravindra and Mor, 2019; Sureshjani *et al.*, 2020).

The review of literature on groundwater quality of will helps to identify the research gaps by providing a detail account on time period of study, selection of sampling locations, impact of industrial wastewater on groundwater quality, and importance of water quality indices on water quality and health risk assessment. The chronologically reviews of related literature are as follows:

Impact of industrial wastewater on groundwater quality

Rameeza *et al.*, (2012) worked on the quality of groundwater in the industrial region of Visakhapatnam. The outcome demonstrated that the quality of water is unfit for consumption. Parihar (2012) analyzed the the groundwater quality in and around Gwalior city. Deepali and Joshi (2012) researched the effects of industrialisation on the District of Haridwar's groundwater quality. Due to the higher amounts of the majority of the distinctive ions, the water in the tested location was deemed unsafe for drinking and domestic use.

Bharti *et al.*, (2013) analyzed the effects of toxic metals on soil and groundwater quality in Panipat City, India. Sirajudeen *et al.*, (2013) investigated the effects of industrial water on groundwater quality, and the water quality indices found that the groundwater was unsuitable for human consumption. The bacteriological approach of surrounding areas in western India has been assessed by Kumar *et al.*, (2013).

Dubey *et al.*, (2014) assessed the quality of groundwater in Dwarka. Dohareet *et al.*, (2014) developed a method to assess the groundwater quality in Indore's industrial region. According to Paul *et al.*, (2014) investigation of the physico-chemical characteristics of groundwater in Kerala, India, the water is excellent and suitable for drinking and irrigation. The water quality index for Smalkhan, Haryana, was examined by Kumari and Rani in 2014. TDS and alkalinity were found to be above desirable levels when Mohan *et al.*, (2014) investigated the physicochemical parameters.

Tyagi *et al.*, (2015) evaluated the quality of groundwater in India's Uttar Pradesh province's Gajraula industrial sector. In the Chrompet industrial area, Karpagam and Ramesh (2015) evaluated the condition of groundwater using the water quality index and came to the conclusion that the groundwater was inappropriate for human consumption. Groundwater quality was crucial, according to research done by Packialakshmie *et al.*, (2015) employing WQI in and around the Sholinganallur area. Vyas (2015) evaluated the physico-chemical and microbial characteristics of drinking water in Junagradh, India. Gogoi (2015) evaluated the impact of industrial effluents on groundwater quality and human health and found that the city suffers from poor sewage management.

Bhutiani *et al.*, (2016) performed research to evaluate the groundwater quality, the source of contaminants, and the health risks associated with heavy metals in groundwater in an industrial region. Kanagaraj and Elango (2016) analyzed hydro-geochemical characteristics in pre and post monsoon sessions and it was found that 80% of groundwater was unfit for human consumption. Groundwater needs to be treated before it can be used, according to Krishna *et al.*, (2016) study on groundwater in the Ballia region of Uttar Pradesh, applying WQI and GIS.

Shrestha *et al.*, (2017) assessed the quality of Selected Industrial Effluents in Nepal industrial area and found that the collected samples were above the national standard of Nepal. Kumar *et al.*, (2017) conducted a study to assess the physico-chemical analysis of groundwater contamination in Tamil Nadu, India and it was discovered that all values were found to be higher than the required Bureau of Indian Standards (BIS).

Sharma and Naithani (2018) researched the impact of industrial effluent on Haridwar's groundwater quality. Akhigbe *et al.*, (2018) assessed the effects of industrial waste on groundwater quality within Trans-Amadi Industrial Layout, Nigeria and found that the Iron (Fe) had levels at some sites over the maximum 0.3 mg/l. Afzal *et al.*, (2018) assessed the impacts of industrial effluents on groundwater quality of Hattar industrial estate in Haripur and found that the effluents were highly polluted due to the presence of heavy metals and hence, contemplated as ill-suited for irrigation and drinking purposes.

Ferrer *et al.*, (2019) assessed how water-dependent industry affects groundwater systems in coastal Kenya and found that the paucity of data on aquifer systems may be partially remedied by combining data from various sources. Popoola *et al.*, (2019) conducted a study to evaluate the physical and chemical properties of industrial and residential groundwater in the Lagos Metropolitan Area. They discovered that substantial effluent discharge from industrial treatment facilities and the dispersion of ionic heavy metals from industrial activities involving large machinery were the causes of increased TDS and EC concentrations recorded for groundwater samples from industrial locations.

Zacchaeus *et al.*, (2020) assessed the impact of industrialization on groundwater quality in Ogun State, Nigeria. Jan *et al.*, (2020) assessed the impact of industrial waste on quality of groundwater in Industrial Estate of Peshawar, Pakistan and revealed that some industries in HIE had the highest levels of Pb and Ni pollution, respectively. Hossain *et al.*, (2020) conducted a study to assess the health risk and groundwater quality: A case study on Setabganj Sugar Mills Limited, Dinajpur, Bangladesh, and discovered that the mean value of Mn content surpassed the drinking water standard. According to Zakir *et al.*, (2020) 95% of the groundwater samples were found unsuitable for drinking purposes. Sureshjani *et al.*, (2020) assessed the effects of industrial wastewater on groundwater in Southwest Iran and found that the highest pollution was related to Zn, P, Cu, and Ba.

Awol *et al.*, (2021) conducted a study to evaluate the impact of industrial effluent on irrigation water quality and farm soil near Kombolcha Town, Ethiopia, it was discovered that greater quantities of heavy metals, except Mn, were also identified during the dry season, however, Cr exhibited significant seasonal change. Ahirwar *et al.*, (2021) assessed the groundwater potential zone in the industrial areas of Hoshangabad and Budhni in Madhya Pradesh. They discovered that the findings were verified by superimposing the health inventory data of observation wells in the identified potential recharge zones using GPS coordinates. The investigation has shown that the observed characteristics regulate the unequal distribution of these groundwater potential zones.

Wekesa *et al.,* (2022) assessed thegroundwater quality using the water quality index from a few springs in the Manga Subcounty of Nyamira County, Kenya has exceptional quality status. The three springs in Manga represent groundwater with a water quality index of less than 30, making it suitable for drinking, agriculture, and industrial usage. According to Meshabaz *et al.,* (2022), the physico-chemical parameters of industrial effluents are higher than the permissible levels of global standard.

Asmamaw *et al.,* (2023) conducted a study to evaluate the groundwater quality in an industrial area for a reliable water supply of water using GIS and WQI in Bahir Dar city, Ethiopia and found that the pH, alkalinity, Escherichia coli, nitrite, manganese, and iron were exceeded than the recommended limits in the groundwater of built-up areas. Rahman *et al.,* (2023), evaluated the groundwater quality for the Baddi-Brotiwala-Nalagarh Industrial Area and discovered that various heavy metals and parameters were found to be exceeding and degrading rather than changing seasonally as the limit specified in BIS 10500. The high level of industrialization, the expansion of companies, and their untreated dumping of contaminated wastewater are the causes of the presence of heavy metals.

Water Quality Status of Uttrakhand State

Bahukhandi and Bartarya (2014) investigated the chemistry of major ions of ground water samples from the Asan Catchment. At a few sample sites, fluoride, potassium, nitrate, and sulphate concentrationsfluoride, potassium, nitrate, and sulphate concentrations were also higher than the BIS requirements for allowed limits for drinking water quality. Another research by Singh *et al.,* (2014) evaluated ground water quality in the Siwalik area of the Doon Valley with the aid of remote sensing and GIS. A considerable amount of WQI, or more than 100, was discovered at the water sample locations, indicating that the water was unfit for human consumption.

Jain *et al.,* (2015) assessed the groundwater quality in Bahadrabad, Haridwar and found that 42.3% of the samples had total dissolved solid concentrations that were higher than the acceptable limit of 500 mg/L, but the values were still well below the permissible limit.

Sharma *et al.,* (2016a) assessed the viable consumption of water resources in Uttarakhand's challenging topography; the state's residents struggle to access safe fresh water because of the state's slope, management issues, urban sprawl, deforestation, and other environmental factors. Sharma *et al.,* (2016b) assessed the groundwater quality of Bhagwanpur Industrial Area of Haridwar in Uttarakhand and found that the arsenic and iron concentrations in several groundwater samples were higher than BIS standard limits.

Gupta *et al.,* (2017) examined the heavy metal contamination for the chosen Dehradun drinking water sources using water quality indices. The study's findings demonstrated that the water quality was good for drinking and that heavy metal pollution was below the threshold levels. The acceptability of water for drinking from several water sources in Dehradun was assessed using physico-chemical and bacteriological methods by Rawat *et al.,* (2017). The study found that the water quality at the analysed locations was good for potable consumption. Chakraborti *et al.,* (2017) conducted a study titled groundwater arsenic pollution and discovered levels of arsenic that were higher than the WHO's drinking water standards in the groundwater (up to 4730 g/L), irrigation water (about 1000 g/L), and food materials (up to 3947 g/kg).

Bhutiani *et al.,* (2018) evaluated the ground water quality of the Laksarblock inHaridwar region of Uttarakhand state and they discovered that the following parameters were found to be below WHO and BIS standards: temperature, TS, EC, pH, DO, total hardness (TH), chloride acidity, and sulphate.

Bhutiani *et al.,* (2019) assessed the suitability of groundwater quality in Haridwar, region on the basis Water Quality Index (WQI) and found that only sample site S-4 is suitable for drinking while sites S-1, S-2, S-3, S-5, and S-6 have meager WQI ratings and are unfit for consumption. According to Bunkar *et al.,* (2019) assessment of the BIS standards for drinking water quality and the WQI for Groundwater Quality Parameters in Udham Singh Nagar District of Uttrakhand, only 50% of groundwater samples were evaluated as excellent. The majority of the remaining 50% of areas had poor groundwater quality.

Kumar *et al.,* (2020) conducted a study of Ground Water of SIDCUL-Rudrapur of Udham Singh Nagar district, Uttarakhand and found that discovered that it is the sole source of drinking water for those who live in the study region, and that there is significant variance in the findings of the chemical analysis of the ground water. Despite past interventions, few areas in the researched region have water quality that is fit for drinking. Singh *et al.,* (2020) analysed the groundwater quality for suitability of irrigation purposes in the Udham Singh Nagar, Uttarakhand and found that all the groundwater samples were observed to be appropriate for irrigation purpose.

Bhutiani *et al.,* (2021) evaluated the adequacy of the ground water quality in Laksar Block, Haridwar in Uttarakhand, India, and discovered that Hardness was generally over the standard level at sample locations. Additionally, it was discovered that Fe was above WHO and BIS thresholds. Sinha *et al.,* (2021) conducted a study of groundwater hydrochemistry of springs in Almoraregion using graphical tools and found that the results of most samples showed that they were not determined to be fit for drinking.

Bhardwaj *et al.,* (2022) evaluate the impact of rainfall on groundwater quality in Uttarakhand state and found that the highest water level was also observed during January and February. From June to August, the water table starts to drop, and typically from August to September, it reaches its lowest point.

Singh *et al.,* (2023) conducted a study to assess the impact of various industries on the drinking water quality and associated health risks in the Upper Gangetic Alluvial Plains of northern India, it was discovered that groundwater samples fell into the unacceptable category. Sharma *et al.,* (2023) evaluated the impacts of industrial wastewater on groundwater quality in Haridwar and found that the groundwater in the vicinity of the industrial region has been contaminated. The study's findings lead to the conclusion that industrial waste water is compromising the region's ground water quality and posing a severethreat to the health of its residents.

Evaluation of water quality using WQI and Multivariate analysis

Giri and Singh (2013) determined the metal concentrations, source apportionment and risk assessment of water quality. The PCA interpretation of this work demonstrated that four factors together explained 76.9% of the variance by innate and anthropogenic activities are contributing factors as source of metal profusion. Risk of metals on human health was then evaluated using Hazard Quotients (HQ) by ingestion and dermal pathways for adult and child and it was indicated that as with $HQ_{ingestion}>1$,was the most important pollutant leading to non-carcinogenic concerns. The largest contributors to chronic risks were as, V and Co, in all the seasons. HQ_{dermal} of all the elements for adultand child were below unity, suggesting that the metals posed little hazards via dermal absorptionindicating that the oral intake was the primary exposure pathway.

Ogwueleka (2015) investigated the spatial and temporal variation and applicability of multivariate statistical techniques such as cluster analysis (CA) and principal component analysis/factor analysis (PCA/FA) of water quality. In addition, Mann-Whitney *U* test was also used to identify the most significant variables responsible for spatial variation.

Singh *et al.,* (2016) evaluated the heavy metal contamination and its indexing technique for the groundwater of the Goa mining region, India and found that the HPI values throughout the monsoon season were 1.5 and 2.1, respectively. Both of these values fall significantly below the critical index limit of 100.

Herojeet *et al.*, (2016) applied the WQIs and chemometric analysis to determine the origins and severity of pollution in Himachal Pradesh, India. The primary contributors to the deterioration of water quality, including industrial effluent discharge, agricultural runoff, and minor natural or geogenic input, were identified through the application of PCA and CA.

Sobhanardakani *et al.*, (2017) assessed the groundwater quality in Toyserkan Plain using the water quality pollution indices, the mean values of the contamination index (C_d), heavy metal pollution index (HPI), and heavy metal evaluation index (HEI) in samples taken during the spring and summer seasons, respectively, were 2.67, 9.51, and 1.32, indicating low contamination levels.

Jareda *et al.*, (2018) conducted a study to assess the Water quality index, heavy metal pollution index and seasonal variation correlation of groundwater and found that it is found that a sizable portion of groundwater samples in the Dantewada district of Chhattisgarh, India, contain iron at toxic levels. Pb concentrations are occasionally discovered to be above the allowable limit. The World Health Organization's (WHO) guidelines were determined to be met by the Al, Cr, and Zn concentration of groundwater.

Ubuoh *et al.*, (2019) used a multivariate and analytical index technique to ze groundwater's cation chemistry near hospital waste dumpsites in Umuahia, Nigeria. They discovered that the Cation Contamination Index (CPI) revealed that cations in boreholes ranged in level of contamination from mild to severely contaminated. The measured water's geo-accumulation of cations ranged from a little polluted state to a substantially contaminated one.

Hossain *et al.*, (2020) researched to evaluate the health risk and groundwater quality: According to a case study on Setabganj Sugar Mills Limited in Dinajpur, Bangladesh, the groundwater is more contaminated near the sugar mill than it is farther away, as shown by the Metal Index, Degree of Contamination, Pollution Load Index, and Heavy Metal Pollution Index. According to the health risk assessment, adults are only at high risk around the sugar mill, whereas infants and young children are very susceptible to polluted groundwater across the study area. Sahoo and Swain (2020) proposed a new modification in heavy metal pollution index (m-HPI) to represent heavy metal pollution. The modified version of HPI represents a Positive Index (PI) and a Negative index (NI) which indicates pollution level exceeding or equal to and less than the maximum desirable limit respectively. This study's results indicated water suitability for domestic and irrigation uses. The m-HPI was found user friendly, robust, flexible and may appraise index value considering any water quality standard.

Islam *et al.,* (2021) appraisal of the groundwater quality and risk assessment of heavy metal pollution in Bangladesh and revealed that the research area's water samples had medium to high levels of contamination. According to the analysis's findings, there was, on average, a low, medium, and increasedrisk of exposure to heavy metals in 32.6%, 15.6%, and 51.8% of the water samples, respectively. According to Krishan *et al.,* (2021) the majority of Punjab's measured groundwater is free of heavy metal contamination and suitable for human consumption.Kadam *et al.,* (2021) found that children are more liable to non-carcinogenic health risk than adults.

Tesseme *et al.,* (2022) evaluated the impact of dumping sites on Ethiopian water quality using leachate and comprehensive pollution indices (CPI), and they discovered that, according to the CPI's grading scale, the water quality was significantly contaminated at every monitoring station. In a study by Souid *et al.,* (2022) that evaluated the risk of groundwater pollution using a calculated contamination index and a geostatistical analysis of the Jerba Island case study, it was discovered that the vast majority of the unconfined aquifer (76%) is located in the significant pollution zone, while the remaining areas (24%) are categorized as areas with low levels of pollution.

Dheeraj *et al.,* (2023) conducted a study to assess the groundwater quality in the Korba coalfield region of India using GIS and Heavy Metal Pollution Index (HPI) approaches. They found that the HPI value determined by mean concentration was calculated to be 21.64, significantly less than the reference pollutant index score of 100. According to the HPI computation, the HPI values for 73.33% of the groundwater samples were low, 6.67% were medium, and 20% were high.

Contamination of Drinking Water and its Effect on Human Health

Wu & Sun (2016) assessed the groundwater quality and health risk in Mid-West China and found that 60% of samples are not fit for drinking and the study area are at high health risk, and nitrate originating mainly from industrial and agricultural pollution is the greatest contributory cause of the health risks. Furthermore, children in this area are at higher health risk than adults, and oral ingestion is the dominate exposure pathway of health risk.

Sener *et al.,* (2017) researched to evaluate the health risks associated with drinking water basin groundwater quality. They discovered that HQ intake >1 was the main contaminant causing non-carcinogenic concerns and found that both adults and children, As and Cr were the main causes of chronic hazards.

Varol *et al.,* (2018) assessed the water quality and health hazard of groundwater resources contaminated by agricultural activities in Korkuteli. The study area is at high risk for adverse health effects, and the water samples are not suitable for use as drinking water due to the presence of heavy metals. The levels of arsenic were higher than allowed. Zhang *et al.*, (2018) found that the health risk of dermal contact is less than that of ingestion, and children are at much more risk than adults. Anh & Giao (2018) assessed the groundwater quality and human health risk assessment in Viet Nam and found that the carcinogenic risk for the human population ranged from medium to high for both children and adults.

Krishna *et al.*, (2019) evaluated the groundwater quality, toxicity, and health risk in an industrial area and discovered that for both seasons all cations and anions were in the following order Ca>Na>HCO_3>Cl>CO_3>SO_4>Mg>NO_3>K>F. According to Xiao *et al.*, 2019 research the total potential non-carcinogenic risk presented by nitrogen and heavy metals (As, Zn, Fe, Mn) through drinking water intake occurs in the southwestern area and a small local area in the centre of the study area, and is higher for children. According to Ramadan *et al.*, (2019), cancer risk for adults in groundwater is as follows: high cancer risk for As; low cancer risk for Cd and very high cancer risk for Cr while that for children is very high cancer risk for As and Cr and then low cancer risk for Cd.

Kamble *et al.*, (2020) assessed the health risks associated with groundwater iron and manganese in Chandrapur District, Central India and found that the WHO's discontinued manganese threshold of 0.4 mg/L was exceeded in severalgroundwater samples. Chronic daily consumption, hazard quotient, and hazard index for these heavy metals were deficient, indicating no immediate or long-term health impact to study region residents from ingesting this groundwater. According to Adimalla *et al.*, (2020) the non-carcinogenic health risk in the research area varied from 0.63 to 5.31 2.59 for adults, 0.85 to 7.18 3.50 for children, and 0.98 to 8.29 4.04 for infants, demonstrating that the health risk was larger in infants and children than in adults. Xiao *et al.*, (2020) analysed the groundwater quality and health risk in a near-suburb area of North China and found that hazard quotient was very high in some local areas.

Chen *et al.*, (2021) analysed the groundwater quality and potential human health risk assessment for drinking and irrigation purposes in the Semiarid Region of North China. They found that the NO_3^-, NO_2^- and F^- surpassed the regulatory limits and may represent possible dangers to human health. According to Varol (2021) the arsenic concentrations in the groundwater above the maximum allowable limits, making it unfit for consumption. Kabir *et al.*, (2021) found that for both adults and children, the mean hazard quotient (HQ) and hazard index (HI) were beyond the acceptable range, having a significant negative influence on human chronic health.

Rahman *et al.,* (2022) evaluated the health risks associated with naturally occurring potentially toxic metals on Hatiya Island in Bangladesh and found that the hazard quotient (HQ) values for each metal (Cu, Zn, Fe, and Mn) did not indicate any harm to oral and dermal health. On the other hand, the hazard index (HI) readings showed no risk for combined metals because none of the values exceeded the safety limit value of 1. According to Gevera *et al.,* (2022) diarrhoea and gastrointestinal disturbances are among the health problems that many (43%) respondents in the region mentioned, and these problems may be related to the area's salty drinking water. Gao *et al.,* (2022) found that is up to the 92.25% of non-carcinogenic risk for children and unsuitable for drinking purposes.

Panneerselvam *et al.,* (2023) evaluated groundwater's quality and health risks for drinking and irrigation purposes by using the entropy water quality and statistical approaches in the semi-arid region of India. They discovered that non-carcinogenic risk evaluation revealed that children aged 6 to 12 are at a higher risk than teenagers, adults, and older people in the study area. Gad *et al.,* (2023) state that the HI shows a potential health risk associated with water consumption, which ismore significantfor children in just one area. However, there was little chance of cutaneous and gastrointestinal contact to the water in any place for kids or adults.

In light of the gaps identified and approved research objectives, it is possible to thoroughly analyse the public drinking water sources for a variety of parameters for spatial and seasonal variations based on the compilation of research work about the water quality of Dehradun presented above. The summary of the literature review under this chapter includes the earlier research findings that assisted in identifying and contrasting changes in the various physical, chemical characteristics of water from different ground water sources in the Dehradun Industrial region of Uttarakhand over two years. The planning to maintain the desired potable water quality, which is declining due to water misuse, uncontrolled population growth, industrialization and urbanisation, etc., can be made and achieved using the compilation above of literature about water quality, along with source identification and health issues. Therefore, the current study is being conducted to look at changes in water quality with spatial and seasonal variations and identify the causes and sources (origin) of pollution entering into Dehradun's drinking water sources so that appropriate, cost-effective measures can be taken.

13

Case Study – Methodology

This chapter describes the methods, techniques, and principle of operation of various instruments employed for the research work. National and International Standards and protocols were followed for sample collection and analysis. Multiparameter potable meter and various instruments were used for measuring the in-situ and ex-situ concentration of industrial wastewater and groundwater samples. Multivariate statistical analysis was done by the use of IBM-SPSS 26.

Research Design

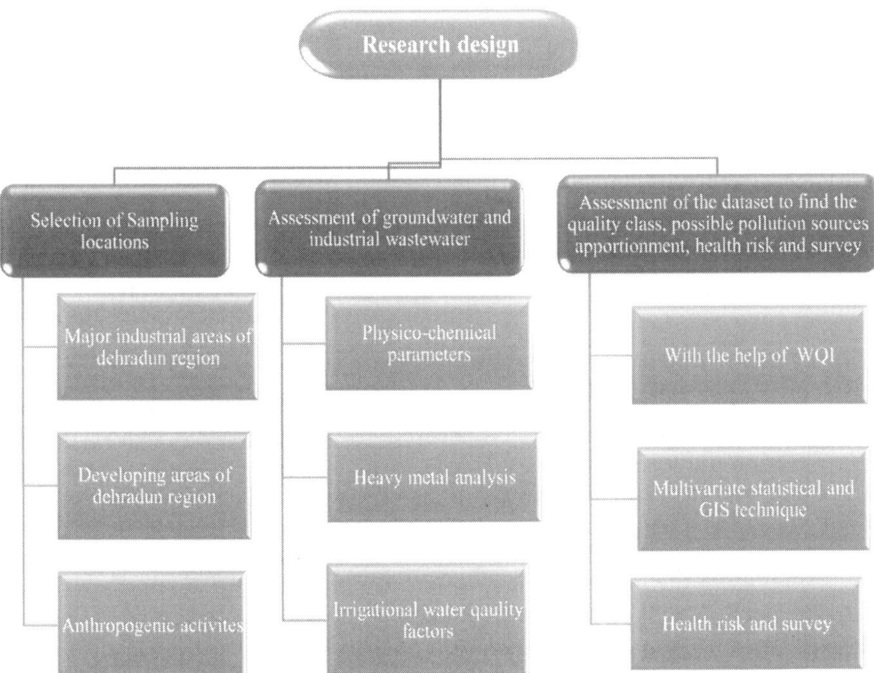

Figure 13.1. Flow chart of the study design

Description of the Research Area

Selection of monitoring locations

The Uttarakhand, nestled in India's northern region, has an elevation of about 300–7000 m above mean sea level (MSL) and is a central Himalayan state (Chauhan, 2010). The Himalaya is known as the third ice polar region (Yao *et al.*, 2012); vast, extensive rivers stretch down to it and constitute the region's geography. Approximately 7.2% of the Indian Himalayan glacier area exists in Uttarakhand (Sangewar and Shukla, 2009). The total forest area of Uttarakhand state is about 34,651 Km^2 which is 63.7 percent of the state's geographical area. The state lies in the highest seismic zone, is prone to geo-hazards, and is exposed to different forms of hydro-meteorological extremes such as glacial lake outburst floods, cloud bursts, floods, flash floods, drought due to water scarcity, hailstorms, landslides, rock falls, forest fires, and debris flow (Kala, 2014; Kumar *et al.*, 2019).

Dehradun District is the State capital of Uttarakhand and is located in the Lower Himalayan Region of Northern India, with latitudes and longitudes of 30.3165° N, 78.0322° E (Figure 3.2). The average elevation of the Dehradun District is around 447 m above sea level, and the total area of the sampling location is 196.48 Km^2. Figure 3.3 shows the satellite view of the proposed industrial area. According to the 2011 national census, the overall population of Dehradun District is 803,983 (Census, 2011). This region may be categorized into four geomorphological components: alluvium, structural and denudational hills, piedmont fan deposits, and residual hills. The climatic condition of the area is humid subtropical and varies from cold to tropical depending upon the elevation of the location. The average yearly rainfall is around 2073.3 mm, and the constant relative humidity ranges from 14.9% to 67.6%. The identification was made for groundwater sampling sites based on the effective use of groundwater impact (expected), majorly near agricultural and industrial areas. These sampling locations' groundwater is used for ingestion, domestic, and irrigation purposes.

The Dehradun District receives drinking water from two significant sources, surface water, and groundwater. According to hydrogeological conditions, the Dehradun is distributed into 3 hydrological components: the Himalayan mountain belt, the Siwalik Zone, and the Doon gravels (CGWB, 2019). In the Himalayan mountain belt unit, groundwater separates local bodies from confined and unconfined conditions. The major rock types in this unit are quartzite, shale, slate, phyllite, limestone, sandstone, and dolomite from Jaunsar. In the Siwalik zone, groundwater is found under confined and unconfined conditions, and this unit's water level is comparatively deep. The part of the intermountain valley in the Dehradun District is sustained through alluvial fan deposits (Bahukhandi & Bartarya 2014). These alluvial fan deposits are known as Doon gravels, Primarily consisting of sandstone, phyllite, and quartzite rocks. The study's impact of industrial wastewater on groundwater quality in the study area was divided into five sampling locations (Figures 3.4 to 3.9) mentioned in the table, and samples were collected for physico-chemical, heavy metal analysis, and health risk assessment (Table 3.1).

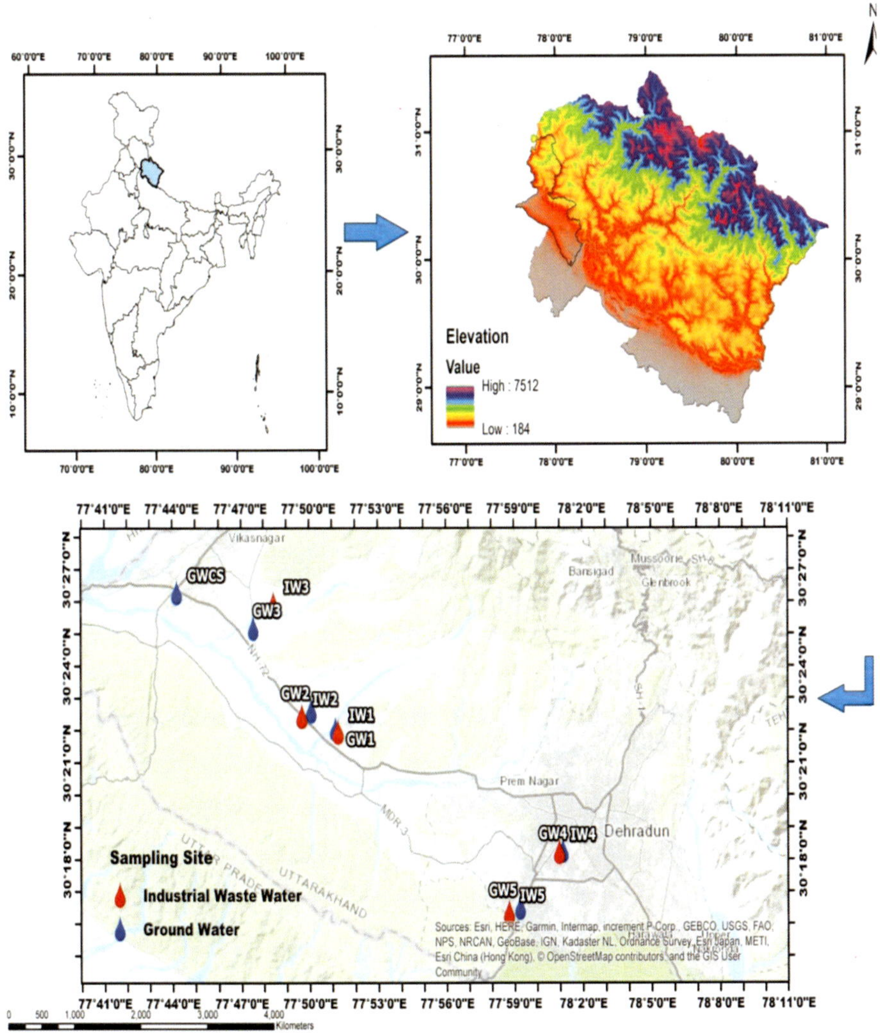

Figure 13.2. Map showing the location and outline of the study area

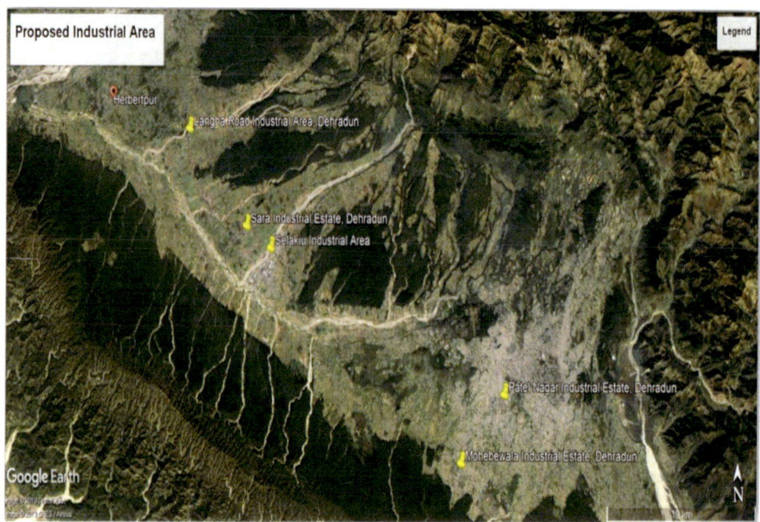

Figure 13.3. Satellite image showing the selected industrial area in Dehradun District in Uttrakhand region (*Source*: Google Earth)

Table 13.1 Description of sampling locations with geo-coordinates

S. N.	**Sampling locations for groundwater/ Industrial waste water**	**Sam-pling code**	**Geo-coor-dinates for ground water samples**	**Source**	**Geo-coordinates for Industrial waste water samples**	**Sam-pling code**	**Ele-vation (m)**
	Herbertpur (Control site)	GWCS	30^{0}43’83.00” N 77^{0}73’66.00” E	Hand pump			
1	Selaqui Industrial estate	GW1	30^{0}22’3.16” “ N 77^{0}51’8.44” “ E	Hand pump	30^{0}21’58” N 77^{0}51’15”E	IW1	525 m
2	Sara Industrial Estate, Dehradun	GW2	30^{0}22’37.87”N 77^{0}50’4.73”E	Hand pump	30^{0}22’27”N 77^{0}49’40”E	IW2	502 m
3	Langha Road, Industrial area Dehradun	GW3	30^{0}25’11.71”N 77^{0}47’32.24”E	Hand pump	30^{0}25’56”N 77^{0}48’26”E	IW3	497 m
4	Patel Nagar, Industrial area Dehradun	GW4	30^{0}18’16.75”N 78^{0} 01’7.44” E	Hand pump	30^{0}30’37.45”N 78^{0} 02’16.63” E	IW4	622 m
5	Mohebewala, Industrial area Dehradun	GW5	30^{0}16’30.90”N 77^{0}59’14.62”E	Hand pump	30^{0}16’21”N 77^{0}58’45”E	IW5	615 m

Figure 13.4. Herbertpur (Control site)

Figure 13.5. Selaqui Industrial Estates (Sampling Site 1)

Figure 13.6. Sara Industrial Estates (Sampling Site 2)

Figure 13.7. Langha Road, Industrial area (Sampling Site 3)

Figure 13.8. Patel Nagar, Industrial area (Sampling Site 4)

Figure 13.9. Mohebewala, Industrial area (Sampling Site 5)

Hydrogeology of Research Area

Initially, Meijerink (1974); Saxena (1974); Kainthola *et al.,* (1988), and Roy, A.K. (1991) provided the initial geo-hydrological framework of the Doon Valley. Latter, Bartarya (1995) gave a detailed hydrogeology of the Doon Valley. Hydrogeological, the Doon Valley has been subdivided (Bartarya 1995) into:

1. Lesser Himalaya Zone
2. Synclinal Central Zone
3. Siwalik Zone

Lesser Himalaya Zone

The steeply sloping northern flank of the valley comprises rocks of the Lesser Himalayan formations, such as quartzite, schist, slates, phyllite, hard sandstone limestone, and dolomite of the Chandpur, Nagthat, Blaini, Krol, and Tal formations of the lesser Himalaya, which have secondary porosity and permeability and are characterized by springs and seepage. Though sedimentary in nature, the rocks have very low intergranular porosity and are characterized by fissures, fractures, and joints. The zone of lineament, fault, and main boundary thrust show pockets of high secondary porosity. The subsurface water in this zone occurs predominantly as disconnected local bodies in favorably perched aquifers under both confined and unconfined conditions and in the zone of jointing, fracturing, and faulting. Relatively flat areas and gently sloping ground are characterized by deep weathering, such as hilltops, ridges, saddles, spurs, and bulges of old landslide debris, river terraces, and fluvial fans from the recharge area, while steeper hill slopes, first and second order streams at slope breaks, and scraps of fans are sites of discharge.

The upper portion of the catchment area is saucer-shaped, giving rise to large underground bodies of perched water. The spring and seepage issue out from all sides of the individual aquifer at different elevations through fractures and joints and from weathered material within short distances. The sand-gravel deposit of fluvial and colluvial origin in the Lesser Himalayan Zone lying in the lower reaches of the stream or near the confluence of two streams in the form of fans and terraces is highly porous and permeable and therefore holds sufficient quantities of water. Reliable groundwater supply from streams in quantities enough to meet the needs of small villages of 150–400 persons is thus available in almost all valleys in which traces of sand and gravel deposits are present, such as Manjgaon, Pipalsain, and Nayar, along the Song river. The springs show wide discharge variability from 1400 to 150700 L/day.

Synclinal Central Zone

The zone classified under the predmont zone and occupied by the Doon gravels, which have primary porosity and permeability, is forming the area's main aquifer. The groundwater is present in multi-tiered aquifers under confined and unconfined conditions. The main water-bearing strata are the coarse sand and gravel underlain by clay beds. The zone is characterized by a high infiltration rate. A number of dug wells and tube wells are primarily confined to this zone. Several water table springs (classified as springs related to fluvial deposits) are present along the valley floor and along the terrace and fan scarp,contributing to the base flow of streams during the lean period of discharge. The discharges of such springs are usually high and vary from 2000 L/day to 250,700 L/day.

The Siwalik Zone

The water is present under confined conditions, and the water table is relatively deep. Though the upper Siwalik's conglomeratic unit is highly porous and permeable, because of steep slopes, fans forming a piedmont zone along the southern fringe of the valley, and moderate to steep dips in beds, the water quickly goes as runoff.

Sub Surface of Geo-hydrology

The lithologs of the borehole used to drill the tube well and the vertical segment that was left exposed in the vicinity provide the best understanding of the subsurface geo-hydrology of the area. The lithologs of the tube well maintained by CGWB, state irrigation, and the Jal Nigam department indicate that the horizons comprising boulders and gravel set in the coarse sandy matrix are the main water-bearing horizons. Such horizons are encountered more than once in the deep drill. Clay horizons separate these horizons; a few are also present in Dharmawala, Raipur, and Karanpur. Such horizons do possess water in their pore spaces, but the yield from such horizons is rather poor or low. The water-yielding horizon occurs at about 20m below ground level in the western part of the valley. A few horizons of perched nature with limited dimension and yield are also present in the valley's northwest, southeast, and eastern parts. Since the Doon gravels are dominantly fan deposits, they generally follow the topographic profile where they have been deposited. The various water-bearing horizons in the present area taper towards the foot of the fans, that is, towards the Asan and Song Rivers and hence the shallow depth of the water table, whereas towards their head region they occur at higher levels.

Sample Collection and Analysis

The sampling plan was prepared according to the selected water quality parameters. The study was conducted for two years, from April 2019–March 2020 to August 2020–July 2021, at all the selected sampling sites in Dehradun District. All the groundwater (bi-monthly) and Industrial effluent (monthly) samples were collected from the studied region for four different seasons (i.e., summer: May to June, monsoon: July to August, autumn: October to November, and winter: January to February) during the study period.

Selection criteria of sites, collection, preservation, and transportation of the water samples

The obtained water samples were put in one-litre pre-acid-washed polypropylene containers for hydro-chemical examination. A GARMIN portable GPSMAP device (Model 60CSx) recorded the geo-coordinates of all the monitoring locations. The equipment used for all analyses was pre-calibrated using standard solutions or as indicated by their company guidelines to preserve accuracy and precision in the observations. Before each in situ observation, all electrodes were thoroughly cleaned with deionised water. The probes were conditioned in the sample before each usage to ensure the best stabilisation time. The pH probe was calibrated using freshly made buffer solutions of two units after being rinsed and dried. Throughout the research period, established protocols for each piece of equipment and chemical analysis were adhered to with adequate safety. Before being transported to the lab, samples were maintained in an ice bucket at 4°C to prevent contamination and the effects of temperature and light. During the study period following Parameters like pH, Electrical conductivity, Water temperature (°C); Turbidity (NTU); Total Dissolved Solid (mg/l); Dissolved Oxygen (mg/l); Biochemical Oxygen Demand (mg/l); Chemical Oxygen demand (mg/l); Acidity; Alkalinity (mg/l); Total Hardness (mg/l); Calcium (mg/l); Magnesium (mg/l); Sodium (mg/l); Potassium (mg/l); Bicarbonates (mg/l); Nitrates (mg/l); Phosphate (mg/l); Chloride (mg/l) using the procedure described in Table 3.2. For dissolved oxygen (DO) and biochemical oxygen demand (BOD), the water samples were collected in a BOD bottle without any bubbles. The BOD bottles were fixed by adding 2 ml of $MnSO_4$ and an alkali azide solution. After that, collected water samples were transported to thelaboratory for further analysis. ome of the major equipment and instruments used in assessment, sampling, and analysis are depicted in Figures 3.10 to 3.14. During the study period, the following parameters, like Zinc, Iron, Cadmium, Copper, Nickel, Manganese, chromium, and lead, were studied in water using a standard procedure. 50 ml of water samples were taken in a beaker, and 5 ml of concentrated HNO_3 was added to them. The sample was slowly boiled on the hot plate until a light color appeared. During the study period, heavy metals in water sample were analysed using flame atomic absorption spectrophotometer (FAAS).

Quality Assurance and Quality Control (QA/QC)

The instruments used for all the analysis were pre-calibrated with standard solutions or as recommended by their company guidelines to maintain accuracy and precision in the observations. All of the electrodes were properly washed with deionized water before each in-situ observation. The probe was conditioned in the sample for the best stabilization time before each use. The rinsed and dry pH probe was calibrated through freshly prepared buffer solutions in two different units. A blank sample was run for each chemical analysis, and at least three observations were taken for each quality parameter. Standard operating protocols were adopted for each instrument and chemical analysis with adequate safety throughout the study period. The 'A' Grade of glassware and chemicals from Borocile, Merch, and Fisher Scientific, USA, were used for all chemical analysis to get accurate information.

Software Used and Statistical Analysis

Data management and WQI calculation were done using Microsoft Excel 2016 (part of the Microsoft Office suite); study area delineation was done using ArcGIS version 10.3; data treatment, all mathematical operations, and graphical presentations were done using Microsoft Excel, IBM SPSS 23, and Origin Pro-2020 b software.

Table 13.2. List of water quality parameters with their adopted methodology.

S.N	Water quality Parameters	Type of Param-eter	Unit	Adopted methodology and instrument used	Standard protocols used of APHA&BIS
	pH	Physical	-	Potable multi-parameter kit HACH(HQ40d)	8156,USEPA, 2018
	Temperature	Physical	^{0}C		
	Electrical conductivity	Physical	(µS/cm)	Digital Conductivity Meter (Model No. ESICO-1601)	2510-B,APHA, 2017
	Total Dissolved Solids	Physical	mg/L	Filtration and Gravimetric	2540 C, APHA,2017
	Turbidity	Physical	NTU	Nephelometric method (Turbidity meter-ESICO-335)	2130B, APHA, 2017
	Dissolved oxygen	Chemical	mg/L	Winkler Titration method	4500O,APHA,2017
	Biochemical Oxygen Demand	Chemical	mg/L	Winkler Titration method	5210B,APHA,2017
	Chemical oxygen demand	Chemical	mg/L	COD analyser (by DR1900: spectrophotometer and DRB200:COD reactor of HACH)	10211, USEPA, 2018
	Hardness (Ca/Mg)	Chemical	mg/L	Titration method (EDTA)	2340C, APHA, 2017
	Alkalinity	Chemical	mg/L	Titration method	2320, APHA, 2017
	Acidity	Chemical	mg/L	Titration method	2320 B, APHA, 2017
	Nitrates	Chemical	mg/L	UV spectrophotometric screening method (Cary 60 UV spectrophotometer)	4500-NO_3^+ B, APHA, 2017
	Phosphate	Chemical	mg/L	Titration method (Cary 60 UV spectrophotometer)	4500-PD, APHA, 2017
	Sulphate	Chemical	mg/L	Turbidimetric method (Cary 60 UV spectrophotometer)	4500-SO_4^{2+} E, APHA, 2017
	Chloride	Chemical	mg/L	Argentometric method	4500-Cl^- B, APHA, 2017
	Potassium (K^+)	Chemical	mg/L	Flame photometric method	3500-K D, APHA, 2017
	Sodium (Na^+)	Chemical	mg/L	Flame photometric method	3500-Na D, APHA, 2017

	Zinc (Zn)	Metal	mg/L	Atomic absorption spectrophotometric method (model: AAS AA 8000 (PC-based)	3111 B, APHA, 2017
20.	Iron (Fe)	Metal	mg/L		
21.	Cadmium (Cd)	Metal	mg/L		
22.	Copper (Cu)	Metal	mg/L		
23.	Nickel (Ni)	Metal	mg/L		
24.	Manganese (Mn)	Metal	mg/L		
25.	Chromium (Cr)	Metal	mg/L		
26.	Lead (Pb)	Metal	mg/L		

Arithmetic mean

It is obtained by sum of the observations divided by the total number of observations. The formula for calculating the Arithmetic Mean as below:

$$X = \frac{n}{N} \quad (1)$$

Where,

X = Arithmetic mean

n = sum of all the observation

N = total number of samples

Standard deviation (S.D.)

It is the square root of the arithmetic mean of the square of the deviation of a given observation from then the arithmetic mean.

The formula for calculating the Standard deviation as below

$$S.\ D. = \sqrt{}(x - X)\ ^\wedge 2/\ (n - 1) \quad (2)$$

Where,

x = individual reading of parameter

X = mean of the sample

n = Number of observations.

Coefficient of correlation (r)

It is defined as the degree of relationship between the two variables called a Coefficient of correlation. The formula for calculating the "r" is given below;

$$r = \sqrt{}\ (\sum (x - X)\ (y - Y) / \sum (x - X)\ ^\wedge 2\ \ \sum (y - Y)\ ^\wedge 2 \quad (3)$$

Where,

x = Individual reading of 1st parameter

X = mean value of X parameters

y = Individual reading of 2nd parameter

Y = Mean of Σ Y parameters

Multivariate Statistical Method

Principal component analysis

PCA is utilized to reduce the dimensionality of the initially observed dataset by defining and standardizing the main responsible water quality variables. A preliminary assumption test (Kolmogorov-Smirnov: KS and Shaphiro-Wilk: SW) was conducted to check the normality of the dataset. The null hypothesis for these normality tests is that the data is normally distributed (H_0) or not normally distributed (H_1). Water quality variables had different scales of measurements and magnitudes; therefore, the data needs to be standardized to fit all parameters' normal distributions (Mishra, 2010; Maji and Chaudhary, 2019). PCA makes the water quality estimation process more feasible and economical, drastically decreasing the effort, cost, and time required for many variables. PCA can obtain information related to a particular variable with the minimum loss of the entire data set (Singh, 2004). PCA discarded redundant and highly correlated variables (Sharma *et al.*, 2015). The suitability or adequacy of the dataset for PCA was determined based on Kaiser-Meyer-Olkin (KMO) and Bertlett's tests. The varimax rotation method is used to recognize the association between data sources and essential factors (Noori *et al.*, 2010). The information is achieved by transforming the original dataset into a normalized version due to unit variation and generating a new set of uncorrelated pseudo-variables called Principal Components (PCs). Those PCs with eigenvalues greater than 1 were recognized by generating screen plots.

Water Quality Index

Groundwater Contamination index (C_d)

Application of the contamination index, which measures water quality by determining the level of contamination in each water sample (Backman *et al.*, 1998). To calculate the maximum allowable concentration (MAC), it was necessary to count the contamination factors of individual components that were higher than the upper permitted limit. Consequently, the Cd takes into account the combined impacts of various water quality factors that are believed to be hazardous to home water. The following equation is used to calculate the Contamination Index (C_d).

$$C_d = \sum_{i=0}^{n} C_{fi} \tag{4}$$

$$C_{fi} = \frac{C_{Ai}}{C_{Ni}} - 1 \tag{5}$$

Where, $C_d < 1$ (Low contamination), $C_d = 1–3$ (Medium contamination), $C_d >$ 3 (High contamination).

Comprehensive Pollution Index

This method is used to categorise the water quality status by many of the research fndings (Mishra *et al.,* 2015; Zhao *et al.,* 2012). The following equations calculate it as:

$$CPI = \frac{1}{n}\sum_{i=1}^{n}\frac{M_i}{S_i} \tag{6}$$

Where, Mi is monitored value for each water quality parameter, Si is standard permissible limit of ith water quality parameter, n is number of parameters. The standard permissible limit of each parameter considered in this work is used from the proposed norms of the diferent governments for a general discharge of an environmental pollutant (BIS 2012; WHO 2011). CPI classifies water quality ≤0.20 as clean; 0.21–0.40is sub-clean; 0.41–1.00 slightly polluted; 1.01–2.0 is moderately polluted; ≥2.01 severely polluted.

Heavy Metal Pollution Indexing (HPI)

A scoring approach for aquatic systems that shows the cumulative impact of toxic metals in water bodies is called the Heavy Metals Pollution Index (HPI) (Mohan *et al.,* 1996). Three types of HPI measurements are used: low (15), medium (15–30), and high (>30) (Edet and Offiong, 2002; Tiwari and Singh, 2016). Water quality for ingesting or other uses may be determined using the water quality index (Mohan *et al.,* 1996; Parsad and Bose, 2001).The following are the steps for calculating HPI:

The first stage involves figuring the amount the ith parameter weighs, the second stage involves zing the water quality for every trace metal, and the third stage combines all sub-indices to get the total index.

The ith parameter's weight is:

$$W_i = k/S_i \tag{7}$$

Wi stands for weight in units

The following equation is used to get S_i =, where k is the proportionality constant

$$K = \frac{1}{\sum_{i=0}^{n} \frac{1}{Si}} \quad (8)$$

The phrase assigns a specific quality score

$$Q_i = \sum_{i=0}^{n} \frac{(M_i - I_i)}{(S_i - I_i)} X100$$

Qi is the ith parameter's sub-index here

Mi measured the ith parameter's value in μg/L, S_i the allowed upper limit, and I_i, the optimum value for the ith parameter. The heavy metal index was determined using the formula:

$$HPI = \sum_{i=}^{n} (Q_i W_i) / \sum_{i=0}^{n} W_i \quad (10)$$

In this case, the ith parameter's sub-index is Q_i, n is the number of parameters taken into consideration, and W_i is a unit weight for the ith parameter. 100 represent the crucial pollution index number (Parsad and Bose, 2001).

Irrigation Groundwater Quality Evaluation

Salinity and Alkalinity Hazard (SAH) Electrical conductivity and US Salinity Laboratory diagram helped in explaining salinity and alkalinity hazard in the study area. The salinity hazard (C) is classified into 4 subzones: very low salinity (C1: 250 μS/cm), medium salinity (C2: 250 to 750 μS/cm), high salinity (C3: 750 to 2250 μS/cm), and very high salinity (C4: above 2250 μS/cm).

Sodium Adsorption Ratio (SAR) SAR was calculated using the equation given by Richards 1954.The concentration of the ions was expressed in meqL^{-}.

$$SAR = \frac{Na^+}{\sqrt{(Ca^{2+} + Mg^+)/2}} \quad (11)$$

Magnesium hazard (MH) MH is used to evaluate groundwater suitability for irrigation by determining the concentration of Ca^{2+} over Mg^{2+} (Ragunath, 1987). Calcium and magnesium are essential nutrients for crops and commonly occur in the soil. Their high water value increases the soil's pH (Joshi *et al.*, 2009). Excess quantity of Mg^{2+} ions in irrigation water adversely impacts soil quality (makes soil alkaline) and reduces crop production (Snousy *et al.*, 2021). The magnesium adsorption ratio (MAR) was calculated by using the following equation (Szabolcs and Darab, 1964):

$$MAR = \frac{Mg^{2+}}{(Ca^{2+} + Mg^{+})} \times 100 \quad (12)$$

Sodium Percentage (Na%) The sodium content of groundwater can be used todetermine its quality for irrigation (Wilcox, 1955). As sodium reacts with soil and reduces its permeability and texture, a high sodium percentage in water causes soil degradation (Karanth, 1987; Naidu, 2021). The sodium percentage was calculated by using the following equation:

$$Na\% = \frac{Na + +K +}{Ca^{2+} + Mg^{2+} + Na^{+} + K^{+}} \times 100 \quad (13)$$

Gibbs Ratio Gibbs (1970) proposed a diagram for interpretation of the mechanism of major ion chemistry of groundwater samples. Gibbs diagram explains the three fields: precipitation dominance, evaporation dominance, and rock dominance. Gibbs ratio was calculated with the help of the following formula:

Gibbs Ratio I (For Anion) $Cl^{-}/(Cl^{-}+HCO_3)$ (14)

Gibbs Ratio II (ForCation) $(Na^{+} + K^{+})/(Na^{+}+K^{+}+Ca^{2+})$ (15)

Human Health Risk assessment

The US Environmental Protection Agency (USEPA) developed a methodology for evaluating potential health risks in 1989. When people regularly consume polluted groundwater, the kind and intensity of negative effects on their bodies are evaluated through health risk assessment (Li *et al.,* 2019; Adimallaa and Qiana, 2019). The USEPA (2004) risk assessment technique was utilised to investigate the health risks brought on by each element. Trace metals in groundwater can be ingested by potable water via the mouth, and they can also be contacted through the skin. The following equations were used to determine regular health and exposure risk monitoring for trace elements linked to groundwater: I chemical daily intake (CDI), which takes into consideration oral ingestion by consuming water (Eq. 15), & (ii) dermal absorbed dose (DAD), that takes into consideration dermal contact of PTEs attached to skin Eq (16)

$$CDI = \frac{[C \times IR \times EF \times ED]}{[BW \times AT]} \quad (16)$$

$$DAD = \frac{C \times SA \times Kp \times ET \times EF \times ED}{BW \times AT} \quad (17)$$

The following formula was used to determine the Hazard Index (HI), Hazard Quotient (HQ), and Chemical Daily Intake (CDI) and Dermal Absorbed Dose (DAD) of metals in groundwater:

$$HQ_{ingestion} = \frac{CDI_{ingestion}}{RFD_{ingestion}} \tag{18}$$

$$HQ_{Dermal} = \frac{DAD_{Dermal}}{RfD_{Dermal}} \tag{19}$$

$$HI\,(Hazard\ Index) = \sum_{i=1}^{n} HQ \tag{20}$$

C for metal level in the groundwater (mg/L), ED for exposure duration (30 for adults and 6 for children), EF for resident ingestion frequency (365 days/year), and BW for body weight are the acronyms for chemical daily intake (CDI) and dermal absorbed dose (DAD) of toxic metals on a per-unit-bodyweight basis (mg/L-day) (70 kg for adults, 15 kg for children) Ingestion rate is denoted by the acronym IR (2 litres per day for adults, 1 litre per day for children). The acronym AT stands for averaging time (365 days per year), SA stands for unprotected skin area (18,000 cm^2 for adults 6600 cm^2 for children), ET stands for exposure time (0.6 h/day for adults, 1 h/day for children), AT stands for averaging time (25550 days for adults and 2190 for children), Kp stands for skin adhesion factor , for Cu, Mn, Fe and Cd the Kp values are 1×10^{-3}, for Zn (6×10^{-4}) $^{, Cr, Ni}$ (2×10^{-3}) and CF is the conversion factor(0.001). RfD stands for reference dose, and the values for Cu, Zn, Mn, Fe, Cd, Cr, and Ni are 0.0371, 0.3, 0.014, 0.7, 0.001, 0.003, 0.02, respectively. N represents the sample count. The hazard index value for HI Total indicates the entire level of non-carcinogenic risk in the human body. Negative health impacts that are not cancerous are proposed if (HQ>1), and none are proposed if (HQ< 1).

Health risk assessment by Survey method

Data collection and analysis

The survey was conducted in the major industrial region of Dehradun District. They are highly populated, with different industries and some locations near the agricultural region. So they might have possibilities for pollution in terms of air, water, noise, plastic pollution, and hazardous contamination. During the ground survey, different age groups were selected near the sampling region to assess the health risk. The data collection techniques, i.e., face-to-face interview, group discussion, and direct observation, were employed using the questionnaires.

Health Risk Assessment Questionnaire

Name	Address
Q1. Age	Q2. Gender
Q3. Occupation	
Research Scholar: Anjali Nayak, Deptt. of zoology and environmental science, GurukulKnagri [Deemed to be University] Haridwar, Uttrakhand Supervisor: Dr. Gagan Matta Deptt. of zoology and environmental science, GurukulKnagri [Deemed to be University] Haridwar, Uttrakhand	

Q.4. Which of the following water resources of drinking water does your household use?

a) Government hand pump/ bore well

b) Surface water

c) House hold water supply

d) Personal hand pump

Q.5. Source depth in feet

a) <100 b) 100 to 150

c) 150 to 200 d) >200 [e] unknown

Q.6. What are the major pollution causing sources in this region?

a) Industrial wastewater b) Agricultural runoff

c) Domestic/Municipal wastewater d) Don't Know

Q.7. In general, what would you say about your health?

a) Excellent b) Good

c) Fair d) Poor

Q.8. Do you have any disease related symptom to?

a) Skin b) Mouth and teeth

c) Stomach or intestine d) Other

e) None

Q.9. In which time duration the water quality is most affected?

a) Winter b) Summer

c) Monsoon d) Post-monsoon

e) Don't Know

Q.10. Does water smell?

a) No smell b) Foul smell

c) Pungent smell

Q.11. What does the water look like?

a) Clear b) Dirty/cloudy

c) Blackish/Brownish d) Yellowish

Q.12. Are you satisfied with your drinking water quality?

a) Yes b) No

Q.13. Have you made a complaint related to your drinking water service to government body or media?

a) Yes b) No

Q. 14. What was the result of the complaint?

a) Prompt action taken v) Delayed action taken

c) No action taken

Q.15. What are the reasons of your dissatisfaction regarding water quality? (in brief)

14

Case Study – Observations and Results

The objective of this research was to identify the level of prominence at which certain physico-chemical and heavy metal indicators were found in groundwater samples used for industrial, agricultural, and home purposes. The contribution of diverse contamination sources was validated using the deterministic indices approach to categorize, identify, and perform cost-effective analysis. The results and interpretation of all huge datasets were made by interpolation, Principal Component Analysis (PCA) and Correlation matrix (CM) were used for pollution source apportionment. The comparison of results with national and international agencies helped to extract more significant variables influencing water quality. The outcomes presented here will support the controlling agencies at the later stage to curb the water pollution at the specific location.

Spatio-temporal variation in physico-chemical parameters and potential toxic elements (PTEs) in Industrial wastewater:

The monthly observed variation in Mean values of hydro-chemical parameters for each monitoring location is represented in Table 4.1 to 4.5 and Table 4.6 to 4.10 for 2019-2020 and 2020-2021, respectively. The annual mean in physico-chemical parameters in groundwater samples of each monitoring location is given in Table 4.11 and Table 4.12 for 2019-2020 and 2020-2021, respectively. The descriptive statistical seasonal variation is shown in Table 4.13 and Table 4.14 for both years.

The description of each variable for minimum, maximum, and mean±sd on a spatial and temporal basis, along the data comparison with a prescribed limit of national agency (CPCB) are represented as follow:

pH

The highest value (8.74) of pH was recorded at sampling location IW1 in the month of May,whereas the lowest pH value (7.10) was measured at IW2 in the month of August for the year 2019-2020 (Table 4.1 to Table 4.10). The pH value was reported below than the acceptance limit for surface water discharge limit prescribed by CPCB which is 9.0. The annual mean of pH values was recorded 7.53±0.15 and 7.47±0.16 for 2019-2020 and 2020-2021, respectively (Table 4.11 and Table 4.12).

The seasonal variation in pH values were estimated as 8.06±0.48, 7.32±0.32, 7.24±0.05 and 7.21±0.05 for SUM, MON, AUT, and WIN season of the year 2019-20. However, the seasonal variation for pH was observed as 7.94±0.49, 7.27±0.22, 7.23±0.06 and 7.20±0.06 for SUM, MON, AUT, and WIN season, respectively, during the year 2020-21 (Table 4.13 and Table 4.14).

DO

In the first year (2019-20) the minimum DO (0.24 mg/l) of wastewater was observed in the month of May, while the maximum DO (0.58 mg/l) was observed in the month of January. The first year annual average was calculated as 0.37 mg/l±0.03. During the second year (2020-21) the minimum DO (0.25 mg/l) was recorded in the month June at site IW1, while the maximum DO (0.46 mg/l) observed in the month of September at site IW2. The second year annual average was calculated as 0.34 mg/l ± 0.02.

The seasonal variation in DO values were estimated 0.31±0.09 mg/L, 0.38±0.07 mg/L, 0.36±0.0.05 mg/L, and 0.41±0.04 mg/L for SUM, MON, AUT and WIN season of the year 2019-20. However, the seasonal variation for DO was observed 0.30±0.05 mg/L, 0.36±0.07 mg/L, 0.33±0.04 mg/L, and 0.37±0.05 mg/L for SUM, MON, AUT, and WIN season respectively during the year 2020-21.

TDS

In the first year (2019-20) the highest value of TDS was recorded 1548.28 mg/L at IW1 in September, whereas the lowest TDS value was measured 741.39 mg/L at IW5 in June. During the second year (2020-21) the maximum valueof TDS was recorded 1399.67 mg/L at IW1 in September, whereas the minimum TDS value was measured 711.31 mg/L at IW5 in June.The annual mean of TDS values was recorded as 1059.46±18.63 mg/L and 1030.97±16.75 mg/L for 2019-20 and 2020-21, respectively.

The seasonal variation in TDS values were estimated 839.85±24.68 mg/L, 1148.22±27.13mg/L, 1236.48±21.36 mg/L, and 1145.48±25.35 mg/L for SUM, MON, AUT and WIN season of the year 2019-20. However, the seasonal variation for TDS was observed 829.02±27.98 mg/L, 1079.53±25.78 mg/L, 1198.32±28.54 mg/L and 1140.11±25.53 mg/L for SUM, MON, AUT and WIN season respectively during the year 2020-21

BOD

In the first year (2019-20) the highest value of BOD was recorded 692.34 mg/L at IW1 in June, whereas the lowest BOD value was measured 224.28 mg/L at

IW5 in December. During the second year (2020-21) the maximum valueof BOD was recorded 587.56 mg/L at IW1 in May, whereas the minimum BOD value was measured 220.19 mg/L at IW5 in December. The annual mean of BOD values was recorded as 370.99±12.87mg/L and 353.45±19.25 mg/L for 2019-20 and 2020-21, respectively.

The seasonal variation in BOD values were estimated 468.54±15.34 mg/L, 380.16±14.76 mg/L, 283.69±12.68 mg/L, and 383.72±14.75 mg/L for SUM, MON, AUT and WIN season of the year 2019-20. However, the seasonal variation for BOD was observed 430.53±27.77 mg/L, 358.76±23.44 mg/L, 273.94±27.65 mg/L, and 378.84±25.36 mg/L for SUM, MON, AUT and WIN season respectively during the year 2020-21.

COD

In the first year (2019-20) the highest value of COD was recorded 1348.39 mg/L at IW1 in August, whereas the lowest COD value was measured 748.39 mg/L at IW3 in April. During the second year (2020-21) the maximum valueof BOD was recorded 1488.78 mg/L at IW1 in September, whereas the minimum BOD value was measured 743.39 mg/L at IW2 in April. The annual mean of COD values was recorded 1084.18±19.62 mg/L and 1057.21±17.46 mg/L for the year 2019-20 and 2020-21 respectively.

The seasonal variation in COD values were estimated 967.14±14.65 mg/L, 1187.25±14.25 mg/L, 1067.84±17.54 mg/L and 1148.04±18.54 mg/L for SUM, MON, AUT and WIN season of the year 2019-20. However, the seasonal variation for COD was observed 37.99±9.05 mg/L, 14.55±3.47 mg/L, 27.61±6.57 mg/L and 27.13±6.45 mg/L for SUM, MON, AUT and WIN season respectively during the year 2020-21.

PO_4^{3-}

In the first year (2019-20) the highest value of PO_4^{3-} was recorded 19.33 mg/L at IW1 in the month of September, whereas the lowest PO_4^{3-} value was measured 2.24 mg/L at sampling location IW2 during the month of April. During the second year (2020-21) the maximum valueof PO_4^{3-} was recorded 17.34 mg/L at IW1 in September, whereas the minimum PO_4^{3-} value was measured 1.85 mg/L at IW2 in April. The annual mean of PO_4^{3-} values was recorded as 9.39±0.61 mg/L and 6.80±0.84 mg/L for 2019-20 and 2020-21, respectively.

The seasonal variation in PO_4^{3-} values were estimated 6.93±2.73 mg/L, 14.15±3.02 mg/L, 9.44±1.81 mg/L, and 7.90±2.10 mg/L for SUM, MON, AUT, and WIN seasons of the year 2019-20. However, the seasonal variation for PO_4^{3-} was observed 4.61±1.70 mg/L, 11.80±2.82 mg/L, 8.17±1.67 mg/L and 3.82±1.42 mg/L for SUM, MON, AUT, and WIN season respectively during the year 2020-21.

Electrical Conductivity (EC)

In the year (2019-20) the minimum conductivity 875.32 μS/cm was observed in April, while the maximum conductivity 1505.66 μS/cm was observed in September. The year annual average was calculated as 1201.03±13.60 μS/cm. During the year (2020-21) the minimumconductivity 868.88 μS/cmwas recorded in April, while the maximum conductivity 1498.53 μS/cm observed inSeptember. The year 2020-21 annual average was calculated as 1177.45± 17.43 μS/cm.

The seasonal variation in EC values were estimated 1139.43±16.78 μS/cm, 1247.73±15.24 μS/cm, 1187.59±14.87 μS/cm and 1245.43±14.44μS/cm for SUM, MON, AUT and WIN season of the year 2019-20. However, the seasonal variation for EC was observed 1119.10±16.33 μS/cm, 1179.41±15.67 μS/cm, 1158.80±13.43 μS/cm and 1265.71±14.22 μS/cm for SUM, MON, AUT and WIN season respectively during the year 2020-21.

PTEs

The order of higher occurrence of these toxic metals/PTEs in wastewater during the study period is Fe>Zn>Mn>Cr>Cu>Ni>Cd>Pb. The highest value of Fe was recorded 9.84mg/L at sampling site IW1 in September (2019-20), whereas the lowest Fe value was measured 2.02 mg/L at sampling location IW3 during March (2019-20). The highest value of Fe was recorded 8.78 mg/L at sampling site IW1 in September-(2020-21) whereas the lowest Fe value was measured 2.01 mg/L at sampling location IW3 during March (2020-21). The annual mean of Fe values was recorded 5.09±1.15 mg/L and 5.02±1.07 mg/L for 2019-20 and 2020-21 respectively. The seasonal variation in Fe values were estimated 4.22±1.48 mg/L, 6.12±2.11 mg/L, 5.57±1.40 mg/L and 4.89±1.27 mg/L for SUM, MON, AUT and WIN season of the year 2019-20. However, the seasonal variation for Fe was observed 4.14±1.45 mg/L, 5.94±2.02 mg/L, 5.42±1.36 mg/L and 5.01±1.21 mg/L for SUM, MON, AUT and WIN season respectively during the year 2020-21. The highest value of Zn was recorded 8.89 mg/L at sampling site IW1 in September (2019-20) whereas the lowest Zn value was measured 2.14 mg/L at site IW3 during March (2019-20). In the year 2020-21, the highest value of Zn was recorded 7.59 mg/L at sampling site IW1 in September, whereas the lowest Zn value was measured 2.09 mg/L at sampling location IW3 in April. The annual mean of Zn values was recorded at 4.75±0.85 mg/L and 4.61±0.80 mg/L for 2019-20 and 2020-21, respectively. The seasonal variation in Zn values were estimated 4.30±1.26 mg/L, 5.87±0.74 mg/L, 5.27±1.40 mg/L and 3.88±0.74 mg/L for SUM, MON, AUT and WIN season of the year 2019-20. However, the seasonal variation for Zn was observed 4.20±1.26 mg/L, 5.65±1.47 mg/L, 5.19±1.24 mg/L and

3.74±0.83 mg/L for SUM, MON, AUT and WIN season respectively during the year 2020-21.

In the year (2019-20) the highest value of Mn was recorded 3.19 mg/L at site IW1 in September whereas the lowest Mn value was measured 0.42 mg/L at site IW3 during April. In 2020-21, the highest value of Mn was recorded 3.13 mg/L at sampling site IW1 in September, whereas the lowest Mn value was measured 0.41 mg/L at sampling location IW3 in April. The annual mean of Mn values was 1.37±0.49 mg/L and 1.35±0.46 mg/L for 2019-20 and 2020-21 respectively. The seasonal variation in Mn values were estimated 1.07±0.43 mg/L, 1.50±0.82 mg/L, 1.46±0.46 mg/L and 1.58±0.51 mg/L for SUM, MON, AUT and WIN season of the year 2019-20. However, the seasonal variation for Mn was observed 1.00±0.40 mg/L, 1.44±0.77 mg/L, 1.50±0.48 mg/L and 1.63±0.48 mg/L for SUM, MON, AUT and WIN season respectively during the year 2020-21. In the year (2019-20) the highest value of Cr was recorded 2.12 mg/L at site IW1 in September whereas the lowest Cr value was measured 0.46 mg/L at site IW3 during April. In 2020-21, the highest value of Cr was recorded 2.08 mg/L at sampling site IW1 in September, whereas the lowest Cr value was measured 0.41 mg/L at sampling location IW3 during the month of April. The annual mean of Cr values was 1.37±0.49 mg/L and 1.35±0.46 mg/L for 2019-20 and 2020-21, respectively. The seasonal variation in Cr values were estimated at 0.78±0.19 mg/L, 0.98±0.38 mg/L, 1.25±0.44 mg/L and 1.14±0.47 mg/L for SUM, MON, AUT and WIN season of the year 2019-20. However, the seasonal variation for Cr was observed 0.78±0.20 mg/L, 0.94±0.38 mg/L, 1.25±0.44 mg/L and 1.23±0.42 mg/L for SUM, MON, AUT and WIN respectively, during the year 2020-21.

In the year (2019-20) the highest value of Cu was recorded 2.11 mg/L at site IW1 in September whereas the lowest Cu value was measured 0.43 mg/L at site IW3 during March. In the year 2020-21, the highest value of Cu was recorded 1.80 mg/L at sampling site IW1 in the month of September, whereas the lowest Cu value was measured 0.32 mg/L at sampling location IW2 during the month of October. The annual mean of Cu values was recorded 0.96±0.24 mg/L and 0.92±0.19 mg/L for the year 2019-20 and 2020-21 respectively. The seasonal variation in Cu values were estimated 0.82±0.27 mg/L, 1.06±0.38 mg/L, 0.83±0.38 mg/L and 1.14±0.26 mg/L for SUM, MON, AUT and WIN season of the year 2019-20. However, the seasonal variation for Cu was observed 0.80±0.24 mg/L, 1.01±0.32 mg/L, 0.83±0.37 mg/L and 1.07±0.19 mg/L for SUM, MON, AUT and WIN season respectively during the year 2020-21.

In the year (2019-20) the highest value of Ni was recorded 1.48 mg/L at site IW1 in September whereas the lowest Ni value was measured 0.12 mg/L at site

IW3 during March. In the year 2020-21, the highest value of Ni was recorded 1.47 mg/L at sampling site IW1 in September, whereas the lowest Ni value was measured 0.12 mg/L at sampling location IW3 during the month of March. The annual mean of Ni values was recorded as 0.50±0.14 mg/L and 0.47±0.13 mg/L for 2019-20 and 2020-21, respectively. The seasonal variation in Ni values were estimated 0.41±0.15 mg/L, 0.67±0.38 mg/L, 0.46±0.13 mg/L and 0.47±0.16 mg/L for SUM, MON, AUT and WIN season of the year 2019-20. However, the seasonal variation for Ni was observed 0.39±0.15 mg/L, 0.65±0.37 mg/L, 0.45±0.15 mg/L and 0.44±0.13 mg/L for SUM, MON, AUT and WIN season respectively during the year 2020-21.

In the year (2019-20) the highest value of Cd was recorded 1.10 mg/L at site IW1 in September whereas the lowest Cd value was measured 0.21 mg/L at site IW3 during April. In the year 2020-21, the highest value of Cd was recorded 1.10 mg/L at sampling site IW1 in the month of September, whereas the lowest Cd value was measured 0.21 mg/L at sampling location IW3 during the month of March. The annual mean of Cd values was recorded as 0.59±0.18 mg/L and 0.57±0.18 mg/L for 2019-20 and 2020-21, respectively. The seasonal variation in Cd values were estimated 0.49±0.20 mg/L, 0.70±0.24 mg/L, 0.52±0.27 mg/L and 0.64±0.21 mg/L for SUM, MON, AUT and WIN season of the year 2019-20. However, the seasonal variation for Cd was observed 0.47±0.19 mg/L, 0.69±0.24 mg/L, 0.52±0.27 mg/L and 0.64±0.21 mg/L for SUM, MON, AUT and WIN season respectively during the year 2020-21. In the year (2019-20) the highest value of Pb was recorded 0.14 mg/L at site IW1 in September whereas the lowest Pb value was measured 0.03 mg/L at site IW3 during April. In 2020-21, the highest value of Pb was recorded 0.14 mg/L at sampling site IW1 in September, whereas the lowest Pb value was measured 0.02 mg/L at sampling location IW3 in April. The annual mean of Pb values was recorded 0.08±0.02 mg/L and 0.07±0.02 mg/L for the year 2019-20 and 2020-21.

Table 4.1 Monthly variation of selected water parameters of Industrial waste water at sampling site 1 (IW1) during the year 2019-2020

Parameters	Month											
	April	**May**	**June**	**July**	**Aug**	**Sep**	**Oct**	**Nov**	**Dec**	**Jan**	**Feb**	**March**
pH	8.68	**8.74**	8.73	7.35	7.11	7.17	7.21	7.23	7.26	7.21	7.15	8.58
DO	0.28	0.24	0.27	0.29	0.31	0.39	0.33	0.35	0.37	0.34	0.38	0.32
TDS	797.65	847.32	957.30	959.32	1268.39	1548.28	1312.48	1412.48	1248.38	1189.39	1229.31	997.65
BOD	581.31	528.32	692.34	304.21	531.38	458.32	478.31	531.38	534.28	531.28	481.18	304.21
COD	827.21	1010.39	1228.31	1334.21	1348.39	1408.28	1133.49	1173.42	1179.38	1429.47	1324.21	1227.21
PO_4^{3-}	3.24	6.24	8.24	10.24	12.24	19.33	10.24	8.24	9.34	8.23	7.10	8.14
EC	975.32	1295.35	1435.44	1445.56	1455.64	1505.66	1205.52	1489.65	1478.43	1436.21	1425.77	1375.56
Zn	6.67	5.87	5.97	5.98	6.78	8.89	7.88	5.78	4.78	4.78	4.46	5.67
Fe	6.15	6.96	7.79	5.80	7.84	9.84	8.77	6.84	6.69	6.98	6.81	4.15
Cd	0.87	0.77	0.89	0.76	1.01	1.10	0.85	0.97	1.02	1.01	0.92	0.77
Cu	1.12	1.21	1.59	1.28	1.31	2.11	0.76	1.58	1.63	1.73	1.54	1.12
Ni	0.53	0.65	0.66	0.56	1.23	1.48	0.51	0.71	0.75	0.72	0.68	0.53
Mr	1.48	1.61	2.10	2.57	2.64	3.19	2.01	2.19	2.33	2.26	2.17	1.48
Cr	1.03	1.01	1.12	1.08	1.23	2.12	2.01	2.02	2.04	2.01	2.04	1.01
Pb	0.08	0.10	0.10	0.09	0.12	0.14	0.11	0.12	0.11	0.12	0.10	0.08

Table 4.2 Monthly variation of selected water parameters of Industrial waste water at sampling site 2 (IW2) during the year 2019-2020

Parameters	Month											
	April	**May**	**June**	**July**	**Aug**	**Sep**	**Oct**	**Nov**	**Dec**	**Jan**	**Feb**	**March**
pH	7.89	7.90	7.93	7.76	7.10	7.19	7.34	7.27	7.23	7.14	7.18	7.89
DO	0.33	0.35	0.38	0.33	0.47	0.42	0.46	0.43	0.55	0.58	0.43	0.33
TDS	782.00	797.22	827.24	729.21	1224.65	1384.50	1221.60	1141.60	1164.80	1042.90	1142.90	904.24
BOD	239.21	344.21	384.21	459.21	288.38	457.21	296.21	423.21	325.28	387.32	331.12	239.31
COD	757.28	990.65	967.65	970.38	1170.38	1234.50	989.21	1096.23	851.21	1327.24	1127.24	917.28
PO_4^{3-}	2.24	5.24	7.24	10.78	11.78	16.33	11.24	8.11	9.89	7.23	7.79	7.58
EC	887.32	1097.32	1197.32	1099.32	1119.32	1219.32	1089.32	1167.25	1161.56	1173.32	1189.32	1087.32
Zn	3.57	3.87	4.91	3.94	5.84	6.24	5.74	4.34	3.84	3.84	3.77	3.58
Fe	3.12	4.77	4.74	3.76	5.87	7.75	5.53	4.34	4.56	4.00	3.87	3.12
Cd	0.33	0.38	0.58	0.38	0.69	0.77	0.47	0.67	0.76	0.67	0.61	0.33
Cu	0.63	0.73	0.68	0.74	0.79	1.11	0.34	1.03	1.03	1.03	1.01	0.63
Ni	0.28	0.39	0.38	0.19	0.73	0.84	0.45	0.39	0.36	0.34	0.33	0.18
Mn	0.78	0.79	0.79	0.79	0.82	1.25	1.03	1.22	1.11	1.14	1.17	0.78
Cr	0.79	0.80	0.81	0.83	0.89	1.02	1.38	1.02	1.02	1.01	1.01	0.79
Pb	0.06	0.07	0.06	0.03	0.08	0.09	0.06	0.08	0.07	0.07	0.08	0.07

Table 4.3 Monthly variation of selected water parameters of Industrial waste water at sampling site 3 (IW3) during the year 2019-2020

Parameters	Month											
	April	**May**	**June**	**July**	**Aug**	**Sep**	**Oct**	**Nov**	**Dec**	**Jan**	**Feb**	**March**
pH	7.56	7.12	7.14	7.19	7.13	7.24	7.23	7.29	7.26	7.18	7.21	7.56
DO	0.34	0.31	0.33	0.36	0.45	0.25	0.37	0.39	0.41	0.44	0.32	0.31
TDS	776.34	787.37	748.31	749.37	1159.34	1328.32	1245.38	1145.38	1131.38	1118.39	1108.29	876.34
BOD	249.31	258.29	348.21	452.26	388.29	434.28	288.38	324.38	328.37	368.48	324.28	279.31
COD	748.39	969.34	989.24	952.14	952.14	1238.32	973.34	1063.28	763.38	1141.48	1041.48	1048.39
PO_4^{3-}	4.34	5.45	8.24	11.24	11.54	15.33	11.45	8.83	9.27	7.48	6.28	7.28
EC	929.46	1067.46	1199.46	989.46	993.46	1221.46	1021.34	1011.67	1008.46	1121.46	1021.46	989.46
Zn	2.15	2.32	4.45	2.47	5.41	6.84	5.53	3.66	2.53	2.86	2.76	2.14
Fe	3.02	2.10	4.18	2.19	4.29	7.78	4.19	3.32	3.32	3.32	3.32	2.02
Cd	0.24	0.28	0.38	0.27	0.58	0.62	0.23	0.33	0.53	0.33	0.28	0.21
Cu	0.63	0.68	0.49	0.50	0.83	1.12	0.47	1.05	1.04	1.01	1.03	0.43
Ni	0.32	0.33	0.44	0.14	0.49	0.89	0.30	0.29	0.29	0.29	0.29	0.12
Mn	0.42	0.49	0.49	0.48	0.52	1.06	1.04	1.02	1.02	1.01	0.89	0.77
Cr	0.46	0.47	0.58	0.47	0.49	0.88	0.79	0.82	0.81	0.79	0.87	0.46
Pb	0.03	0.04	0.04	0.05	0.06	0.07	0.05	0.06	0.06	0.06	0.06	0.04

Table 4.4. Monthly variation of selected water parameters of Industrial waste water at sampling site 4 (IW4) during the year 2019-2020

Parameters	Month											
	April	**May**	**June**	**July**	**Aug**	**Sep**	**Oct**	**Nov**	**Dec**	**Jan**	**Feb**	**March**
pH	8.20	8.67	7.89	7.91	7.11	7.21	7.19	7.22	7.29	7.22	7.24	8.20
DO	0.31	0.32	0.34	0.33	0.28	0.32	0.34	0.36	0.52	0.54	0.29	0.30
TDS	772.34	798.39	818.39	829.37	1282.29	1488.48	1228.48	1318.48	1231.48	1148.67	1008.17	992.34
BOD	343.24	359.38	425.38	469.38	465.45	581.47	329.48	484.28	426.38	426.38	423.27	243.24
COD	814.87	992.21	992.21	1181.29	1181.29	1388.48	1089.34	1158.39	1098.37	1338.48	1238.48	1144.87
PO_4^{3-}	4.75	6.75	10.75	11.75	17.75	16.98	11.75	7.75	12.44	9.34	6.24	9.75
EC	924.43	1294.43	1278.67	1288.43	1298.43	1328.43	1228.43	1318.11	1288.34	1428.43	1278.21	1224.43
Zn	4.21	4.66	5.77	4.79	6.54	7.24	6.54	4.11	4.67	4.41	4.59	4.21
Fe	4.17	4.33	5.44	4.48	6.89	8.58	6.79	5.58	5.61	5.54	5.43	3.17
Cd	0.49	0.49	0.50	0.49	0.79	1.01	0.28	0.58	0.59	0.58	0.51	0.49
Cu	0.87	0.90	0.90	0.89	1.22	1.23	0.54	1.12	1.09	1.10	1.11	0.87
Ni	0.44	0.58	0.55	0.38	0.51	0.96	0.61	0.43	0.48	0.59	0.57	0.34
Mn	1.02	1.04	1.10	1.12	1.14	2.22	1.67	2.01	1.87	2.00	2.12	1.02
Cr	0.86	0.77	0.67	0.89	0.90	1.21	1.10	1.11	1.02	0.85	0.89	0.87
Pb	0.07	0.08	0.07	0.07	0.09	0.12	0.07	0.09	0.09	0.09	0.09	0.07

Table 4.5. Monthly variation of selected water parameters of Industrial waste water at sampling site 5 (IW5) during the year 2019-2020

Parameters	Month											
	April	**May**	**June**	**July**	**Aug**	**Sep**	**Oct**	**Nov**	**Dec**	**Jan**	**Feb**	**March**
pH	8.13	8.16	8.17	8.08	7.12	7.29	7.23	7.21	7.27	7.24	7.22	8.17
DO	0.32	0.37	0.36	0.43	0.34	0.41	0.44	0.42	0.47	0.49	0.43	0.38
TDS	767.43	792.33	741.39	788..39	1159.34	1372.11	1119.48	1219.48	1151.38	1118.48	1148.23	967.43
BOD	315.35	382.41	372.48	369.48	352.48	452.47	317.48	478.48	224.28	326.49	328.21	245.35
COD	804.39	989.34	999.24	1091.34	1091.34	1252.11	953.29	1048.38	945.24	1257.48	1157.48	924.39
PO_4^{3-}	5.36	8.28	14.22	13.12	16.37	17.47	10.29	6.45	8.34	6.28	3.24	5.21
EC	875.32	1199.18	1280.21	1204.34	1223.48	1323.58	1223.27	1121.34	1125.87	1278.21	1266.34	1175.21
Zn	3.98	4.01	4.10	4.12	5.78	7.12	5.78	3.34	3.47	3.66	3.74	3.98
Fe	3.88	3.97	4.06	4.14	5.28	7.28	5.18	5.11	5.01	4.89	4.01	3.18
Cd	0.42	0.48	0.52	0.47	0.70	0.86	0.21	0.61	0.58	0.61	0.67	0.42
Cu	0.77	0.65	0.69	0.79	0.83	1.21	0.55	0.88	0.91	0.92	0.93	0.80
Ni	0.28	0.49	0.50	0.29	0.44	0.89	0.47	0.47	0.47	0.47	0.47	0.28
Mn	1.23	1.33	1.39	1.38	1.48	1.89	1.09	1.34	1.46	1.56	1.63	1.23
Cr	0.79	0.80	0.79	0.80	0.81	1.10	1.15	1.09	0.87	0.94	0.88	0.79
Pb	0.05	0.06	0.06	0.06	0.09	0.09	0.09	0.09	0.09	0.09	0.09	0.05

Table 4.6. Monthly variation of selected water parameters of Industrial waste water at sampling site 1(IW1) during the year 2020-2021

Parameters	**Month**											
	July	**Aug**	**Sep**	**Oct**	**Nov**	**Dec**	**Jan**	**Feb**	**March**	**April**	**May**	**June**
pH	7.25	7.24	7.12	7.24	7.28	7.22	7.19	7.12	8.43	8.55	8.55	8.55
DO	0.40	0.29	0.34	0.37	0.34	0.31	0.33	0.26	0.33	0.31	0.25	0.30
TDS	899.32	1148.28	1397.45	1234.45	1379.44	1234.32	1178.32	1245.35	976.77	786.77	839.66	943.66
BOD	541.31	414.56	387.46	298.34	458.32	519.44	528.22	476.12	302.88	511.45	587.56	422.48
COD	1138.22	1248.33	1488.78	1043.33	1164.39	1174.45	1423.42	1312.78	1215.67	819.66	1007.76	1188.78
PO_4^{3-}	9.87	10.12	14.24	10.43	8.35	2.45	2.21	2.23	5.89	2.78	3.88	5.55
EC	1245.56	1367.88	1498.53	1189.32	1445.67	1439.49	1389.78	1501.78	1356.93	969.48	1248.67	1415.46
Zn	5.58	6.78	7.59	6.88	5.67	4.89	4.65	4.34	5.87	6.88	5.74	5.23
Fe	5.65	7.74	8.78	7.72	6.81	6.79	6.79	6.76	4.09	6.09	6.87	7.64
Cd	0.66	1.02	1.10	0.84	1.02	0.97	1.02	1.01	0.75	0.86	0.69	0.86
Cu	1.17	1.28	1.80	0.75	1.54	1.53	1.44	1.11	1.04	1.09	1.18	1.43
Ni	0.54	1.15	1.47	0.51	0.71	0.53	0.57	0.69	0.52	0.51	0.64	0.62
Mn	2.54	2.14	3.13	2.01	2.17	2.34	2.11	2.10	1.38	1.39	1.54	1.89
Cr	1.04	1.12	2.08	2.04	2.02	2.03	2.01	2.02	1.12	1.02	0.99	1.11
Pb	0.08	0.12	0.14	0.11	0.12	0.10	0.12	0.11	0.08	0.08	0.09	0.10

Table 4.7. Monthly variation of selected water parameters of Industrial waste water at sampling site 2 (IW2) during the year 2020-2021

Parameters	Month											
	July	**Aug**	**Sep**	**Oct**	**Nov**	**Dec**	**Jan**	**Feb**	**March**	**April**	**May**	**June**
pH	7.56	7.18	7.11	7.31	7.24	7.24	7.12	7.11	7.76	7.78	7.23	7.87
DO	0.39	0.44	0.46	0.41	0.39	0.43	0.42	0.35	0.29	0.28	0.26	0.27
TDS	719.29	1178.45	1267.45	1212.67	1132.38	1157.79	1038.87	1137.94	897.78	775.42	788.18	816.14
BOD	329.21	256.32	529.32	328.45	419.19	321.24	382.29	321.11	331.64	319.44	239.15	355.22
COD	870.38	1130.29	1346.34	975.35	1086.21	848.19	1318.45	1134.78	911.88	743.39	976.83	942.88
PO_4^{3-}	8.34	12.35	13.22	9.45	8.43	2.24	1.85	3.78	3.13	2.89	2.77	4.66
EC	1077.23	1045.66	1181.23	1012.34	1138.23	1148.28	1176.67	1167.46	1066.78	882.98	1078.88	1131.64
Zn	3.74	5.34	6.21	5.65	4.32	3.81	3.76	3.65	3.48	3.44	3.77	4.55
Fe	3.26	5.77	7.67	5.50	4.29	4.51	4.05	4.34	3.04	3.10	4.58	4.67
Cd	0.39	0.67	0.77	0.45	0.66	0.73	0.58	0.64	0.32	0.31	0.33	0.53
Cu	0.72	0.73	1.11	0.32	1.02	1.02	1.01	0.97	0.62	0.62	0.72	0.71
Ni	0.17	0.72	0.83	0.30	0.33	0.32	0.32	0.35	0.17	0.23	0.37	0.38
Mn	0.77	0.82	1.34	1.02	1.21	1.18	1.17	1.18	0.77	0.75	0.74	0.76
Cr	0.72	0.80	1.01	1.23	1.01	1.01	1.01	1.01	0.78	0.72	0.77	0.81
Pb	0.02	0.08	0.08	0.06	0.04	0.05	0.07	0.08	0.06	0.06	0.07	0.07

Table 4.8. Monthly variation of selected water parameters of Industrial waste water at sampling site 3 (IW3) during the year 2020-2021

Parameters	**Month**											
	July	**Aug**	**Sep**	**Oct**	**Nov**	**Dec**	**Jan**	**Feb**	**March**	**April**	**May**	**June**
pH	7.19	7.11	7.10	7.14	7.30	7.27	7.13	7.18	7.43	7.49	7.08	7.10
DO	0.34	0.42	0.28	0.38	0.35	0.39	0.38	0.36	0.28	0.27	0.29	0.27
TDS	718.45	1046.56	1312.34	1124.45	1132.41	1128.33	1112.33	1101.88	863.65	764.45	781.29	737.28
BOD	428.26	367.34	414.25	237.45	317.31	323.34	368.48	319.18	336.49	352.55	247.77	338.67
COD	852.14	932.45	1138.43	955.39	1053.21	758.31	1132.89	1038.34	1037.99	743.56	951.69	955.54
PO_4^{3-}	8.45	10.56	12.32	8.56	4.75	3.75	3.78	3.87	6.69	3.89	2.88	2.46
EC	889.56	953.56	1219.34	1012.45	1025.48	1018.39	1118.55	1017.89	976.59	922.69	1053.78	1162.21
Zn	2.27	5.38	6.76	5.49	3.54	2.62	2.61	2.59	2.11	2.09	2.24	4.34
Fe	2.11	4.21	7.71	4.11	3.28	3.29	3.39	3.28	2.01	3.01	2.09	4.12
Cd	0.24	0.55	0.62	0.21	0.32	0.52	0.32	0.38	0.21	0.22	0.27	0.34
Cu	0.48	0.81	1.11	0.47	1.05	1.06	1.03	1.01	0.43	0.60	0.66	0.46
Ni	0.13	0.48	0.88	0.29	0.28	0.28	0.28	0.22	0.12	0.31	0.31	0.41
Mn	0.37	0.51	1.01	1.03	1.01	1.01	1.01	1.01	0.42	0.41	0.46	0.48
Cr	0.44	0.47	0.82	0.78	0.87	0.86	0.87	0.87	0.46	0.41	0.43	0.57
Pb	0.03	0.05	0.06	0.04	0.05	0.05	0.05	0.05	0.03	0.03	0.03	0.04

Table 4.9. Monthly variation of selected water parameters of Industrial waste water at sampling site 4 (IW4) during the year 2020-2021

Parameters	Month											
	July	Aug	Sep	Oct	Nov	Dec	Jan	Feb	March	April	May	June
pH	7.77	7.15	7.19	7.12	7.24	7.28	7.20	7.13	8.18	8.24	8.55	7.75
DO	0.31	0.30	0.34	0.33	0.45	0.43	0.32	0.27	0.32	0.28	0.29	0.27
TDS	789.37	1162.34	1399.67	1167.45	1308.11	1221.43	1139.67	1003.76	983.59	765.56	786.35	811.27
BOD	449.29	424.45	563.55	280.56	471.21	423.32	419.59	418.22	439.55	438.35	348.88	422.78
COD	1071.17	1082.44	1356.28	998.38	1142.31	1087.31	1328.84	1216.87	1138.81	811.79	983.66	967.78
PO_4^{3-}	9.28	14.87	17.34	6.45	9.75	4.75	5.89	3.89	5.94	5.69	6.89	8.58
EC	1118.59	1148.36	1312.45	1178.45	1237.23	1312.48	1421.89	1318.76	1214.88	916.69	1283.78	1253.77
Zn	4.59	6.39	7.19	6.63	4.64	4.59	4.53	4.62	4.10	4.18	4.57	5.62
Fe	4.34	6.76	8.43	6.68	5.53	5.49	5.53	5.43	3.11	4.11	4.17	5.33
Cd	0.46	0.77	1.01	0.28	0.57	0.59	0.58	0.48	0.48	0.48	0.48	0.47
Cu	0.78	1.10	1.21	0.53	1.11	1.11	1.09	1.08	0.85	0.86	0.89	0.81
Ni	0.38	0.50	0.93	0.61	0.53	0.54	0.53	0.54	0.34	0.43	0.55	0.52
Mn	1.14	1.12	2.10	1.55	2.09	1.89	2.08	2.12	1.01	1.01	1.01	1.12
Cr	0.88	0.86	1.18	1.10	1.19	1.21	1.12	1.17	0.85	0.83	0.85	0.89
Pb	0.07	0.09	0.12	0.07	0.08	0.09	0.07	0.09	0.07	0.07	0.07	0.07

Table 4.10. Monthly variation of selected water parameters of Industrial waste water at sampling site 5 (IW5) during the year 2020-2021

Parameters	Month											
	July	**Aug**	**Sep**	**Oct**	**Nov**	**Dec**	**Jan**	**Feb**	**March**	**April**	**May**	**June**
pH	7.67	7.14	7.22	7.21	7.23	7.23	7.27	7.24	8.08	8.13	8.10	7.88
DO	0.39	0.28	0.43	0.31	0.33	0.42	0.41	0.31	0.31	0.36	0.32	0.31
TDS	776.34	1119.29	1323.45	1089.46	1202.33	1148.33	1118.45	1134.89	943.55	754.29	788.47	711.31
BOD	339.27	319.44	418.56	276.49	471.39	220.19	319.48	322.45	227.89	221.25	275.98	367.89
COD	981.27	987.34	1252.89	932.45	1034.31	941.19	1241.98	1143.76	912.86	801.55	967.98	988.88
PO_4^{3-}	8.21	12.56	15.24	8.34	7.21	6.21	5.39	4.97	3.66	4.89	5.69	3.37
EC	1133.44	1167.34	1332.44	1134.56	1214.27	1318.34	1319.27	1316.67	1168.17	868.88	1187.88	1221.89
Zn	4.11	5.66	7.11	5.88	3.16	3.10	3.16	3.14	3.87	3.88	3.97	4.09
Fe	4.18	5.23	7.21	5.11	5.21	5.22	5.17	5.17	3.08	3.76	3.88	4.11
Cd	0.48	0.68	0.86	0.22	0.62	0.61	0.62	0.58	0.41	0.42	0.46	0.44
Cu	0.79	0.81	1.21	0.55	0.93	0.91	0.81	0.87	0.79	0.74	0.81	0.72
Ni	0.28	0.41	0.88	0.47	0.46	0.46	0.46	0.46	0.27	0.23	0.48	0.46
Mn	1.29	1.43	1.83	1.08	1.81	1.77	1.65	1.76	1.19	1.18	1.31	1.21
Cr	0.75	0.79	1.12	1.09	1.12	1.11	1.12	1.09	0.79	0.79	0.71	0.78
Pb	0.06	0.08	0.09	0.08	0.08	0.09	0.08	0.08	0.05	0.05	0.04	0.06

Table 4.11. Annual average of selected water parameters of Industrial waste water at sampling sites during the year 2019-2020 (Mean ± S.D)

Parameters	IW1		IW2		IW3		IW4		IW5		Mean	SD	CPCB standardfor surface discharge
	Mean	SD	Mean	SD	Mean	SD	Mean	SD	Mean	SD			
pH	7.69	0.73	7.49	0.35	7.25	0.16	7.61	0.53	7.60	0.47	7.53	0.15	9.0
DO	0.34	0.06	0.42	0.09	0.36	0.06	0.35	0.09	0.39	0.06	0.37	0.03	-
TDS	1147.33	13.12	1030.24	11.54	1014.52	11.78	1076.41	13.65	1028.79	11.76	1059.46	18.63	-
BOD	478.56	16.81	327.55	11.56	314.49	11.67	406.44	18.87	327.91	17.76	370.99	12.87	30
COD	1219.85	17.34	1033.27	16.34	990.08	14.67	1134.86	16.76	1042.84	13.76	1084.18	19.62	250
PO_4^{3-}	9.24	3.89	8.79	3.57	8.89	3.09	10.50	4.00	9.55	4.71	9.39	0.61	5
EC	1377.01	11.37	1124.00	17.57	1047.88	18.92	1264.90	15.54	1191.36	16.54	1201.03	13.60	-
Zn	6.13	1.29	4.46	0.97	3.59	1.59	5.15	1.08	4.42	1.16	4.75	0.85	5.0
Fe	7.05	1.44	4.62	1.30	3.59	1.54	5.50	1.44	4.67	1.05	5.09	1.15	3.0
Cd	0.91	0.11	0.55	0.17	0.35	0.15	0.57	0.18	0.55	0.17	0.59	0.18	2.0
Cu	1.42	0.35	0.81	0.23	0.77	0.27	0.99	0.20	0.83	0.17	0.96	0.24	3.0
Ni	0.75	0.30	0.41	0.20	0.35	0.20	0.54	0.16	0.46	0.16	0.50	0.14	3.0
Mn	2.17	0.50	0.97	0.20	0.77	0.27	1.53	0.49	1.42	0.21	1.37	0.49	2.0
Cr	1.56	0.51	0.95	0.17	0.66	0.18	0.93	0.15	0.90	0.14	1.00	0.30	2.0
Pb	0.11	0.02	0.07	0.01	0.05	0.01	0.09	0.02	0.08	0.02	0.08	0.02	0.1

Table 4.12. Annual average of selected water parameters of Industrial waste water at sampling sites during the year 2020-2021 (Mean ± S.D)

Parameters	IW1		IW2		IW3		IW4		IW5				CPCB standard for surface discharge
	Mean	SD	Mean	SD	Mean	SD	Mean	SD	Mean	SD	Mean	SD	
pH	7.65	0.65	7.38	0.29	7.21	0.14	7.57	0.51	7.53	0.41	7.47	0.16	9.0
DO	0.33	0.04	0.38	0.06	0.34	0.05	0.32	0.06	0.35	0.05	0.34	0.02	-
TDS	1105.32	11.34	1010.20	19.18	985.29	19.34	1044.88	22.13	1009.18	20.15	1030.97	16.75	-
BOD	452.25	10.50	304.07	18.34	304.26	16.18	391.65	17.24	315.02	78.07	353.45	12.25	30
COD	1185.46	16.45	1023.75	16.65	962.50	12.45	1098.80	15.22	1015.54	20.13	1057.21	17.46	250
PO_4^{3-}	6.50	4.03	6.09	4.07	6.00	3.26	8.28	4.09	7.15	3.56	6.80	0.84	5
EC	1339.05	16.47	1092.28	18.23	1030.87	14.28	1226.44	12.32	1198.60	12.56	1177.45	17.43	-
Zn	5.84	1.01	4.31	0.93	3.50	1.60	5.14	1.05	4.26	1.28	4.61	0.80	5.0
Fe	6.81	1.19	4.57	1.30	3.55	1.52	5.41	1.42	4.78	1.06	5.02	1.07	3.0
Cd	0.90	0.14	0.53	0.17	0.35	0.14	0.55	0.18	0.53	0.16	0.57	0.18	2.0
Cu	1.28	0.28	0.80	0.23	0.76	0.28	0.95	0.20	0.83	0.16	0.92	0.19	3.0
Ni	0.70	0.30	0.37	0.20	0.33	0.20	0.53	0.15	0.44	0.16	0.47	0.13	3.0
Mn	2.06	0.49	0.97	0.23	0.73	0.30	1.52	0.50	1.46	0.28	1.35	0.46	2.0
Cr	1.55	0.51	0.91	0.16	0.65	0.20	1.01	0.16	0.94	0.18	1.01	0.30	2.0
Pb	0.10	0.02	0.06	0.02	0.04	0.01	0.08	0.02	0.07	0.02	0.07	0.02	0.1

Table 4.13. Seasonal variation of selected water parameters of Industrial waste water during the year 2019-2020 (Mean ± S.D)

Parameters	Season												CPCB standard for surface discharge
	Summer			Monsoon			Autumn			Winter			
	Min	Max	Mean ± S.D	Min	Max	Mean ± S.D	Min	Max	Mean ± S.D	Min	Max	Mean ± S.D	
pH	7.12	8.74	8.06± 0.48	7.10	8.08	7.32±0.32	7.19	7.34	7.24±0.05	7.11	7.29	7.21±0.05	9.0
DO	0.24	0.38	0.31±0.09	0.24	0.47	0.38±0.07	0.31	0.46	0.36±0.05	0.30	0.58	0.41±0.04	-
TDS	749.37	997.65	839.85± 14.68	729.21	1548.28	1148.22 ±17.13	1119.48	1412.48	1236.48 ±11.36	1008.17	1248.38	1145.48 ±15.35	-
BOD	344.21	592.34	468.54± 15.34	288.38	531.38	380.16 ±14.76	239.31	478.31	283.69 ±12.68	224.28	534.28	383.72 ±14.75	30
COD	748.39	1228.31	967.14± 14.65	952.14	1488.78	1187.25 ±14.25	953.29	1173.42	1067.84 ±17.54	763.38	1429.47	1148.04 ±18.54	250
PO_4^{3-}	2.24	14.22	6.93±2.73	10.24	19.33	14.15± 3.02	6.45	11.75	9.44±1.81	3.24	12.44	7.90±2.10	5
EC	875.32	1435.44	1139.43 ±16.78	989.46	1505.66	1247.73 ±15.24	1011.67	1489.65	1187.59 ±14.87	1008.46	1478.43	1245.43 ±14.44	-
Zn	2.14	6.67	4.30±1.26	2.47	8.89	5.87±1.57	3.34	7.88	5.27±1.40	2.53	4.78	3.88±0.74	5.0
Fe	2.02	7.79	4.22±1.48	2.19	9.84	6.12±2.11	3.32	8.77	5.57±1.57	3.32	6.98	4.89±1.27	3.0
Cd	0.21	0.89	0.49±0.20	0.27	1.10	0.70±0.24	0.23	0.97	0.52±0.27	0.28	1.02	0.64±0.21	2.0
Cu	0.43	1.59	0.82±0.27	0.50	2.11	1.06±0.38	0.34	1.58	0.83±0.38	0.91	1.73	1.14±0.26	3.0
Ni	0.12	0.66	0.41±0.15	0.14	1.48	0.67±0.38	0.29	0.71	0.46±0.13	0.29	0.75	0.47±0.16	3.0
Mn	0.42	2.10	1.07±0.43	0.48	3.19	1.50±0.82	1.02	2.19	1.46±0.46	0.89	2.33	1.58±0.51	2.0
Cr	0.46	1.12	0.78±0.19	0.47	2.12	0.98±0.38	0.79	2.02	1.25±0.44	0.79	2.04	1.14±0.47	2.0
Pb	0.03	0.10	0.06±0.02	0.03	0.14	0.08±0.03	0.05	0.12	0.08±0.02	0.06	0.12	0.08±0.02	0.1

Table 4.14. Seasonal variation of selected water parameters of Industrial waste water during 2020-2021 (Mean ± S.D)

Parameters	Season												CPCB standard for surface discharge
	Summer			Monsoon			Autumn			Winter			
	Min	Max	Mean ± S.D	Min	Max	Mean ± S.D	Min	Max	Mean ± S.D	Min	Max	Mean ± S.D	
pH	7.08	8.55	7.94±0.49	7.10	7.77	7.27±0.22	7.12	7.31	7.23±0.06	7.11	7.28	7.20±0.06	9.0
DO	0.25	0.36	0.30±0.05	0.26	0.46	0.36±0.07	0.28	0.41	0.33±0.04	0.28	0.46	0.37 ±0.05	-
TDS	737.28	983.59	829.02± 17.98	711.31	1399.67	1079.53 ±15.78	1089 .46	1379 .44	1198.32 ±18.54	1003 .76	1245 .35	1140.11 ±15.53	-
BOD	319.44	587.56	430.53± 17.77	256.32	529.32	358.76 ±13.44	237 .45	458 .32	273.94 ±17.65	220 .19	528 .22	378.84 ±15.36	30
COD	743.39	1215.67	953.45± 12.68	852.14	1488.78	1125.[illegible]2 ±18.35	932 .45	1164 .39	1038.53 ±16.67	758 .31	1423 .42	1140.10 ±14.23	250
PO_4^{3-}	2.46	8.58	4.61±1 .70	8.21	17.34	11.80 ±2.82	4.75	10.43	8.17±1.67	1.85	6.21	3.82±1.42	5
EC	868.88	1415.46	1119.10± 16.33	889.56	1498.53	1179.41 ±15.67	1012 .34	1445 .67	1158.80 ±13.43	1017 .89	1501 .78	1265.71 ±14.22	-
Zn	2.09	6.88	4.20±1.26	2.27	7.59	5.65±1.47	3.16	6.88	5.19±1.24	2.59	4.89	3.74±0.83	5.0
Fe	2.01	7.64	4.14±1.45	2.11	8.78	5.94±2.02	3.28	7.72	5.42±1.36	3.28	6.79	5.01±1.21	3.0
Cd	0.21	0.86	0.47±0.19	0.24	1.10	0.69±0.24	0.21	1.02	0.52±0.27	0.32	1.02	0.64±0.21	2.0
Cu	0.43	1.43	0.80±0.24	0.48	1.80	1.01±0.32	0.32	1.54	0.83±0.37	0.81	1.53	1.07±0.19	3.0
Ni	0.12	0.64	0.39±0.15	0.13	1.47	0.65±0.37	0.28	0.71	0.45±0.15	0.22	0.69	0.44±0.13	3.0
Mn	0.41	1.89	1.00±0.40	0.37	3.13	1.44±0.77	1.01	2.17	1.50±0.48	1.01	2.34	1.63±0.48	2.0
Cr	0.41	1.12	0.78±0.20	0.44	2.08	0.94 ±0.38	0.78	2.04	1.25±0.44	0.86	2.03	1.23±0.42	2.0
Pb	0.03	0.10	0.06±0.02	0.02	0.14	0.08±0.03	0.04	0.12	0.07±0.03	0.05	0.12	0.08±0.02	0.1

The seasonal variation in Pb values was estimated 0.06±0.02 mg/L, 0.08±0.03 mg/L, 0.08±0.02 mg/L, and 0.08±0.02 mg/L for SUM, MON, AUT and WIN season of the year 2019-20. However, the seasonal variation for Pb was observed as 0.06±0.02 mg/L, 0.08±0.03 mg/L, 0.07±0.03 mg/L, and 0.08±0.02 mg/L for SUM, MON, AUT and WIN season respectively during the year 2020-21.

Comparison in annual average values of all the variables in wastewater for both the years (from interpolated figures)

Many academics employ GIS-based methodologies for spatial interpolation, analysis, and the extraction of many thematic layers to achieve desired results. To create the appropriate output maps, spatial distributions of Industrial wastewater quality elements will be estimated using a GIS-based spatial analysis tool in ArcGIS. Spatial analysis was carried out in this work using ArcGIS 10.3 software. Each sampling station is given a location in the Dehradun region based on its corresponding coordinates using the geographic coordinate system of the World Geodetic System (WGS)—1984. This is done to assess the critical areas in the region that are affected by one or more water quality issues. The actual data values of each physico-chemical and potential toxic elements variable are assigned to each of 5 monitoring locations and the data is spatially interpolated using GIS Inverse Distance Weight (IDW) method.

Figure 4.1 indicated a comparable spatial variation in pH during the years 2019-2020 and 2020-2021 using the IDW method. The pH spatial variation ranged from 7.25 to 7.69 and 7.21 to 7.65 during 2019-2020 and 2020-2021 respectively. Figure 4.2, DO interpolation using IDW ranged from 0.34 to 0.42 mg/L during 2019-2020 whereas, the DO variation during 2020-2021 ranged from 0.32 to 0.38 mg/L. Figure 4.3, shows the interpolation of TDS using IDW ranged from 1014.52 to 1147.33 mg/L during the year 2019-2020 whereas the spatial variation of TDS during 2020-2021 was ranged from 985.29 to 1105.32 mg/L. Figure 4.4 shows the interpolation of BOD using IDW ranged from 314.49 to 478.56 mg/L during the year 2019-2020 whereas the spatial variation of BOD during 2020-2021 was ranged from 304.07 to 452.25 mg/L. Figure 4.5 represents the spatial variation of COD which ranged from 990.08 to 1219.85 mg/L and 962.50 to 1185.46 mg/L for 2019-2020 and 2020-2021 respectively. A comparable spatial variation in Phosphate during the year 2019-2020 and 2020-2021 using the IDW method is depicted in Figure 4.6. The Phosphate spatial variation ranged from 8.79 to 10.50 mg/L and 6.0 to 8.28 mg/L during the year 2019-2020 and 2020-2021 respectively. Figure 4.7 represents the EC spatial variation ranging from 1047.88 to 1377.01μS/cm and 1030.87 to 1339.05μS/cm during 2019-2020 and 2020-2021, respectively. Figure 4.8 represents the Zinc interpolation using IDW ranged from 3.59 to 6.13 mg/L during 2019-2020 whereas, Zinc variation during 2020-2021 ranged from 3.50 to 5.84 mg/L.

Figure 4.9 indicated a comparable spatial variation in Cadmium during the years 2019-2020 and 2020-2021 using the IDW method. The Cadmium spatial variation ranged from 0.35 to 0.91 mg/L and 0.35 to 0.90 mg/L during 2019-2020 and 2020-2021, respectively. Figure 4.10 that the Iron interpolation using IDW ranged from 3.59 to 7.05 mg/L during 2019-2020, whereas the Iron variation during 2020-2021 ranged from 3.55 to 6.81 mg/L. Figure 4.11 shows the interpolation of Nickel using IDW ranged from 0.35 to 0.75 mg/L during the year 2019-2020 whereas the spatial variation of Nickel during 2020-2021 was ranged from 0.33 to 0.70 mg/L. Figure 4.12 shows the interpolation of Copper using IDW ranged from 0.77 to 1.42 mg/L during the year 2019-2020 whereas the spatial variation of Copper during 2020-2021 was ranged from 0.76 to 1.28 mg/L. A comparable spatial variation in Manganese during the year 2019-2020 and 2020-2021using the IDW method is depicted in Figure 4.13. The Manganese spatial variation ranged from 0.77 to 2.17 mg/L and 0.73 to 2.06 mg/L during the year 2019-2020 and 2020-2021 respectively. Figure 4.14 represents the Chromium interpolation using IDW ranged from 0.66 to 1.56 mg/L during the year 2019-2020 whereas, the Chromium variation during 2020-2021 ranged from 0.65 to 1.55 mg/L. Figure 4.15 represents the lead interpolation using IDW ranged from 0.05 to 0.11 mg/L during 2019-2020 whereas the lead variation during 2020-2021 ranged from 0.04 to 0.10 mg/L.

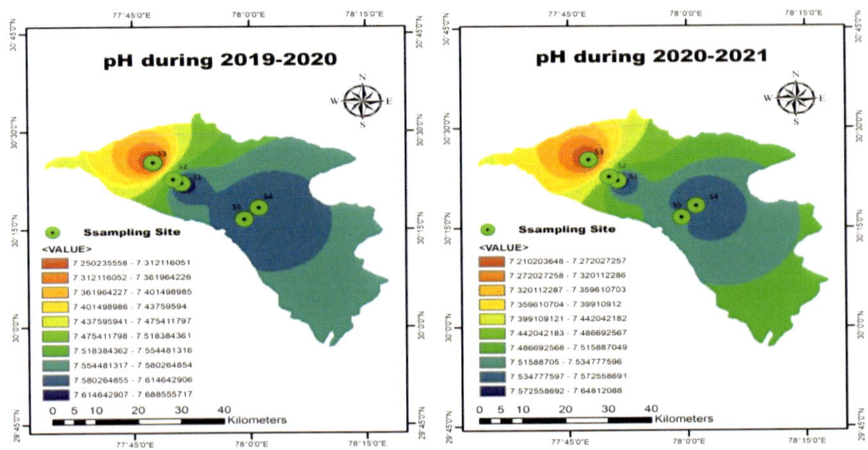

Figure 14.1. Spatio-temporal variations in pH during both the years of the study in Industrial wastewater

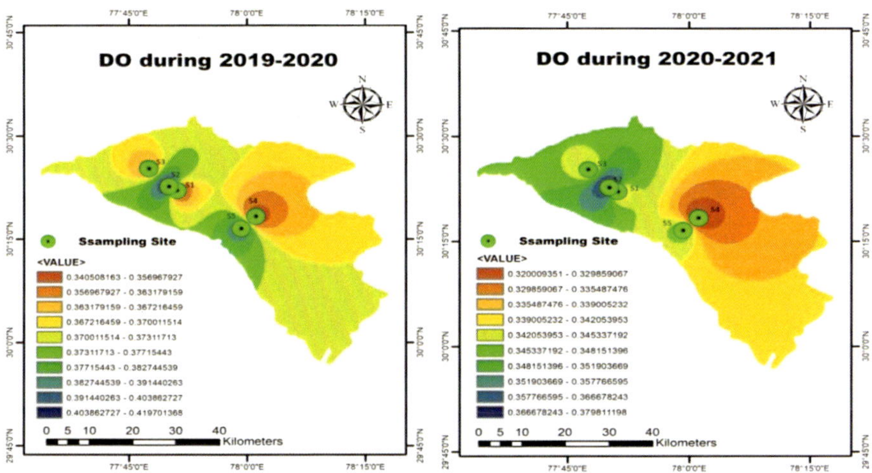

Figure 14.2. Spatio-temporal variations in DO during both the years of the study in Industrial wastewater

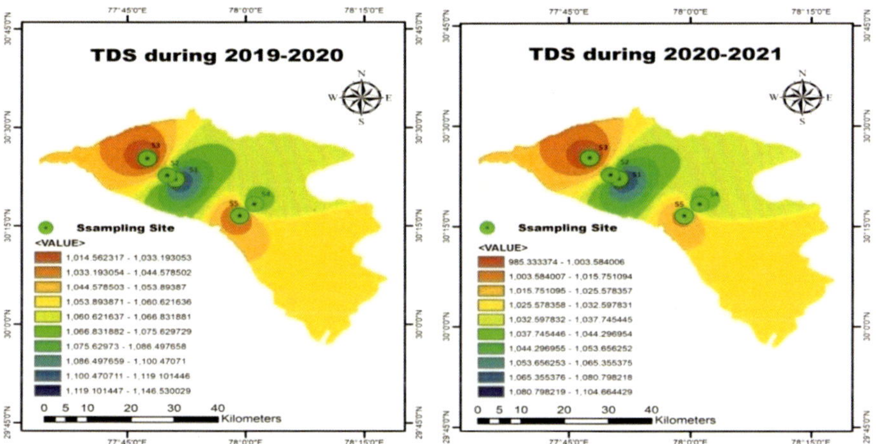

Figure 14.3. Spatio-temporal variations in TDS during both the years of the study in Industrial wastewater

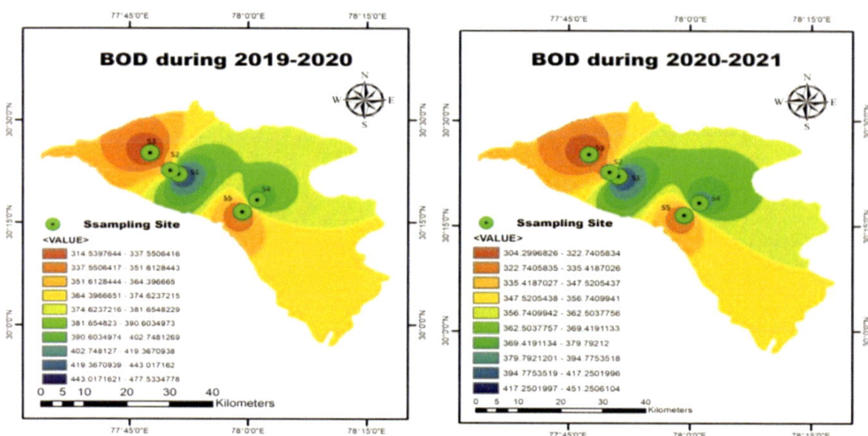

Figure 14.4. Spatio-temporal variations in BOD during both the years of the study in Industrial wastewater

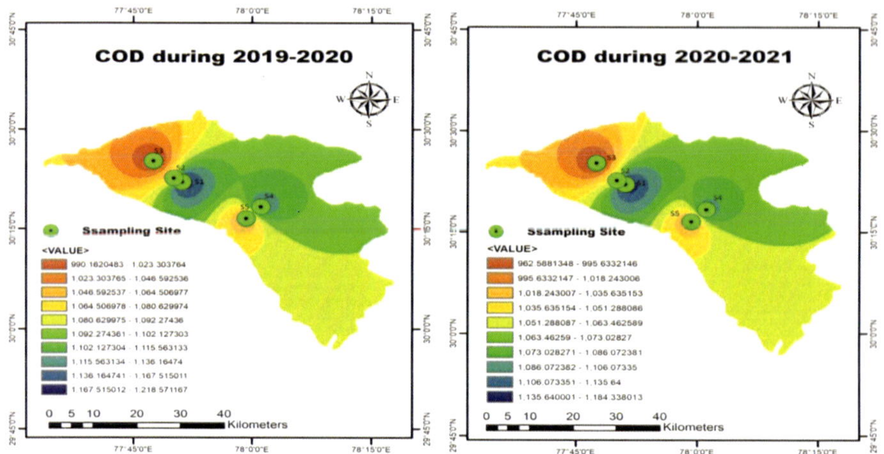

Figure 14.5. Spatio-temporal variations in COD during both the years of the study in Industrial wastewater

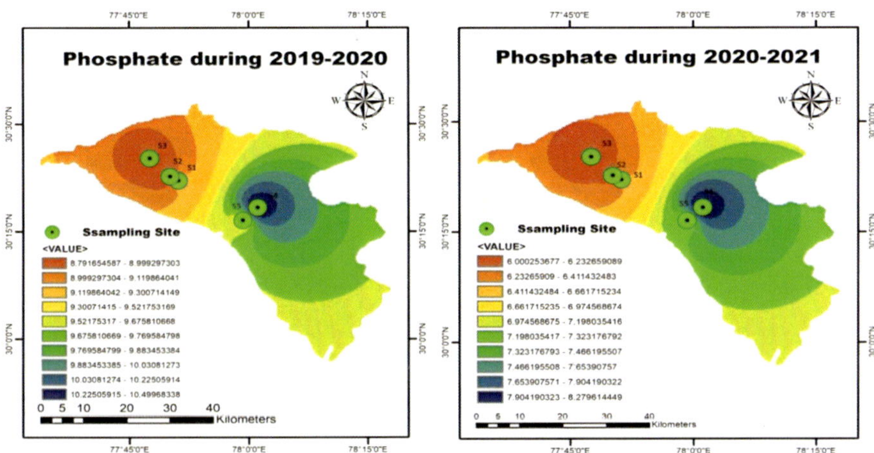

Figure 14.6. Spatio-temporal variations in Phosphate during both the years of the study in Industrial wastewater

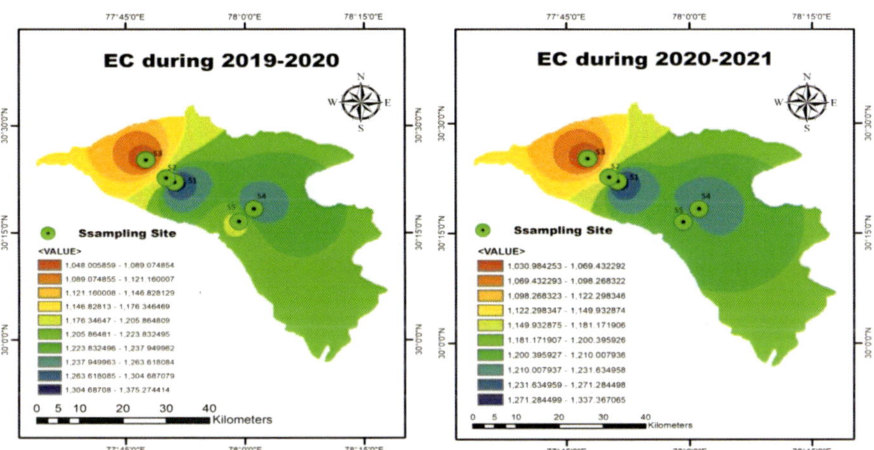

Figure 14.7. Spatio-temporal variations in EC during both the years of the study in Industrial wastewater

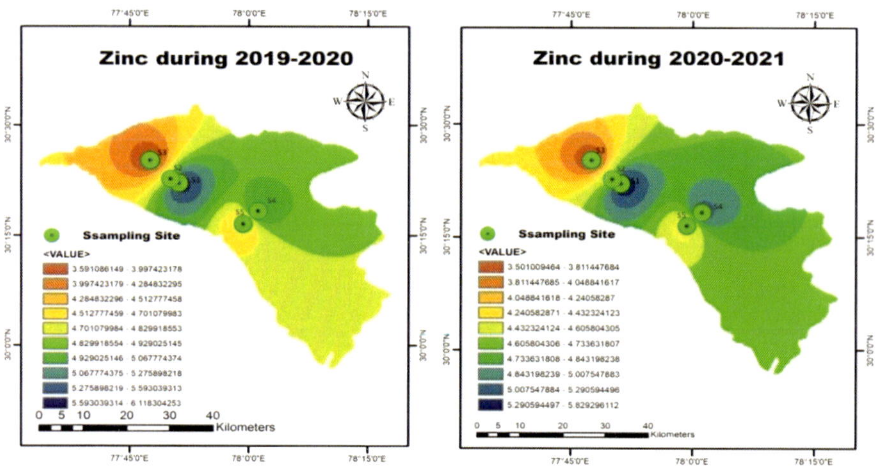

Figure 14.8. Spatio-temporal variations in Zinc during both the years of the study in Industrial wastewater

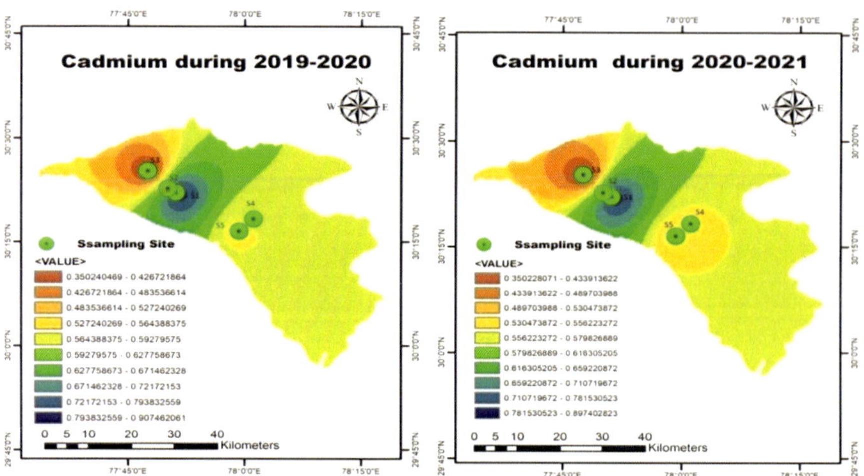

Figure 14.9. Spatio-temporal variations in Cadmium during both the years of the study in Industrial wastewater

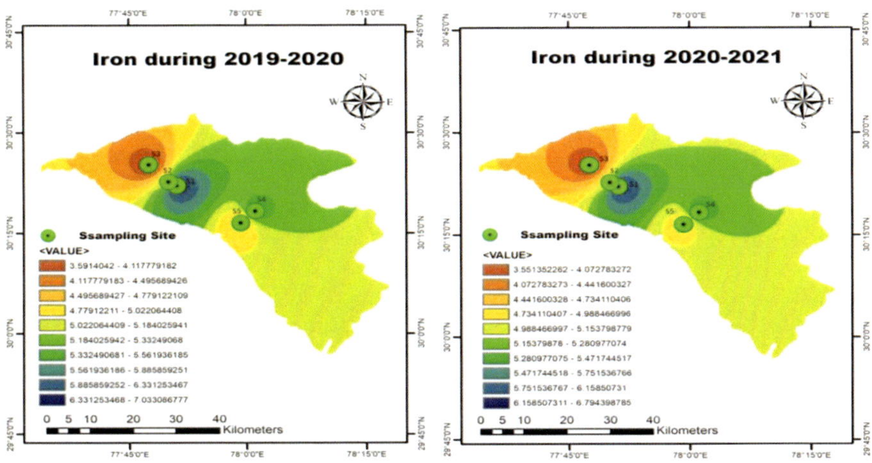

Figure 14.10. Spatio-temporal variations in Iron during both the years of the study in Industrial wastewater

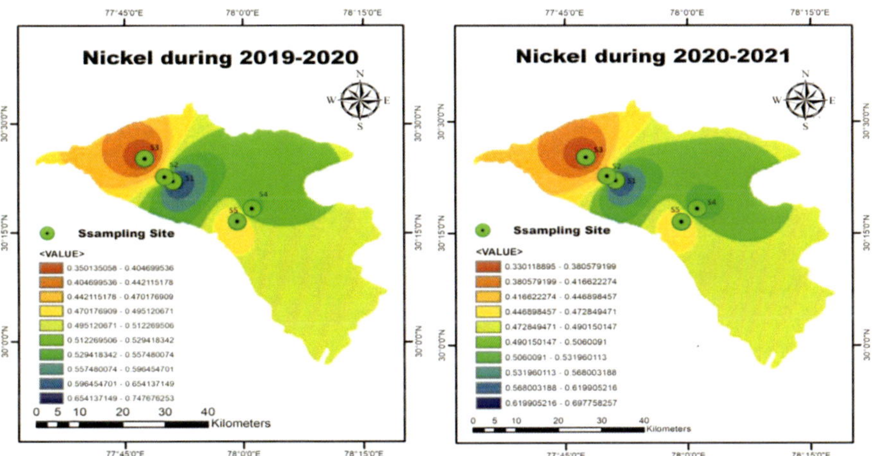

Figure 14.11. Spatio-temporal variations in Nickel during both the years of the study in Industrial wastewater

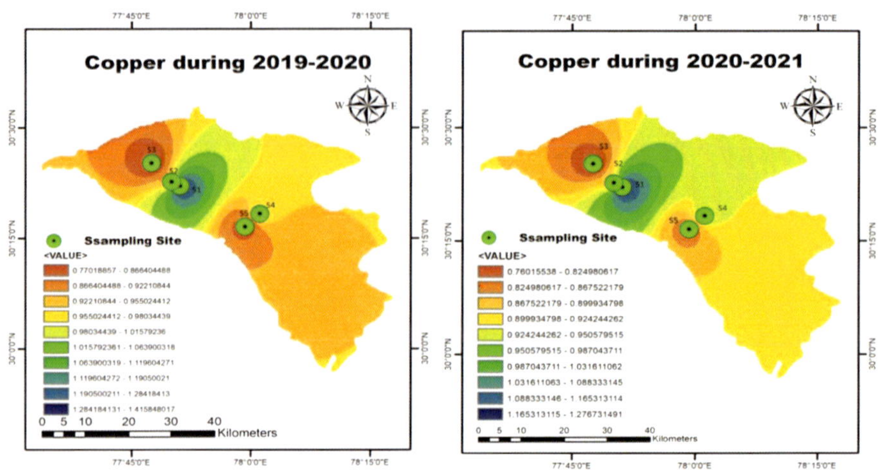

Figure 14.12. Spatio-temporal variations in Copper during both the years of the study in Industrial wastewater

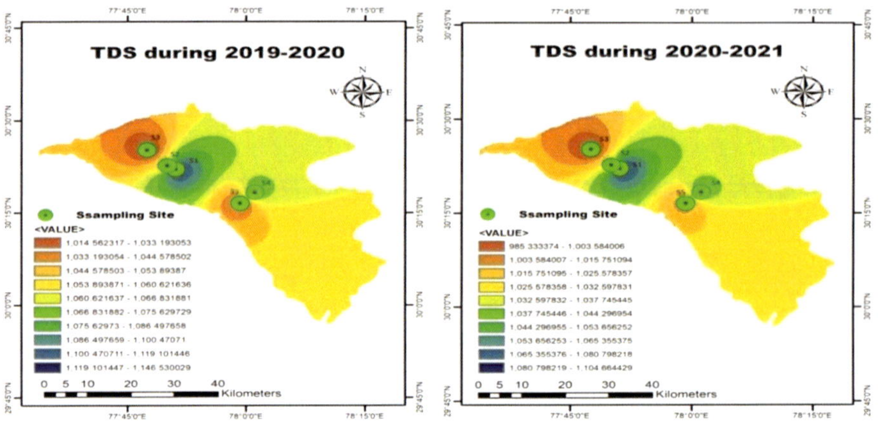

Figure 14.13. Spatio-temporal variations in Manganese during both the years of the study in Industrial wastewater

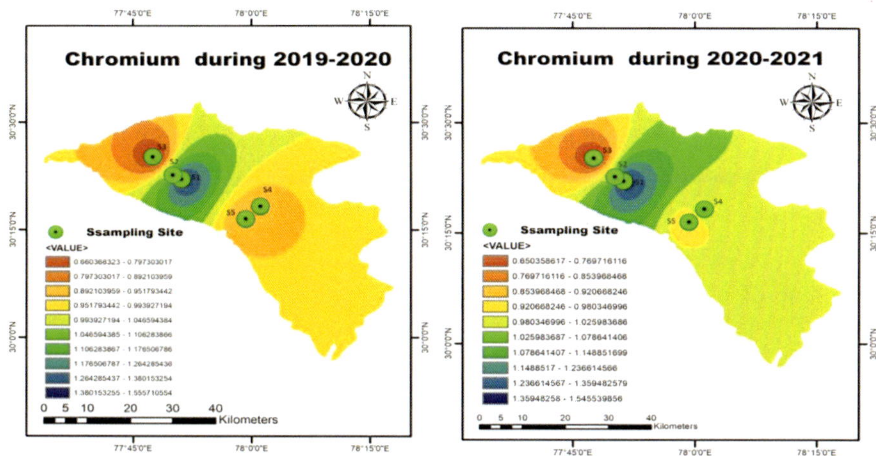

Figure 14.14. Spatio-temporal variations in Chromium during both the years of the study in Industrial wastewater

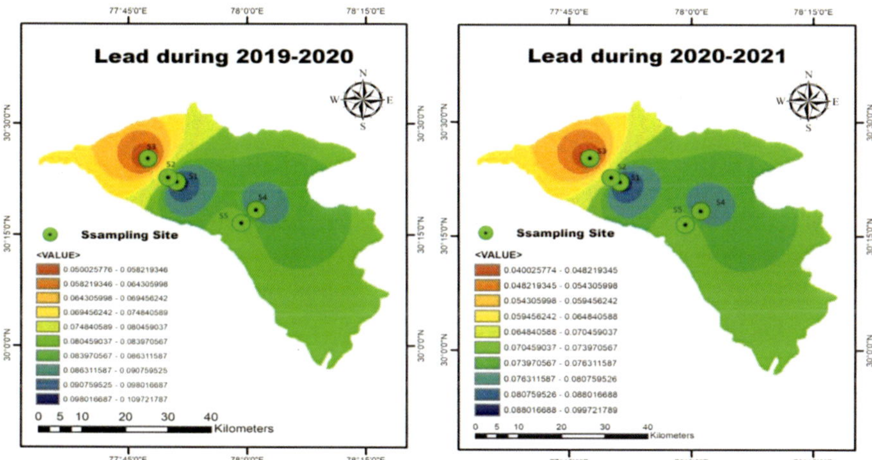

Figure 14.15. Spatio-temporal variations in Lead during both the years of the study in Industrial wastewater

Spatio-temporal variation in physico-chemical parameters and potential toxic elements (PTEs) in groundwater

The monthly observed variation in Mean values of hydro-chemical parameters for each monitoring location is represented in Table 4.15 to 4.20 and Table 4.21 to 4.26 for 2019-2020 and 2020-2021, respectively. The annual mean in physico-chemical parameters in groundwater samples of each monitoring location is given in Table 4.27 and Table 4.28 for 2019-2020 and 2020-2021, respectively. Table 4.29 and Table 4.30 show the descriptive statistical seasonal variation for both years.

The description of each variable for minimum, maximum, and mean±sd on a spatial and temporal basis, along with the data comparison with a prescribed limit of national and international agencies,are represented as follows:

pH

The highest value (8.01) of pH was recorded at sampling location GW1 in the month of June,whereas the lowest pH value (7.02) was measured at GW2 in the month of December for the year 2019-2020 (Table 4.16 and Table 4.17). The pH value was reported below the acceptance limit for groundwater prescribed by BIS, which is 8.5. The annual mean of pH values was recorded as 7.35±0.13 and 7.34±0.08 for 2019-2020 and 2020-2021, respectively (Table 4.27 and Table 4.28).

The seasonal variation in pH values were estimated as 7.31±0.25, 7.37±0.31, 7.51±0.33 and 7.21±0.08 for SUM, MON, AUT, and WIN season of the year 2019-20. However, the seasonal variation for pH was observed as 7.27±0.10, 7.39±0.09, 7.41±0.16 and 7.30±0.16 for SUM, MON, AUT, and WIN season, respectively, during the year 2020-21 (Table 4.29 and Table 4.30).

Electrical Conductivity (EC)

In the year (2019-20) the minimum conductivity of 271.36 µmhos/cm was observed in January at site GW2, while the maximum conductivity of 495.35 µS/cm was observed in August at site GW4. The year annual average was calculated as 405.32±26.46 µS/cm. During the year (2020-21) the minimumconductivity 281.28 µS/cmwas recorded in May at site GWCS, while the maximum conductivity 428.57 µS/cm was observed in August at site GW1. The 2020-21 annual average was 383.66± 21.29µS/cm.

The seasonal variation in EC values were estimated as 377.89±38.89 µS/cm, 439.93±29.34 µS/cm, 432.33±17.54 µS/cm, and 371.13±46.73µS/cm for SUM, MON, AUT and WIN season of the year 2019-20. However, the seasonal variation for EC was observed as 361.39±29.60 µS/cm, 410.08±16.48 µS/cm, 393.28±19.44 µS/cm, and 369.88±27.97µS/cm for SUM, MON, AUT and WIN season respectively during the year 2020-21.

Temperature

In the first year (2019-20), the highest value of Temperature was recorded 33.21 °C at GW5 in the month of June, whereas the lowest temperature value was measured 14.18 °C at GW3 during the month of January. During the second year (2020-21), the maximum value of Temperature was recorded

33.24 °C at GW4 in the month of June, whereas the minimum temperature value was measured 13.33 at GW5 during the month of January. The annual mean of temperature values was recorded as 23.87±0.11°C and 23.19±0.06 °C for 2019-20 and 2020-21, respectively.

The seasonal variation in temperature values were estimated as 30.38±2.37 °C, 28.30±1.03 °C, 21.53±2.29 °C, and 15.27±1.05 °C for SUM, MON, AUT, and WIN season of the year 2019-20. However, the seasonal variation for Temperature was observed as 29.68±5.43 °C, 27.27±1.00 °C, 20.18±2.05 °C, and 15.65±4.37°C for SUM, MON, AUT, and WIN seasons, respectively, during the year 2020-21.

TDS

In the first year (2019-20), the highest value of TDS was recorded 433.42 mg/L at GW4 in August, whereas the lowest TDS value was measured 273.39 mg/L at GWCS during January. During the second year (2020-21), the maximum value of TDS was recorded 411.68 mg/L at GW1 in August, whereas the minimum value of TDS was measured 263.53 mg/L at GWCS in January. The annual mean of TDS values was recorded as 351.09±22.51 mg/L and 337.34±18.46 mg/L for 2019-20 and 2020-21, respectively.

The seasonal variation in TDS values were estimated 348.96±26.72 mg/L, 385.86±36.37 mg/L, 330.18±14.77 mg/L, and 339.35±34.22 mg/L for SUM, MON, AUT, and WIN season of the year 2019-20. However, the seasonal variation for TDS was observed 337.35±28.94 mg/L, 371.52±29.27mg/L, 337.19±16.97 mg/L, and 303.30±24.24 mg/L for SUM, MON, AUT, and WIN season respectively during the year 2020-21.

TU

In the first year (2019-20), the highest value of TU was recorded 2.14 NTU at GW3 in August, whereas the lowest TU value was measured 0.65 mg/L at GWCS during November. During the second year (2020-21), the maximum value of TU was recorded at 1.45 mg/L at GW1 in August, whereas the minimum TU value was measured 0.71 NTU at GWCS during June. The annual mean of TU values was recorded 1.09±0.15 NTU and 1.00±0.09 NTU for 2019-20 and 2020-21, respectively.

The seasonal variation in TU values were estimated 1.09±0.15 NTU, 1.50±0.45NTU, 0.92±0.13 NTU, and 0.86±0.13 NTUfor SUM, MON, AUT, and WIN season of the year 2019-20. However, the seasonal variation for TU was observed 1.02±0.18 NTU, 1.18±0.15 NTU, 0.97±0.07 NTU, and 0.85±0.11 mg/L for SUM, MON, AUT and WIN season respectively during the year 2020-21.

TH

In the first year (2019-20), the highest value of TH was recorded 264.21 mg/L at GW4 in August, whereas the lowest TH value was measured 110.38 mg/L at GW3 in January. During the second year (2020-21), the maximum value of TH was recorded 184.21 mg/L at GW4 in August, whereas the minimum TH value was measured 111.34 mg/L at GWCS during May.The annual mean of TH values was recorded as 148.77±24.77 mg/L and 135.36±13.30 mg/L for the year 2019-20 and 2020-21, respectively.

The seasonal variation in TH values were estimated 134.01+16.64 mg/L, 170.07±44.18 mg/L, 170.94±43.40 mg/L, and 120.06±7.51 mg/L for SUM, MON, AUT, and WIN season of the year 2019-20. However, the seasonal variation for TH was observed 128.09±13.82 mg/L, 149.38±17.67 mg/L, 144.73±19.48 mg/L, and 119.26±5.54 mg/L for SUM, MON, AUT, and WIN seasons respectively during the year 2020-21.

Alkalinity

In the first year (2019-20) the highest alkalinity value was recorded 159.41 mg/L at GW4 in July, whereas the lowest alkalinity value was measured 107.43 mg/L at GWCS in December. During the second year (2020-21), the maximum valueof Alkalinity was recorded 156.38 mg/L at GW4 in August, whereas the minimum alkalinity value was measured 105.38 mg/L at GW3 during May. The annual mean of alkalinity values was recorded as 128.59±9.12 mg/L and 126.85±8.67 mg/L for the year 2019-20 and 2020-21, respectively.

The seasonal variation in alkalinity values were estimated 124.03±9.84 mg/L, 141.10±13.77 mg/L, 134.10±10.07 mg/L, and 115.15±6.70 mg/L for SUM, MON, AUT, and WIN season of the year 2019-20. However, the seasonal variation for Alkalinity was observed 118.92±10.13 mg/L, 137.60±12.40 mg/L, 133.16±9.60 mg/L, and 117.70±6.80 mg/L for SUM, MON, AUT, and WIN season respectively during the year 2020-21.

Acidity

In the first year (2019-20) the highest value of Acidity was recorded 41.89 mg/L at GW1 in August, whereas the lowest acidity value was measured 16.32 mg/L at GWCS during November. During the second year (2020-21), the maximum value of Acidity was recorded at 41.68 mg/L at GW1 in August, whereas the minimum acidity value was measured 16.38 mg/L at GWCS in December.The annual mean of acidity values was 32.88±6.16 mg/L and 30.72±5.24 mg/L for 2019-20 and 2020-21, respectively.

The seasonal variation in acidity values were estimated 34.49±5.11 mg/L, 35.27±6.46 mg/L, 31.56±7.04 mg/L, and 30.17±6.11 mg/L for SUM, MON,

AUT, and WIN season of the year 2019-20. However, the seasonal variation for Acidity was observed 30.48±4.57 mg/L, 33.97±5.66 mg/L, 30.89±5.66 mg/L, and 27.53±5.34 mg/L for SUM, MON, AUT, and WIN season respectively during the year 2020-21.

PO_4^{3-}

In the first year (2019-20) the highest value of PO_4^{3-} was recorded 0.29 mg/L at GW3 in June, whereas the lowest PO_4^{3-} value was measured 0.10 mg/L at GWCS during January. During the second year (2020-21) the maximum value of PO_4^{3-} was recorded 0.27 mg/L at GW4 in July, whereas the minimum PO_4^{3-} value was measured 0.11 mg/L at GWCS during May.The annual mean of PO_4^{3-} values was recorded 0.18±0.02 mg/L and 0.18±0.03 mg/L for 2019-20 and 2020-21, respectively.

The seasonal variation in PO_4^{3-} values were estimated 0.20±0.06 mg/L, 0.20±0.04 mg/L, 0.19±0.03 mg/L, and 0.14±0.03 mg/L for SUM, MON, AUT, and WIN season of the year 2019-20. However, the seasonal variation for PO_4^{3-} was observed 0.17±0.05 mg/L, 0.22±0.04 mg/L, 0.18±0.02mg/L and 0.14±0.01 mg/L for SUM, MON, AUT, and WIN season respectively during the year 2020-21.

HCO^-_3

In the first year (2019-20) the highest value of HCO^-_3 was recorded at 225.38 mg/L at GW4 in October, whereas the lowest HCO^-_3 value was measured 98.34 mg/L at GWCS during January. During the second year (2020-21), the maximum value of HCO^-_3 was recorded at 154.24 mg/L at GW4 in October, whereas the minimum HCO^-_3 value was measured 97.18 mg/L at GWCS during January.The annual mean of HCO^-_3 values was recorded as 131.79±19.64 mg/L and 118.21±9.19 mg/L for 2019-20 and 2020-21, respectively.

The seasonal variation in HCO^-_3 values were estimated at 118.58±10.19 mg/L, 148.60±33.41 mg/L, 152.68±40.33 mg/L, and 107.30±6.37 mg/L for SUM, MON, AUT, and WIN season of the year 2019-20. However, the seasonal variation for HCO^-_3 was observed at 114.30 ±7.25 mg/L, 128.65±13.09 mg/L, 124.07±13.84 mg/L, and 105.82±6.12 mg/L for SUM, MON, AUT, and WIN season respectively during the year 2020-21.

NO_3^-

In the first year (2019-20), the highest value of NO_3^- was recorded at 19.33 mg/L at GW4 in October, whereas the lowest NO_3^- value was 8.21 mg/L at GWCS in January. During the second year (2020-21), the maximum value of NO_3^- was recorded at 18.34 mg/L at GW4 in October, whereas the minimum

NO_3^- value was 8.21 mg/L at GWCS in January.The annual mean of NO_3^- values was recorded as 14.32±1.81 mg/L and 13.60±1.62 mg/L for 2019-20 and 2020-21, respectively.

The seasonal variation in NO_3^- values was estimated at 13.30±1.33 mg/L, 16.10±1.97 mg/L, 15.95±2.61 mg/L, and 11.94±1.99 mg/L for SUM, MON, AUT, and WIN season of the year 2019-20. However, the seasonal variation for NO_3^- was observed at 12.35 ±1.04 mg/L, 14.64±1.71 mg/L, 14.84±2.55 mg/L, and 12.54±2.12 mg/L for SUM, MON, AUT, and WIN season respectively during the year 2020-21.

SO_4^{2-}

In the first year (2019-20) the highest value of SO_4^{2-} was recorded at 19.27 mg/L GW4 in October, whereas the lowest SO_4^{2-} value was measured at 7.24 mg/L at GWCS during January. During the second year (2020-21) the maximum valueof SO_4^{2-} was recorded at 19.21mg/L at GW4 in October, whereas the minimum SO_4^{2-} value was 7.24 mg/L at GWCS in January.The annual mean of SO_4^{2-} values was recorded as 13.61±1.72 mg/L and 13.18±1.72 mg/L for 2019-20 and 2020-21, respectively.

The seasonal variation in SO_4^{2-} values were estimated 12.78±1.77 mg/L, 15.75±2.10 mg/L, 15.17±2.40 mg/L, and 10.72±1.84 mg/L for SUM, MON, AUT, and WIN season of the year 2019-20. However, the seasonal variation for SO_4^{2-} was observed 11.92 ±1.62 mg/L, 14.95±2.13 mg/L, 14.74±2.47 mg/L, and 11.09±2.00 mg/L for SUM, MON, AUT and WIN season respectively during the year 2020-21.

Cl^-

In the first year (2019-20), the highest value of Cl^- was recorded 21.80 mg/L at GW5 in August, whereas the lowest was measured 10.85 mg/L at GWCS in January. During the second year (2020-21), the maximum valueof Cl^- was recorded 20.34 mg/L at GW5 in August, whereas the minimum Cl^- value was measured 9.37 mg/L at GWCS during January. The annual mean of Cl^- values was recorded as 16.02±2.41 mg/L and 15.37±2.02 mg/L for 2019-20 and 2020-21, respectively.

The seasonal variation in Cl^- values were estimated 15.46±3.41 mg/L, 17.62±2.26 mg/L, 17.20±2.23 mg/L, and 13.82±3.39 mg/L for SUM, MON, AUT, and WIN season of the year 2019-20. However, the seasonal variation for Cl^- was observed 13.97 ±2.40 mg/L, 17.16±1.60 mg/L, 16.73±2.14 mg/L, and 13.60±3.02 mg/L for SUM, MON, AUT, and WIN seasons respectively during the year 2020-21.

K^+

In the first year (2019-20) the highest value of K^+ was recorded 23.21 mg/L at GW1 in August, whereas the lowest K^+ value was measured 8.23 mg/L at GWCS during January. During the second year (2020-21) the maximum value of K^+ was recorded 11.78 mg/L at GW1 in August, whereas the minimum K^+ value was measured 2.38 mg/L at GW2 during January.

The seasonal variation in K^+ values was estimated 14.28±2.45 mg/L, 16.91±2.99 mg/L, 16.30±2.80 mg/L, and 13.03±3.42 mg/L for SUM, MON, AUT, and WIN season of the year 2019-20. However, the seasonal variation for K^+ was observed 9.68 ±0.99 mg/L, 10.99±0.54 mg/L, 9.59±1.00 mg/L, and 7.51±2.38 mg/L for SUM, MON, AUT, and WIN seasons respectively during the year 2020-21.

Na^+

In the first year (2019-20), the highest value of Na^+ was recorded 22.48 mg/L at GW1 in August, whereas the lowest Na^+ value was measured 9.37mg/L at GWCS during January. During the second year (2020-21), the maximum valueof Na^+ was recorded 21.48 mg/L at GW4 in October, whereas the minimum Na^+ value was measured 8.78 mg/L at GWCS during January. The annual mean of Na^+ values was recorded 15.94±2.41 mg/L and 15.11±2.37 mg/L for 2019-20 and 2020-21, respectively.

The seasonal variation in Na^+ values was estimated 14.69±2.91 mg/L, 18.94±3.00 mg/L, 17.20±3.66 mg/L, and 12.92±2.69 mg/L for SUM, MON, AUT, and WIN season of the year 2019-20. However, the seasonal variation for Na^+ was observed 13.62 ±2.76 mg/L, 17.65±2.80 mg/L, 16.50±3.16 mg/L, and 12.67±2.53 mg/L for SUM, MON, AUT, and WIN seasons respectively during the year 2020-21.

Ca^{2+}

In the first year (2019-20), the highest value of Ca^{2+} was recorded 128.45 mg/L at GW4 in August, whereas the lowest Ca^{2+} value was measured 39.45 mg/L at GW3 during January. During the second year (2020-21) the maximum valueof Ca^{2+} was recorded 98.34 mg/L at GW4 in August, whereas the minimum Ca^{2+} value was measured 36.85 mg/L at GW3 during January.The annual mean of Ca^{2+} values was recorded 74.42±21.62 mg/L and 66.43±13.41 mg/L for 2019-20 and 2020-21, respectively.

The seasonal variation in Ca^{2+} values was estimated 68.51±15.12 mg/L, 87.23±26.13 mg/L, 81.76±27.12 mg/L, and 60.18±15.62 mg/L for SUM, MON, AUT, and WIN season of the year 2019-20. However, the seasonal variation for Ca^{2+} was observed 63.36 ±16.09 mg/L, 76.20±12.27 mg/L,

68.95±10.61 mg/L, and 57.18±14.56 mg/L for SUM, MON, AUT, and WIN season respectively during the year 2020-21.

Mg^{2+}

In the first year (2019-20), the highest value of Mg^{2+} was recorded 56.44 mg/L at GW4 in July, whereas the lowest Mg^{2+} value was measured 12.56 mg/ at GWCS during January. During the second year (2020-21), the maximum valueof Mg^{2+} was recorded 61.38 mg/L at GW4 in August, whereas the minimum Mg^{2+} value was measured 12.34 mg/L at GWCS during January.The annual mean of Mg^{2+} values was recorded as 32.05+9.80 mg/L and 31.10±11.04 mg/L for 2019-20 and 2020-21, respectively.

The seasonal variation in Mg^{2+} values was estimated as 27.80±9.85 mg/L, 40.18±11.11 mg/L, 35.19±8.22 mg/L, and 25.03±12.01 mg/L for SUM, MON, AUT, and WIN season of the year 2019-20. However, the seasonal variation for Mg^{2+} was observed as 25.34±9.46 mg/L, 40.69±14.47 mg/L, 33.57±8.99 mg/L, and 24.78±11.03 mg/L for SUM, MON, AUT, and WIN season respectively during the year 2020-21.

PTEs

The order of higher occurrence of these toxic metals/PTEs in wastewater during the study period is Zn>Fe>Ni>Mn>Cr>Cu>Cd>Pb. The highest value of Zn was recorded 0.489 mg/L at sampling site GW1 in November (2019-20), whereas the lowest Zn value was measured at 0.252 mg/L at site GWCS during March (2019-20). In 2020-21, the highest value of Zn was recorded at 0.478 mg/L at sampling site GW1 in August, whereas the lowest value was measured 0.291 mg/L at sampling location GWCS during January. The annual mean of Zn values was recorded as 0.348±0.05 mg/L and 0.342±0.04 mg/L for 2019-20 and 2020-21, respectively. The seasonal variation in Zn values was estimated 0.333±0.044 mg/L, 0.383±0.06 mg/L, 0.357±0.065 mg/L, and 0.321±0.027 mg/L for SUM, MON, AUT and WIN season of the year 2019-20. However, the seasonal variation for Zn was observed 0.325 ±0.04 mg/L, 0.371±0.05 mg/L, 0.360±0.05 mg/L, and 0.313±0.02 mg/L for SUM, MON, AUT and WIN season respectively during the year 2020-21.

The highest value of Fe was recorded 0.411 mg/L at sampling site GW1 in July (2019-20), whereas the lowest Fe value was measured 0.191 mg/L at sampling location GWCS during November (2019-20). The highest value of Fe was recorded 0.392 mg/L at sampling August (2020-21), whereas the lowest was measured 0.191 mg/L at sampling location GW1. The annual mean of Fe values was observed 0.278±0.08 mg/L and 0.285±0.07 mg/L during the year 2019-20 and 2020-21, respectively. The seasonal variation in Fe values was estimated 0.269±0.074 mg/L, 0.296±0.07 mg/L, 0.281±0.074 mg/L, and 0.263±0.076 mg/L for SUM, MON, AUT, and WIN season of the year 2019-20. However, the seasonal variation for Fe was observed 0.259 ±0.07 mg/L, 0.308±0.07 mg/L, 0.304±0.07 mg/L, and 0.271±0.06 mg/L for SUM, MON, AUT and WIN seasons respectively, during the year 2020-21. The presence of Cu, Cr, Cd concentration was reported on a few monitoring locations and varied between the ranges of 0.001-0.006, 0.001-0.003, 0.001-0.002 mg/L during the study period.

In the year (2019-20) the highest value of Mn was recorded 0.008 mg/L at site GW1 in August whereas the lowest Mn value was measured 0.001 mg/L at site GWCS during May. In 2020-21, the highest value of Mn was recorded 0.006 mg/L at sampling site GW1 in August, whereas the lowest Mn value was measured 0.001 mg/L at site GWCS during May. The seasonal variation in Mn values were estimated 0.003±0.00 mg/L, 0.005±0.00 mg/L, 0.005±0.0 mg/L, and 0.002±0.00 mg/L for SUM, MON, AUT and WIN season of the year 2019-20. However, the seasonal variation for Mn was observed 0.003±0.00 mg/L, 0.005±0.00 mg/L, 0.005±0.00 mg/L, and 0.002±0.00 mg/L for SUM, MON, AUT and WIN seasons, respectively, during the year 2020-21.

In the year (2019-20) the highest value of Ni was recorded 0.013 mg/L at site GW1 in August, whereas the lowest value 0.001 mg/L was measured at various sampling locations. In 2020-21, the highest value of Ni was recorded 0.011 mg/L at sampling site GW1 in August, whereas the lowest Ni value was measured 0.001 mg/L at various sampling locations. The annual mean of Ni values was recorded 0.004±0.00 mg/L and 0.005±0.00 mg/L for 2019-20 and 2020-21, respectively. The seasonal variation in Ni values were estimated 0.004±0.00 mg/L, 0.006±0.00 mg/L, 0.005±0.00 mg/L, and 0.003±0.00 mg/L for SUM, MON, AUT and WIN season of the year 2019-20. However, the seasonal variation for Ni was observed as 0.004±0.00 mg/L, 0.007±0.00 mg/L, 0.005±0.00 mg/L and 0.003±0.00 mg/L for SUM, MON, AUT and WIN season respectively during the year 2020-21. The lead (Pb) concentrations were not detected for any ground water sample of any monitoring location during the study period due to the less detection limit of FAAS used to analyse PTEs.

Table 14.15. Monthly variation of selected water parameters of Groundwater water at Control site (GWCS) during the year 2019-2020

Parameters	Month							
	May	**June**	**July**	**Aug**	**Oct**	**Nov**	**Dec**	**Jan**
pH	7.12	7.15	7.14	7.11	7.51	7.34	7.12	7.18
EC	279.23	342.34	384.25	442.34	431.45	421.67	280.74	271.36
Temp	28.23	31.36	29.18	27.38	23.38	19.38	16.36	14.36
TDS	281.12	317.39	323.21	347.27	322.31	312.31	276.28	273.39
TU	0.87	0.78	0.95	1.01	0.74	0.65	0.81	0.68
TH	116.45	121.57	125.34	133.43	124.45	122.45	114.67	110.42
Alkalinity	114.19	114.39	121.32	121.49	119.38	111.43	107.43	109.23
Acidity	23.21	25.78	21.37	23.27	17.47	16.32	18.34	17.78
PO_4^{3-}	0.11	0.13	0.19	0.18	0.16	0.14	0.12	0.10
HCO_3^-	108.23	110.38	119.62	128.21	111.18	114.18	104.28	98.34
NO_3^-	11.13	11.23	12.21	12.45	11.23	11.01	9.28	8.21
SO_4^{2-}	9.07	11.45	12.7	13.23	12.32	11.23	8.24	7.24
Cl^-	11.22	13.7	14.12	15.42	13.34	12.65	11.11	10.85
K^+	11.48	13.18	11.41	13.28	13.24	11.13	10.38	8.23
Na^+	10.23	12.22	12.54	15.21	11.24	10.53	10.12	9.37
Ca^{2+}	48.12	56.55	59.34	67.55	61.34	56.34	44.67	41.65
Mg^{2+}	15.34	18.25	21.25	24.44	21.13	18.23	14.22	12.56
Zn	0.295	0.297	0.304	0.312	0.267	0.252	0.294	0.293
Fe	0.198	0.201	0.203	0.204	0.198	0.191	0.199	0.198
Cd	0.000	0.000	0.001	0.001	0.001	0.000	0.000	0.000
Cu	0.001	0.001	0.002	0.002	0.001	0.001	0.001	0.000
Ni	0.000	0.001	0.001	0.001	0.000	0.000	0.000	0.000
Mn	0.001	0.001	0.002	0.002	0.001	0.002	0.001	0.001
Cr	0.000	0.000	0.001	0.001	0.000	0.000	0.000	0.000
Pb	0.000	0.000	0.000	0.000	0.000	0.000	0.000	0.000

Table 14.16. Monthly variation of selected water parameters of groundwater at sampling site 1 (GW1) during the year 2019-2020

Parameters	Month							
	May	June	July	Aug	Oct	Nov	Dec	Jan
pH	7.65	8.01	7.55	7.88	7.55	7.42	7.14	7.22
EC	421.12	433.12	439.38	463.12	454.24	413.36	421.14	398.38
Temp	28.2	32.38	29.38	27.21	23.18	19.18	16.38	14.38
TDS	374.21	384.37	381.42	422.38	353.21	313.21	364.83	354.38
TU	1.23	1.25	1.31	2.12	1.04	0.97	1.02	0.75
TH	129.79	131.34	145.48	161.48	153.47	146.67	121.28	117.39
Alkalinity	123.89	129.37	135.78	157.49	143.31	139.65	118.28	111.37
Acidity	39.22	36.49	37.28	41.89	38.23	34.23	32.34	31.59
PO_4^{3-}	0.18	0.25	0.15	0.21	0.22	0.19	0.15	0.12
HCO^-_3	114.35	120.37	127.82	139.34	131.33	125.33	109.49	101.34
NO_3^-	14.23	13.7	17.2	15.70	16.65	14.37	13.21	12.36
SO_4^{2-}	11.23	12.45	14.25	19.60	14.21	13.24	10.34	9.78
Cl^-	12.46	15.8	17.72	18.60	17.42	17.38	12.13	12.33
K^+	14.89	13.21	18.16	23.21	17.27	15.13	13.49	10.24
Na^+	15.23	19.55	21.12	22.48	15.24	14.22	14.89	11.32
Ca^{2+}	64.28	67.66	73.66	76.74	72.12	68.24	62.67	59.45
Mg^{2+}	23.95	28.22	34.43	41.41	36.48	33.23	21.38	18.22
Zn	0.353	0.452	0.487	0.489	0.478	0.464	0.351	0.324
Fe	0.395	0.399	0.411	0.398	0.383	0.367	0.391	0.393
Cd	0.002	0.002	0.002	0.002	0.002	0.002	0.002	0.002
Cu	0.002	0.002	0.004	0.006	0.005	0.003	0.002	0.001
Ni	0.011	0.011	0.012	0.013	0.012	0.009	0.010	0.011
Mn	0.003	0.005	0.007	0.008	0.006	0.006	0.003	0.002
Cr	0.002	0.002	0.003	0.002	0.002	0.002	0.002	0.002
Pb	0.000	0.000	0.000	0.000	0.000	0.000	0.000	0.000

Table 14.17 Monthly variation of selected water parameters of Ground water at sampling site 2 (GW2) during the year 2019-2020

Parameters	Month							
	May	**June**	**July**	**Aug**	**Oct**	**Nov**	**Dec**	**Jan**
pH	7.17	7.45	7.84	7.18	7.19	7.12	7.02	7.27
EC	377.45	384.23	414.25	454.23	423.35	411.23	377.45	396.23
Temp	28.12	32.28	29.48	27.34	24.39	19.39	16.28	14.28
TDS	358.32	351.38	368.41	415.42	322.34	314.34	350.23	387.48
TU	1.01	1.15	1.21	1.78	0.97	0.83	0.94	0.73
TH	124.75	137.67	142.32	156.31	151.37	142.78	119.55	118.38
Alkalinity	118.39	124.28	141.27	143.58	138.22	132.48	112.54	118.47
Acidity	33.35	37.48	34.46	38.45	33.89	32.34	30.48	35.21
PO_4^{3-}	0.15	0.22	0.16	0.25	0.21	0.17	0.14	0.10
HCO^-_3	112.34	123.35	131.78	139.45	139.21	128.23	107.23	104.25
NO_3^-	12.6	13.6	16.34	15.90	16.47	15.23	11.29	13.85
SO_4^{2-}	11.24	14.1	16.23	14.90	15.34	14.24	10.58	10.24
Cl^-	11.62	16.4	17.46	19.90	17.23	18.68	10.34	11.45
K^+	12.68	14.28	16.29	15.27	21.13	18.13	12.33	11.39
Na^+	13.68	17.69	18.51	20.24	18.38	17.45	13.54	12.28
Ca^{2+}	59.55	63.56	68.34	72.55	68.34	63.24	55.34	53.11
Mg^{2+}	19.35	27.41	31.12	37.22	40.77	38.34	19.21	17.45
Zn	0.321	0.381	0.417	0.423	0.369	0.362	0.321	0.381
Fe	0.311	0.312	0.344	0.385	0.371	0.361	0.309	0.31
Cd	0.002	0.001	0.002	0.002	0.002	0.002	0.002	0.002
Cu	0.001	0.002	0.003	0.004	0.002	0.002	0.001	0.001
Ni	0.001	0.001	0.005	0.006	0.003	0.002	0.001	0.001
Mn	0.002	0.003	0.005	0.005	0.005	0.005	0.002	0.003
Cr	0.002	0.002	0.002	0.002	0.002	0.002	0.002	0.002
Pb	0.000	0.000	0.000	0.000	0.000	0.000	0.000	0.000

Table 14.18 Monthly variation of selected water parameters of Ground water at sampling site 3 (GW3) during the year 2019-2020

Parameters	Month							
	May	**June**	**July**	**Aug**	**Oct**	**Nov**	**Dec**	**Jan**
pH	7.18	7.29	7.16	7.55	7.18	7.18	7.19	7.29
EC	368.45	373.12	426.34	443.12	433.46	413.12	368.45	375.67
Temp	28	33.18	29.12	27.41	23.47	19.47	16.18	14.18
TDS	346.39	359.21	368.34	417.34	334.27	324.27	339.87	321.24
TU	1.04	1.22	1.11	2.14	1.01	0.89	1.01	0.79
TH	120.46	131.89	147.32	178.21	159.37	142.37	112.28	110.38
Alkalinity	110.22	119.78	123.38	139.32	133.49	133.27	107.38	112.48
Acidity	33.79	36.77	34.47	41.59	36.21	34.43	31.29	32.39
PO_4^{3-}	0.18	0.29	0.21	0.19	0.18	0.16	0.14	0.16
HCO^-_3	110.24	123.29	128.43	143.78	137.11	127.11	105.27	99.37
NO_3^-	12.8	13.9	15.44	16.80	17.69	17.34	11.34	14.28
SO_4^{2-}	12.7	13.7	16.4	15.80	16.13	14.24	11.39	10.37
Cl^-	11.67	16.8	14.36	17.50	16.45	16.83	11.24	13.45
K^+	11.81	13.78	15.39	18.39	19.28	18.32	11.73	10.29
Na^+	12.39	13.48	17.57	19.56	19.38	18.34	12.24	10.28
Ca^{2+}	51.44	61.22	71.11	69.47	61.45	58.34	49.28	39.45
Mg^{2+}	17.23	29.24	39.89	44.74	34.41	31.43	17.19	16.88
Zn	0.317	0.312	0.348	0.351	0.336	0.328	0.314	0.292
Fe	0.284	0.287	0.322	0.311	0.294	0.287	0.281	0.283
Cd	0.002	0.001	0.002	0.002	0.002	0.002	0.002	0.002
Cu	0.003	0.002	0.002	0.004	0.003	0.002	0.001	0.002
Ni	0.003	0.002	0.004	0.005	0.003	0.003	0.001	0.001
Mn	0.002	0.004	0.006	0.005	0.005	0.005	0.002	0.003
Cr	0.002	0.001	0.001	0.001	0.001	0.001	0.002	0.001
Pb	0.000	0.000	0.000	0.000	0.000	0.000	0.000	0.000

Table 14.19. Monthly variation of selected water parameters of Ground water at sampling site 4 (GW4) during the year 2019-2020

Parameters	Month							
	May	June	July	Aug	Oct	Nov	Dec	Jan
pH	7.58	7.16	7.11	7.21	7.43	7.37	7.16	7.31
EC	388.45	395.35	425.24	495.35	465.67	435.35	388.45	395.48
Temp	28.3	33.21	29.18	27.32	23.37	19.37	16.21	14.21
TDS	349.29	358.28	361.44	433.42	352.37	332.47	361.29	358.48
TU	1.06	1.23	1.28	2.03	1.03	0.94	1.02	0.75
TH	135.69	179.56	181.91	264.21	241.48	213.67	130.39	129.34
Alkalinity	129.57	142.37	159.41	158.38	148.23	139.28	125.37	128.47
Acidity	40.39	36.44	37.36	38.43	34.26	33.45	38.48	32.89
PO_4^{3-}	0.19	0.27	0.18	0.27	0.21	0.17	0.14	0.17
HCO^-_3	119.37	143.38	147.39	222.39	225.38	193.45	118.23	115.28
NO_3^-	15.51	13.6	18.34	17.60	19.33	17.24	14.29	13.56
SO_4^{2-}	14.9	14.4	15.3	18.70	19.27	17.25	13.28	11.38
Cl^-	20.1	16.3	16.95	19.70	18.32	18.42	19.34	17.29
K^+	18.68	14.12	17.39	17.78	15.49	13.94	18.35	15.36
Na^+	19.45	13.21	16.38	21.41	22.39	19.34	18.54	16.48
Ca^{2+}	87.35	91.45	121.88	128.45	121.11	118.34	83.28	79.44
Mg^{2+}	47.69	34.36	56.44	49.88	44.15	41.45	45.87	41.11
Zn	0.321	0.308	0.361	0.368	0.359	0.347	0.341	0.301
Fe	0.211	0.221	0.267	0.265	0.251	0.243	0.201	0.203
Cd	0.001	0.001	0.001	0.001	0.001	0.001	0.001	0.001
Cu	0.001	0.002	0.005	0.006	0.004	0.003	0.001	0.001
Ni	0.002	0.003	0.007	0.01	0.009	0.008	0.002	0.002
Mn	0.003	0.004	0.007	0.006	0.006	0.006	0.003	0.002
Cr	0.002	0.002	0.002	0.002	0.002	0.002	0.002	0.002
Pb	0.000	0.000	0.000	0.000	0.000	0.000	0.000	0.000

Table 14.20. Monthly variation of selected water parameters of Ground water at sampling site 5 (GW5) during the year 2019-2020

Parameters	Month							
	May	**June**	**July**	**Aug**	**Oct**	**Nov**	**Dec**	**Jan**
pH	7.21	7.12	7.27	7.19	7.81	7.78	7.12	7.34
EC	389.45	382.35	419.21	472.35	452.68	432.35	389.45	392.34
Temp	28.13	33.21	29.34	27.23	24.38	19.38	16.21	14.21
TDS	350.22	357.28	362.32	429.34	348.37	332.68	347.39	337.39
TU	1.09	1.17	1.19	1.89	1.07	0.88	1.01	0.83
TH	131.23	147.68	153.39	251.44	236.38	216.76	127.29	129.38
Alkalinity	122.49	139.39	148.27	143.48	138.29	132.13	117.29	113.48
Acidity	34.59	36.42	37.48	37.22	35.46	32.45	31.45	29.78
PO_4^{3-}	0.17	0.28	0.16	0.22	0.23	0.21	0.15	0.19
HCO^-_3	109.38	128.28	141.48	213.48	214.29	185.34	114.28	110.29
NO_3^-	12.4	14.9	17.34	17.90	18.48	16.34	11.24	10.38
SO_4^{2-}	14.8	13.3	14.3	17.60	18.34	16.23	13.49	12.35
Cl^-	21.7	17.7	17.89	21.80	19.78	19.84	18.23	18.12
K^+	19.49	13.78	17.17	19.12	17.28	15.24	19.32	15.27
Na^+	15.38	13.78	20.78	21.49	20.48	19.35	13.38	12.59
Ca^{2+}	81.45	89.45	116.89	120.74	118.87	113.43	78.87	74.89
Mg^{2+}	40.67	31.88	54.23	47.14	41.88	40.82	38.55	37.66
Zn	0.326	0.316	0.353	0.388	0.367	0.358	0.326	0.311
Fe	0.199	0.212	0.229	0.21	0.213	0.21	0.192	0.198
Cd	0.001	0.001	0.001	0.001	0.001	0.001	0.001	0.001
Cu	0.002	0.002	0.002	0.002	0.002	0.002	0.001	0.002
Ni	0.004	0.004	0.004	0.003	0.003	0.003	0.002	0.002
Mn	0.003	0.005	0.006	0.005	0.005	0.005	0.003	0.002
Cr	0.002	0.001	0.001	0.002	0.001	0.001	0.002	0.001
Pb	0.000	0.000	0.000	0.000	0.000	0.000	0.000	0.000

Table 14.21. Monthly variation of selected water parameters of Groundwater water at Control site (GWCS) during the year 2020-2021

Parameters	Month							
	July	**August**	**Oct**	**Nov**	**Dec**	**Jan**	**May**	**June**
pH	7.23	7.32	7.21	7.34	7.12	7.16	7.23	7.21
EC	378.65	384.57	362.34	358.77	349.23	297.68	281.28	332.35
Temp	28.21	26.38	22.12	18.22	15.32	13.36	29.24	32.33
TDS	319.35	326.78	320.38	317.57	272.44	263.53	272.22	312.28
TU	0.87	0.95	0.87	0.81	0.79	0.76	0.78	0.71
TH	123.57	131.76	121.35	120.23	111.53	112.43	111.34	118.45
Alkalinity	118.35	121.68	118.34	110.68	106.67	107.87	108.23	111.48
Acidity	22.57	23.31	21.34	17.46	16.38	17.13	21.23	23.34
PO_4^{3-}	0.14	0.18	0.17	0.15	0.13	0.14	0.11	0.12
HCO_3^-	115.57	117.45	114.18	108.23	97.34	97.18	105.37	109.43
NO_3^-	12.12	11.33	11.13	10.23	9.15	8.21	11.02	11.17
SO_4^{2-}	11.35	12.34	11.78	11.21	8.11	7.24	9.04	10.34
Cl^-	14.56	15.76	13.21	11.78	10.34	9.37	11.01	12.34
K^+	10.41	11.34	10.11	9.34	8.84	7.21	10.14	9.38
Na^+	12.32	13.25	11.17	10.43	9.35	8.78	10.11	11.23
Ca^{2+}	59.34	63.24	60.38	55.28	41.38	39.86	41.34	45.63
Mg^{2+}	20.43	22.45	20.44	18.11	14.18	12.34	14.21	17.32
Zn	0.301	0.311	0.308	0.302	0.294	0.291	0.293	0.295
Fe	0.205	0.211	0.194	0.192	0.199	0.198	0.195	0.198
Cd	0.001	0.001	0.000	0.000	0.000	0.000	0.000	0.000
Cu	0.002	0.002	0.001	0.001	0.001	0.000	0.001	0.001
Ni	0.001	0.001	0.001	0.001	0.000	0.001	0.000	0.001
Mn	0.002	0.002	0.001	0.002	0.001	0.001	0.001	0.001
Cr	0.001	0.001	0.000	0.000	0.000	0.000	0.000	0.000
Pb	0.000	0.000	0.000	0.000	0.000	0.000	0.000	0.000

Table 14.22. Monthly variation of selected water parameters of groundwater at sampling site 1(GW1) during the year 2020-2021

Parameters	Month							
	July	**August**	**Oct**	**Nov**	**Dec**	**Jan**	**May**	**June**
pH	7.44	7.43	7.43	7.38	7.31	7.52	7.76	7.21
EC	412.57	428.57	414.24	389.78	384.37	417.28	381.12	374.76
Temp	28.23	26.21	22.14	18.21	15.34	29.17	32.23	13.37
TDS	382.13	411.68	362.45	309.46	298.54	362.23	376.34	296.65
TU	1.23	1.45	1.01	1.05	0.98	1.12	1.19	0.72
TH	138.24	155.78	141.34	139.32	120.54	120.34	127.36	123.41
Alkalinity	131.44	147.45	138.68	137.67	119.27	118.34	123.45	110.97
Acidity	34.63	41.68	37.38	33.45	31.33	32.48	33.38	29.38
PO_4^{3-}	0.19	0.22	0.2	0.17	0.15	0.13	0.18	0.13
HCO^-_3	117.23	128.28	120.28	115.34	103.24	108.48	111.34	111.28
NO_3^-	15.23	14.25	13.38	12.34	13.58	13.17	13.45	12.36
SO_4^{2-}	13.24	17.33	15.38	12.38	11.23	10.38	12.32	9.78
Cl^-	16.77	18.34	17.44	16.46	13.49	12.23	13.45	12.22
K^+	11.08	11.78	10.38	9.48	9.21	10.11	10.87	11.38
Na^+	20.23	21.36	16.38	14.28	13.56	14.24	18.34	11.28
Ca^{2+}	73.66	76.24	74.29	64.28	58.27	60.24	61.51	55.45
Mg^{2+}	34.34	38.48	35.45	32.37	22.48	21.36	25.18	19.67
Zn	0.448	0.478	0.452	0.448	0.343	0.348	0.434	0.311
Fe	0.387	0.392	0.378	0.371	0.368	0.387	0.375	0.353
Cd	0.002	0.002	0.002	0.002	0.002	0.002	0.002	0.002
Cu	0.003	0.004	0.003	0.002	0.002	0.002	0.002	0.001
Ni	0.01	0.011	0.009	0.01	0.007	0.008	0.009	0.005
Mn	0.007	0.006	0.006	0.005	0.003	0.002	0.005	0.002
Cr	0.002	0.002	0.002	0.002	0.002	0.002	0.002	0.002
Pb	0.000	0.000	0.000	0.000	0.000	0.000	0.000	0.000

Table 14.23. Monthly variation of selected water parameters of Ground water at sampling site 2 (GW2) during the year 2020-2021

Parameters	Month							
	July	**August**	**Oct**	**Nov**	**Dec**	**Jan**	**May**	**June**
pH	7.47	7.28	7.27	7.43	7.25	7.24	7.23	7.33
EC	389.65	411.67	387.47	381.68	374.76	365.89	351.92	374.45
Temp	28.23	26.34	22.13	18.23	15.31	13.36	29.37	32.24
TDS	358.34	389.71	334.45	327.67	311.68	301.76	352.24	349.24
TU	1.18	1.24	0.98	0.92	0.86	0.81	1.02	1.13
TH	138.45	146.31	138.48	135.68	117.24	118.34	119.48	133.47
Alkalinity	138.34	141.22	135.35	132.11	122.37	116.78	114.28	121.37
Acidity	33.24	35.68	32.48	31.56	29.34	28.37	28.37	31.45
PO_4^{3-}	0.18	0.21	0.19	0.16	0.13	0.12	0.11	0.17
HCO_3^-	116.24	128.34	117.47	119.34	102.45	112.29	109.37	118.24
NO_3^-	13.24	15.24	16.38	14.55	13.45	13.85	11.53	12.35
SO_4^{2-}	14.26	16.33	14.28	13.28	11.84	10.24	10.34	13.45
Cl^-	17.24	18.34	17.42	17.66	11.34	11.93	11.23	15.25
K^+	11.24	10.36	9.38	8.11	3.45	2.38	7.68	9.47
Na^+	18.21	19.35	17.36	16.23	14.54	12.18	12.23	15.35
Ca^{2+}	68.34	71.22	69.28	60.28	53.48	51.87	53.45	61.34
Mg^{2+}	30.45	33.45	31.34	30.37	20.38	17.54	18.12	24.41
Zn	0.378	0.384	0.355	0.35	0.318	0.304	0.310	0.337
Fe	0.357	0.368	0.368	0.361	0.298	0.289	0.315	0.304
Cd	0.002	0.002	0.001	0.002	0.002	0.002	0.002	0.001
Cu	0.002	0.002	0.002	0.002	0.001	0.001	0.001	0.002
Ni	0.006	0.007	0.006	0.004	0.001	0.001	0.004	0.004
Mn	0.005	0.005	0.005	0.004	0.002	0.003	0.002	0.003
Cr	0.002	0.002	0.002	0.002	0.002	0.002	0.002	0.002
Pb	0.000	0.000	0.000	0.000	0.000	0.000	0.000	0.000

Table 14.24. Monthly variation of selected water parameters of Ground water at sampling site 3 (GW3) during the year 2020-2021

Parameters	Month							
	July	**August**	**Oct**	**Nov**	**Dec**	**Jan**	**May**	**June**
pH	7.37	7.58	7.22	7.31	7.28	7.24	7.29	7.21
EC	416.54	423.46	413.45	384.76	371.45	362.84	354.45	368.46
Temp	28.22	26.41	22.15	18.2	15.33	13.35	29.31	33.22
TDS	361.22	379.68	342.48	331.54	313.78	298.54	341.38	348.22
TU	1.16	1.22	0.93	1.01	0.81	0.75	1.02	1.18
TH	142.35	148.21	133.48	130.86	114.65	114.85	116.35	128.46
Alkalinity	121.35	135.68	138.45	132.24	112.38	112.48	105.38	117.38
Acidity	33.24	37.83	34.28	33.65	27.34	26.38	31.48	32.45
PO_4^{3-}	0.26	0.22	0.2	0.17	0.14	0.16	0.15	0.24
HCO^-_3	124.25	120.23	115.27	114.34	104.24	99.34	110.34	112.48
NO_3^-	14.23	16.21	17.34	16.33	12.48	14.28	11.76	12.87
SO_4^{2-}	14.24	16.23	17.38	13.84	12.39	10.37	12.45	12.45
Cl^-	15.23	17.11	17.34	16.23	12.48	12.34	11.43	14.26
K^+	10.21	11.11	9.22	7.57	6.32	8.89	9.81	10.56
Na^+	17.21	16.24	18.45	17.34	11.59	10.28	11.21	12.45
Ca^{2+}	71.11	66.36	67.23	55.23	47.26	36.85	48.45	59.45
Mg^{2+}	37.23	36.25	30.48	31.12	17.11	16.22	15.19	24.23
Zn	0.356	0.361	0.342	0.339	0.309	0.294	0.312	0.309
Fe	0.311	0.315	0.359	0.343	0.279	0.272	0.276	0.272
Cd	0.002	0.002	0.001	0.002	0.002	0.002	0.002	0.001
Cu	0.002	0.003	0.003	0.002	0.003	0.003	0.003	0.002
Ni	0.007	0.006	0.007	0.006	0.003	0.003	0.004	0.005
Mn	0.006	0.005	0.005	0.004	0.002	0.003	0.002	0.004
Cr	0.001	0.001	0.001	0.001	0.002	0.001	0.002	0.001
Pb	0.000	0.000	0.000	0.000	0.000	0.000	0.000	0.000

Table 14.25. Monthly variation of selected water parameters of Ground water at sampling site 4(GW4) during the year 2020-2021

Parameters	Month							
	July	**August**	**Oct**	**Nov**	**Dec**	**Jan**	**May**	**June**
pH	7.39	7.41	7.76	7.48	7.36	7.34	7.41	7.22
EC	418.45	423.79	414.38	407.56	384.56	375.48	376.87	389.32
Temp	28.23	26.32	22.14	18.22	15.32	13.34	29.11	33.24
TDS	364.46	408.73	361.48	342.75	310.65	295.14	348.38	353.38
TU	1.18	1.25	0.98	1.02	0.89	0.74	1.04	1.19
TH	178.45	184.21	178.37	169.46	129.76	124.25	129.45	164.47
Alkalinity	152.41	156.38	143.29	138.46	127.37	126.84	121.38	139.37
Acidity	36.53	38.11	33.48	32.57	31.34	31.38	38.42	33.48
PO_4^{3-}	0.27	0.26	0.22	0.19	0.16	0.17	0.19	0.25
HCO_3^-	141.34	154.24	151.38	138.33	115.28	112.28	114.38	131.45
NO_3^-	15.25	17.11	18.34	16.38	14.89	13.56	14.38	12.45
SO_4^{2-}	16.24	18.21	19.21	16.38	14.29	11.38	13.23	13.45
Cl^-	16.22	18.47	18.78	17.45	18.24	17.22	18.38	15.36
K^+	11.23	11.55	10.42	9.68	8.23	7.89	8.68	9.34
Na^+	16.46	20.45	21.48	18.33	17.53	14.28	18.32	15.36
Ca^{2+}	91.88	98.34	89.24	81.23	79.38	75.86	82.47	87.32
Mg^{2+}	58.34	61.38	49.34	44.28	44.29	39.75	42.47	33.24
Zn	0.364	0.365	0.368	0.351	0.337	0.297	0.321	0.343
Fe	0.342	0.347	0.321	0.311	0.288	0.219	0.207	0.213
Cd	0.001	0.002	0.002	0.001	0.001	0.001	0.002	0.001
Cu	0.003	0.004	0.001	0.001	0.001	0.001	0.002	0.002
Ni	0.009	0.009	0.005	0.004	0.003	0.001	0.008	0.003
Mn	0.007	0.007	0.006	0.006	0.003	0.002	0.003	0.004
Cr	0.002	0.002	0.002	0.002	0.002	0.002	0.002	0.002
Pb	0.000	0.000	0.000	0.000	0.000	0.000	0.000	0.000

Table 14.26. Monthly variation of selected water parameters of Ground water at sampling site 5 (GW5) during the year 2020-2021

Parameters	Month							
	July	**August**	**Oct**	**Nov**	**Dec**	**Jan**	**May**	**June**
pH	7.34	7.38	7.65	7.42	7.26	7.22	7.23	7.17
EC	413.36	419.68	409.27	395.68	382.67	372.34	373.28	378.38
Temp	28.24	26.23	22.15	18.21	15.31	13.33	29.23	33.23
TDS	358.35	397.75	354.28	341.78	312.67	298.65	345.38	352.45
TU	1.17	1.21	1.01	1.01	0.87	0.81	1.05	1.15
TH	148.45	156.78	169.37	158.78	125.75	121.49	125.36	139.48
Alkalinity	141.24	145.68	141.48	131.22	123.34	118.67	119.38	134.31
Acidity	34.46	36.32	31.38	31.66	29.56	29.38	31.38	31.45
PO_4^{3-}	0.24	0.23	0.21	0.18	0.15	0.14	0.15	0.23
HCO_3^-	132.34	148.24	145.38	129.34	110.38	107.29	113.48	124.45
NO_3^-	16.25	15.23	16.38	15.28	13.48	10.38	11.44	13.45
SO_4^{2-}	13.25	16.36	17.39	14.37	13.21	12.35	13.84	12.34
Cl^-	17.54	20.34	18.38	18.65	17.11	17.11	16.35	16.36
K^+	10.24	11.36	11.14	10.22	9.36	8.28	9.49	9.38
Na^+	17.23	19.46	18.37	18.22	14.28	11.38	14.38	13.21
Ca^{2+}	86.33	88.38	78.35	72.34	71.38	70.35	79.34	84.56
Mg^{2+}	56.23	59.28	39.45	40.13	35.78	35.89	40.67	29.35
Zn	0.341	0.367	0.359	0.349	0.322	0.301	0.320	0.31
Fe	0.229	0.227	0.231	0.221	0.246	0.211	0.191	0.214
Cd	0.002	0.001	0.001	0.001	0.001	0.001	0.001	0.001
Cu	0.002	0.002	0.002	0.002	0.002	0.002	0.002	0.002
Ni	0.008	0.008	0.007	0.003	0.002	0.002	0.006	0.004
Mn	0.006	0.005	0.005	0.005	0.003	0.002	0.003	0.005
Cr	0.001	0.002	0.001	0.001	0.002	0.001	0.002	0.001
Pb	0.000	0.000	0.000	0.000	0.000	0.000	0.000	0.000

Table 14.27. Annual average of selected water parameters of Ground water at sampling sites during the year 2019-2020 (Mean±S.D)

Parameters	**GWCS**		**GW1**		**GW2**		**GW3**		**GW4**		**GW5**		**Mean**	**SD**	**BIS (2012)**
	Mean	**SD**	**Mean**	**SD**	**Mean**	**SD**	**Mean**	**SD**	**Mean**	**SD**	**Mean**	**SD**			
pH	7.21	0.14	7.55	0.30	7.28	0.26	7.25	0.13	7.44	0.38	7.36	0.28	7.35	0.13	8.5
EC	356.48	73.09	430.48	21.43	404.80	26.49	400.22	31.96	423.67	39.85	416.27	33.59	405.32	26.46	-
Temp	23.70	6.35	23.79	6.56	23.95	6.57	23.88	6.72	23.90	6.76	24.01	6.75	23.87	0.11	-
TDS	306.66	26.72	368.50	31.30	358.49	32.85	351.37	31.15	363.38	29.85	358.12	30.33	351.09	22.51	500
TU	0.81	0.13	1.21	0.41	1.08	0.32	1.15	0.42	1.17	0.39	1.14	0.33	1.09	0.15	1
TH	121.10	7.17	138.36	15.75	136.64	14.35	137.79	23.75	184.53	51.87	174.19	51.87	148.77	24.77	200
Alkalinity	114.86	5.42	132.39	14.81	128.65	11.81	122.42	12.00	141.39	13.25	131.85	12.76	128.59	9.12	200
Acidity	20.44	3.43	36.41	3.52	34.46	2.60	35.12	3.17	36.46	2.70	34.36	2.84	32.88	6.16	-
PO_4^{3-}	0.14	0.03	0.18	0.04	0.18	0.05	0.19	0.05	0.20	0.05	0.20	0.04	0.18	0.02	-
HCO_3^-	111.80	9.17	121.17	12.37	123.23	13.89	121.83	15.59	160.61	46.47	152.10	45.39	131.79	19.64	300
NO_3^-	10.84	1.43	14.68	1.69	14.41	1.88	14.95	2.27	16.18	2.24	14.87	3.16	14.32	1.81	45
SO_4^{2-}	10.69	2.22	13.14	3.10	13.36	2.32	13.84	2.24	15.56	2.70	15.05	2.14	13.61	1.72	150
Cl^-	12.80	1.64	15.48	2.74	15.39	3.69	14.79	2.47	18.30	1.37	19.38	1.67	16.02	2.41	250
K^+	11.54	1.74	15.70	3.91	15.19	3.27	14.87	3.49	16.39	1.89	17.08	2.16	15.13	1.93	12
Na^+	11.43	1.87	16.76	3.85	16.47	2.89	15.405	3.693	18.40	2.97	17.15	3.73	15.94	2.41	200
Ca^{2+}	54.45	8.87	68.10	5.87	63.00	6.73	57.720	10.61	103.91	20.29	99.32	19.93	74.42	21.62	75
Mg^{2+}	18.18	4.01	29.67	8.06	28.86	9.43	23.87	10.87	45.12	6.59	41.60	6.67	32.05	9.80	30
Zn	0.289	0.02	0.425	0.07	0.372	0.04	0.325	0.02	0.338	0.03	0.340	0.03	0.348	0.05	5
Fe	0.199	0.00	0.392	0.01	0.338	0.03	0.294	0.01	0.233	0.03	0.210	0.01	0.278	0.08	0.2
Cd	0.001	0.00	0.002	0.00	0.002	0.00	0.002	0.00	0.001	0.00	0.001	0.00	0.002	0.00	0.003
Cu	0.001	0.00	0.003	0.00	0.002	0.00	0.002	0.00	0.003	0.00	0.002	0.00	0.002	0.00	2

Ni	0.001	0.00	0.011	0.00	0.003	0.00	0.003	0.00	0.005	0.00	0.003	0.00	0.004	0.00	0.02
Mn	0.001	0.00	0.005	0.00	0.004	0.00	0.004	0.00	0.005	0.00	0.004	0.00	0.004	0.00	0.1
Cr	0.001	0.00	0.002	0.00	0.002	0.00	0.001	0.00	0.002	0.00	0.001	0.00	0.002	0.00	0.05
Pb	0.000	0.00	0.000	0.00	0.000	0.00	0.000	0.00	0.000	0.00	0.000	0.00	0.000	0.00	0.01

Table 14.28. Annual average of selected water parameters of Ground water at sampling sites during the year 2020-2021(Mean ± S.D)

Parameters	GWCS		GW1		GW2		GW3		GW4		GW5		Mean	SD	BIS (2012)
	Mean	SD	Mean	SD	Mean	SD	Mean	SD	Mean	SD	Mean	SD			
pH	7.23	0.07	7.44	0.16	7.31	0.09	7.31	0.12	7.42	0.16	7.33	0.15	7.34	0.08	8.5
EC	343.11	37.10	400.34	20.06	379.69	17.73	386.93	27.09	398.80	19.45	393.08	19.03	383.66	21.29	-
Temp	23.15	6.97	23.11	6.93	23.15	6.97	23.27	7.15	23.24	7.13	23.24	7.14	23.19	0.06	-
TDS	300.57	26.25	349.94	43.05	340.64	28.00	339.61	25.57	348.12	34.61	345.16	30.01	337.34	18.46	500
TU	0.82	0.08	1.09	0.21	1.02	0.15	1.01	0.17	1.04	0.17	1.04	0.14	1.00	0.09	1
TH	118.83	7.06	133.29	12.50	130.93	11.06	128.65	12.74	157.30	25.19	143.18	17.92	135.36	13.30	200
Alkalinity	112.91	5.72	128.41	12.40	127.73	10.29	121.92	12.21	138.19	12.47	131.92	10.56	126.85	8.67	200
Acidity	20.35	2.90	34.21	3.82	31.31	2.55	32.08	3.73	34.41	2.88	31.95	2.35	30.72	5.24	-
PO_4^{3-}	0.14	0.02	0.17	0.03	0.16	0.04	0.19	0.04	0.21	0.04	0.19	0.04	0.18	0.03	-
HCO_3^-	108.09	7.79	114.43	7.70	115.47	7.65	112.56	8.06	132.34	16.80	126.36	15.46	118.21	9.19	300
NO_3^-	10.55	1.28	13.47	0.95	13.82	1.56	14.44	2.02	15.30	1.92	13.99	2.21	13.60	1.62	45
SO_4^{2-}	10.18	1.85	12.76	2.54	13.00	2.09	13.67	2.27	15.30	2.67	14.14	1.84	13.18	1.72	150
Cl^-	12.30	2.15	15.05	2.46	15.05	3.07	14.55	2.29	17.52	1.21	17.73	1.35	15.37	2.02	250
K^+	9.60	1.23	10.54	0.91	7.76	3.21	9.21	1.60	9.63	1.35	9.93	1.02	9.45	0.94	12
Na^+	10.83	1.48	16.21	3.51	15.68	2.63	14.35	3.28	17.78	2.43	15.82	2.89	15.11	2.37	200
Ca^{2+}	50.81	9.74	65.49	8.08	61.16	7.78	56.49	11.79	85.72	7.37	78.88	7.07	66.43	13.41	75
Mg^{2+}	17.44	3.60	28.67	7.30	25.76	6.44	25.98	9.04	46.64	9.39	42.10	10.34	31.10	11.04	30
Zn	0.299	0.007	0.408	0.063	0.342	0.03	0.328	0.02	0.343	0.02	0.334	0.02	0.342	0.04	5
Fe	0.198	0.006	0.376	0.013	0.333	0.03	0.303	0.03	0.281	0.06	0.221	0.02	0.285	0.07	0.2
Cd	0.001	0.00	0.002	0.00	0.002	0.00	0.002	0.00	0.001	0.00	0.001	0.00	0.002	0.00	0.003
Cu	0.001	0.00	0.002	0.001	0.002	0.00	0.003	0.00	0.002	0.00	0.002	0.00	0.002	0.00	2

Ni	0.001	0.00	0.009	0.002	0.004	0.00	0.005	0.00	0.005	0.00	0.005	0.00	0.005	0.00	0.02
Mn	0.001	0.00	0.005	0.002	0.004	0.00	0.004	0.00	0.005	0.00	0.004	0.00	0.004	0.00	0.1
Cr	0.001	0.00	0.002	0.00	0.002	0.00	0.001	0.00	0.002	0.00	0.001	0.00	0.002	0.00	0.05
Pb	0.000	0.00	0.000	0.00	0.000	0.00	0.000	0.00	0.000	0.00	0.000	0.00	0.000	0.00	0.01

Table 14.29. Seasonal variation of selected water parameters of Groundwater water during the year of 2019-2020

Parameters	Season												BIS (2012)
	Summer			Monsoon			Autumn			Winter			
	Min	Max	Mean ± S.D	Min	Max	Mean ± S.D	Min	Max	Mean ± S.D	Min	Max	Mean ± S.D	
pH	7.12	8.01	7.31± 0.25	7.11	7.88	7.37± 0.31	7.12	8.01	7.51± 0.33	7.02	7.34	7.21± 0.08	8.5
EC	279.23	433.12	377.89± 38.89	384.25	495.35	439.93± 29.34	411.23	465.67	432.33± 17.54	271.36	421.12	371.13± 46.73	-
Temp	28	33.21	30.38± 2.37	27.21	29.48	28.30± 1.03	19.18	24.39	21.53± 2.29	14.18	16.38	15.27± 1.05	-
TDS	281.12	384.37	348.96± 26.72	323.21	433.42	385.86± 36.37	312.31	353.21	330.18± 14.77	273.39	387.48	339.35± 34.22	500
TU	0.78	1.25	1.09±0.15	0.95	2.14	1.50± 0.45	0.65	1.07	0.92± 0.13	0.68	1.02	0.86± 0.13	1
TH	116.45	179.56	134.01± 16.64	125.34	264.21	170.07± 44.18	122.45	241.48	170.94± 43.40	110.38	130.39	120.06± 7.51	200
Alkalinity	110.22	142.37	124.03±9.84	121.32	159.41	141.10± 13.77	111.43	148.23	134.10± 10.07	107.38	128.47	115.15± 6.70	200
Acidity	23.21	40.39	34.49±5.11	21.37	41.89	35.27± 6.46	16.32	38.23	31.56± 7.04	17.78	38.48	30.17± 6.11	-
PO_4^{3-}	0.11	0.29	0.20±0.06	0.15	0.27	0.20± 0.04	0.14	0.23	0.19± 0.03	0.10	0.19	0.14± 0.03	-
HCO_3^-	108.23	143.38	118.58± 10.19	119.62	222.39	148.60± 33.41	111.18	225.38	152.68± 40.33	98.34	118.23	107.30± 6.37	300
NO_3^-	11.13	15.51	13.30±1.33	12.21	18.34	16.10± 1.97	11.01	19.33	15.95± 2.61	8.21	14.29	11.94± 1.99	45

SO_4^{2-}	9.07	14.9	12.78±1.77	12.7	19.6	15.75± 2.10	11.23	19.27	15.17± 2.40	7.24	13.49	10.72± 1.84	150
Cl^-	11.22	21.7	15.46±3.41	14.12	21.8	17.62± 2.26	12.65	19.84	17.20± 2.23	10.34	19.34	13.82± 3.39	250
K^+	11.48	19.49	14.28±2.45	11.41	23.21	16.91± 2.99	11.13	21.13	16.30± 2.80	8.23	19.32	13.03± 3.42	12
Na^+	10.23	19.55	14.69±2.91	12.54	22.48	18.94± 3.00	10.53	22.39	17.20± 3.66	9.37	18.54	12.92± 2.69	200
Ca^{2+}	48.12	91.45	68.51±15.12	59.34	128.45	87.23± 26.13	56.34	121.11	81.76± 27.12	39.45	83.28	60.18± 15.62	75
Mg^{2+}	15.34	47.69	27.80±9.85	21.25	56.44	40.18± 11.11	18.23	44.15	35.19± 8.22	12.56	45.87	25.03± 12.01	30
Zn	0.295	0.452	0.333±0.044	0.304	0.489	0.383± 0.06	0.252	0.478	0.357± 0.065	0.292	0.381	0.321± 0.027	5
Fe	0.198	0.399	0.269±0.074	0.203	0.411	0.296± 0.07	0.191	0.383	0.281± 0.074	0.192	0.393	0.263± 0.076	0.2
Cd	0.000	0.002	0.001±0.00	0.001	0.002	0.002± 0.00	0.001	0.002	0.002± 0.00	0.000	0.002	0.001± 0.00	0.003
Cu	0.001	0.003	0.002±0.00	0.002	0.006	0.004± 0.00	0.001	0.005	0.003± 0.00	0.000	0.002	0.001± 0.00	2
Ni	0.000	0.011	0.004±0.00	0.001	0.013	0.006± 0.00	0.001	0.012	0.005± 0.00	0.000	0.011	0.003± 0.00	0.02
Mn	0.001	0.005	0.003±0.00	0.002	0.008	0.005± 0.00	0.001	0.006	0.005± 0.0	0.001	0.003	0.002± 0.00	0.1
Cr	0.000	0.002	0.002±0.00	0.001	0.003	0.002± 0.00	0.001	0.002	0.002± 0.00	0.000	0.002	0.002± 0.00	0.05
Pb	0.000	0.000	0.00±0.00	0.000	0.000	0.00± 0.00	0.000	0.000	0.00± 0.00	0.000	0.000	0.00±0 .00	0.01

Table 14.30. Seasonal variation of selected water parameters of Groundwater water during 2020-2021

Parameters	Season												BIS (2012)
	Summer			Monsoon			Autumn			Winter			
	Min	Max	Mean ± S.D	Min	Max	Mean ± S.D	Min	Max	Mean ± S.D	Min	Max	Mean ± S.D	
pH	7.17	7.76	7.27 ±0.10	7.23	7.58	7.39± 0.09	7.21	7.76	7.41± 0.16	7.12	7.52	7.30± 0.16	8.5
EC	281.28	389.32	361.39 ±29.60	378.65	428.57	410.08± 16.48	358.77	414.38	393.28± 19.44	297.68	417.28	369.88± 27.97	-
Temp	13.37	33.24	29.68 ±5.43	26.21	28.24	27.27± 1.00	18.2	22.15	20.18± 2.05	13.33	29.17	15.65± 4.37	-
TDS	272.22	376.34	337.35 ±28.94	319.35	411.68	371.52± 29.27	309.46	362.45	337.19± 16.97	263.53	362.23	303.30± 24.24	500
TU	0.71	1.19	1.02 ±0.18	0.87	1.45	1.18± 0.15	0.81	1.05	0.97± 0.07	0.74	1.12	0.85± 0.11	1
TH	111.34	164.47	128.09 ±13.82	123.57	184.21	149.38± 17.67	120.23	178.37	144.73± 19.48	111.53	129.76	119.26± 5.54	200
Alkalinity	105.38	139.37	118.92 ±10.13	118.35	156.38	137.60± 12.40	110.68	143.29	133.16± 9.60	106.67	127.37	117.70± 6.80	200
Acidity	21.23	38.42	30.48 ±4.57	22.57	41.68	33.97± 5.66	17.46	37.38	30.89± 5.66	16.38	32.48	27.53± 5.34	-
PO_4^{3-}	0.11	0.25	0.17 ±0.05	0.14	0.27	0.22± 0.04	0.15	0.22	0.18± 0.02	0.12	0.17	0.14± 0.01	-
HCO^-_3	105.37	131.45	114.30 ±7.25	115.57	154.24	128.65± 13.09	108.23	151.38	124.07± 13.84	97.18	115.28	105.82± 6.12	300
NO_3^-	11.02	14.38	12.35 ±1.04	11.33	17.11	14.64± 1.71	10.23	18.34	14.84± 2.55	8.21	14.89	12.54± 2.12	45

SO_4^{2-}	9.04	13.84	11.92 ±1.62	11.35	18.21	14.95± 2.13	11.21	19.21	14.74± 2.47	7.24	14.29	11.09± 2.00	150
Cl^-	11.01	18.38	13.97 ±2.40	14.56	20.34	17.16± 1.60	11.78	18.78	16.73± 2.14	9.37	18.24	13.60± 3.02	250
K^+	7.68	11.38	9.68 ±0.99	10.21	11.78	10.99± 0.54	7.57	11.14	9.59± 1.00	2.38	10.11	7.51± 2.38	12
Na^+	10.11	18.34	13.62 ±2.76	12.32	21.36	17.65± 2.80	10.43	21.48	16.50± 3.16	8.78	17.53	12.67± 2.53	200
Ca^{2+}	41.34	87.32	63.36 ±16.09	59.34	98.34	76.20± 12.27	55.23	89.24	68.95± 10.61	36.85	79.38	57.18± 14.56	75
Mg^{2-}	14.21	42.47	25.34 ±9.46	20.43	61.38	40.69± 14.47	18.11	49.34	33.57± 8.99	12.34	44.29	24.78± 11.03	30
Zn	0.293	0.434	0.325 ±0.04	0.301	0.478	0.371± 0.05	0.302	0.452	0.360± 0.05	0.291	0.348	0.313± 0.02	5
Fe	0.191	0.375	0.259 ±0.07	0.205	0.392	0.308± 0.07	0.192	0.378	0.304± 0.07	0.198	0.387	0.271± 0.06	0.2
Cd	0.000	0.002	0.001 ±0.00	0.001	0.002	0.002± 0.00	0.001	0.002	0.001± 0.00	0.000	0.002	0.001± 0.00	0.003
Cu	0.001	0.003	0.002 ±0.00	0.002	0.004	0.003± 0.00	0.001	0.003	0.002± 0.00	0.000	0.003	0.002± 0.00	2
Ni	0.000	0.009	0.004 ±0.00	0.001	0.011	0.007± 0.00	0.001	0.01	0.005± 0.00	0.000	0.008	0.003± 0.00	0.02
Mn	0.001	0.005	0.003 ±0.00	0.002	0.007	0.005± 0.00	0.001	0.006	0.005± 0.00	0.001	0.003	0.002± 0.00	0.1
Cr	0.000	0.002	0.002 ±0.00	0.001	0.002	0.002± 0.00	0.001	0.002	0.002± 0.00	0.000	0.002	0.002± 0.00	0.05
Pb	0.000	0.000	0.000 ±0.00	0.000	0.000	0.000 ± 0.00	0.000	0.00	0 ±0.00	0.000	0.000	0 ±0.00	0.01

Comparison in annual average values of all the variables for both the years (from interpolated figures)

Many academics employ GIS-based methodologies for spatial interpolation, analysis, and the extraction of many thematic layers to achieve desired results. To create the appropriate output maps, spatial distributions of groundwater quality elements will be estimated using a GIS-based spatial analysis tool in ArcGIS. Spatial analysis was carried out in this work using ArcGIS 10.3 software. Each sampling station is given a location in the Dehradun region based on its corresponding coordinates using the geographic coordinate system of the World Geodetic System (WGS)—1984. This is done to assess the critical areas in the region that are affected by one or more water quality issues.The actual data values of each physico-chemical variable and potentially toxic elements are assigned to each of 6 monitoring locations and the data is spatially interpolated using GIS Inverse Distance Weight (IDW) method.

Figure 4.16 indicated a comparable spatial variation in pH during the years 2019-2020 and 2020-2021 using the IDW method. The pH spatial variation ranged from 7.21 to 7.55 and 7.23 to 7.44 during 2019-2020 and 2020-2021, respectively. Figure 4.17, EC interpolation using IDW ranged from 356.48 to 430.48 µS/cm during the year 2019-2020, whereas, the EC variation during 2020-2021 ranged from 343.11 to 400.34 µS/cm. Figure 4.18, Temperature interpolation using IDW ranged from 23.70 to 24.01°C during the year 2019-2020, whereas, the Temperature variation during 2020-2021 ranged from 23.11 to 23.27 °C. Figure 4.19, shows the interpolation of TDS using IDW ranged from 306.66 to 368.50 mg/L during the year 2019-2020 whereas the spatial variation of TDS during 2020-2021 was ranged from 300.57 to 349.94 mg/L. Figure 4.20 shows the interpolation of TU using IDW ranged from 0.81 to 1.21 NTU during the year 2019-2020 whereas, the spatial variation of TU during 2020-2021 was ranged from 0.82 to 1.09 NTU. Figure 4.21 represents the spatial variation of TH ranging from 121.10 to 184.53 mg/L and 118.83 to 157.30 mg/L for 2019-2020 and 2020-2021 respectively. A comparable spatial variation in Alkalinity during the year 2019-2020 and 2020-2021 using the IDW method is depicted in Figure 4.22. The alkalinity spatial variation ranged from 114.86 to 141.39 mg/L and 112.91 to 138.19 mg/L during the year 2019-2020 and 2020-2021, respectively. Figure 4.23 represents the acidity interpolation using IDW ranged from 20.44 to 36.46 mg/L during 2019-2020 whereas the acidity variation during 2020-2021 ranged from 20.35 to 34.41 mg/L.

Figure 4.24 indicated a comparable spatial variation in Phosphate during the years 2019-2020 and 2020-2021 using the IDW method. The phosphate spatial variation ranged from 0.14 to 0.20 mg/L and 0.14 to 0.21 mg/L during the year 2019-2020 and 2020-2021 respectively. Figure 4.25 showed that the bicarbonate interpolation using IDW ranged from 111.80 to 160.61 mg/L during 2019-2020 whereas the bicarbonate variation during 2020-2021 ranged from 108.09 to 132.34 mg/L. Figure 4.26 shows the interpolation of nitrate using IDW ranged from 10.84 to 16.18 mg/L during the year 2019-2020, whereas, the spatial variation of nitrate during 2020-2021 was ranged from 10.55 to 15.30 mg/L. Figure 4.27 shows the interpolation of sulphate using IDW ranged from 10.69 to 15.56 mg/L during the year 2019-2020 whereas, the spatial variation of sulphate during 2020-2021 was ranged from 10.18 to 15.30 mg/L.Figure 4.28 shows the interpolation of chloride using IDW ranged from 12.80 to 19.38 mg/L during the year 2019-2020, whereas, the spatial variation of chloride during 2020-2021 was ranged from 12.30 to 17.73 mg/L. A comparable spatial variation in potassium between 2019-2020 and 2020-2021 using the IDW method is depicted in Figure 4.29. The potassium spatial variation ranged from 11.54 to 17.08 mg/L and 9.60 to 10.54 mg/L during 2019-2020 and 2020-2021 respectively. Figure 4.30 represents the sodium interpolation using IDW ranged from 11.43 to 18.40 mg/L during 2019-2020 whereas, the sodium variation during 2020-2021 ranged from 10.83 to 17.78 mg/L. Figure 4.31 shows the interpolation of calcium using IDW ranged from 54.45 to 103.91mg/L during the year 2019-2020 whereas, the spatial variation of calcium during 2020-2021 was ranged from 50.81 to 85.72 mg/L. Figure 4.32 shows the interpolation of magnesium using IDW ranged from 18.18 to 45.12 mg/L during 2019-2020 whereas the spatial variation of magnesium during 2020-2021 ranged from 17.44 to 46.64 mg/L.

Figure 4.33 showed a comparable spatial variation in Zn during 2019-2020 and 2020-2021 using the IDW method. The Zn spatial variation ranged from 0.289 to 0.425 mg/L and 0.299 to 0.408 mg/L during the year 2019-2020 and 2020-2021 respectively. Figure 4.34 showed the Fe interpolation using IDW ranged from 0.199 to 0.392 mg/L during 2019-2020 whereas the Fe variation during 2020-2021 ranged from 0.198 to 0.376 mg/L. Figure 4.35 showed the Cd interpolation using IDW ranged from 0.001 to 0.002 mg/L during 2019-2020 whereas the Cd variation during 2020-2021 ranged from 0.001 to 0.002 mg/L. Figure 4.36 shows the interpolation of Cu using IDW ranged from 0.001 to 0.003 mg/L during the year 2019-2020 whereas, the spatial variation of Cu during 2020-2021 was ranged from 0.001 to 0.003 mg/L. Figure 4.37 shows the interpolation of Ni using IDW ranged from 0.001 to 0.011 mg/L during the year 2019-2020 whereas, the spatial variation of Ni during 2020-2021 was ranged from 0.001 to 0.009 mg/L. Figure 4.38 shows the interpolation of Mn using IDW ranged from 0.001 to 0.005 mg/L during 2019-2020 whereas, the spatial variation of Mn during 2020-2021 ranged from 0.001 to 0.005 mg/L. Figure 4.39 shows the interpolation of Cr using IDW ranged from 0.001 to 0.002 mg/L during the year 2019-2020 whereas, the spatial variation of Cr during 2020-2021 was ranged from 0.001 to 0.002 mg/L. Similarly, Pb was not detected in any ground water sample therefore, the spatial distribution of these PTEs was not considered in this study.

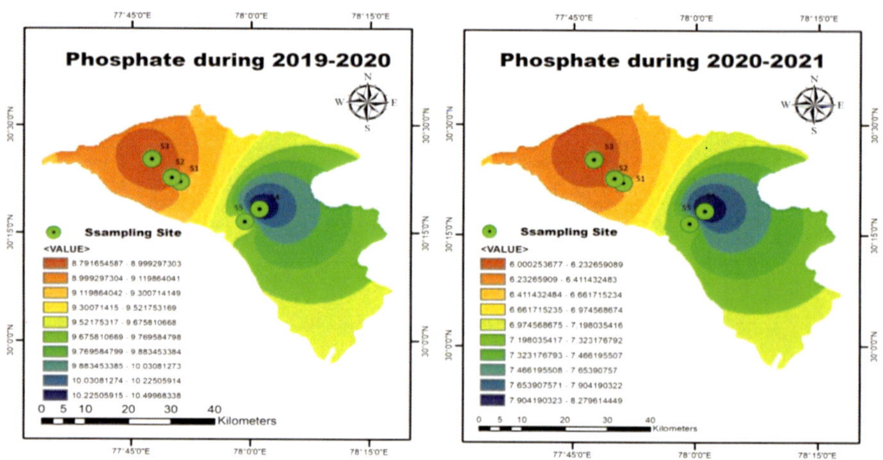

Figure 14.16. Spatio-temporal variations in pH during both the years of the study in groundwater

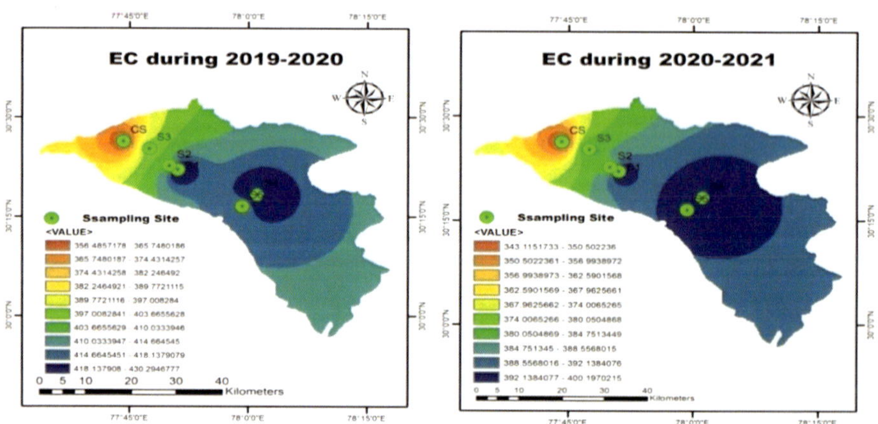

Figure 14.17. Spatio-temporal variations in EC during both the years of the study in groundwater

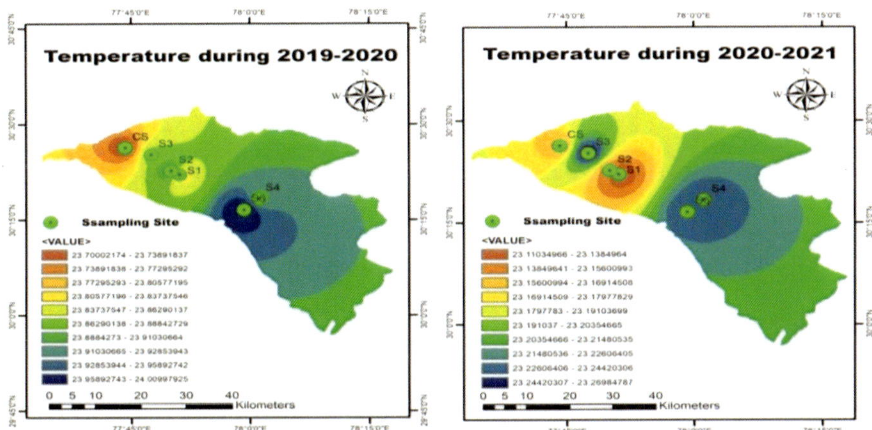

Figure 14.18. Spatio-temporal variations in Temperature during both the years of the study in groundwater

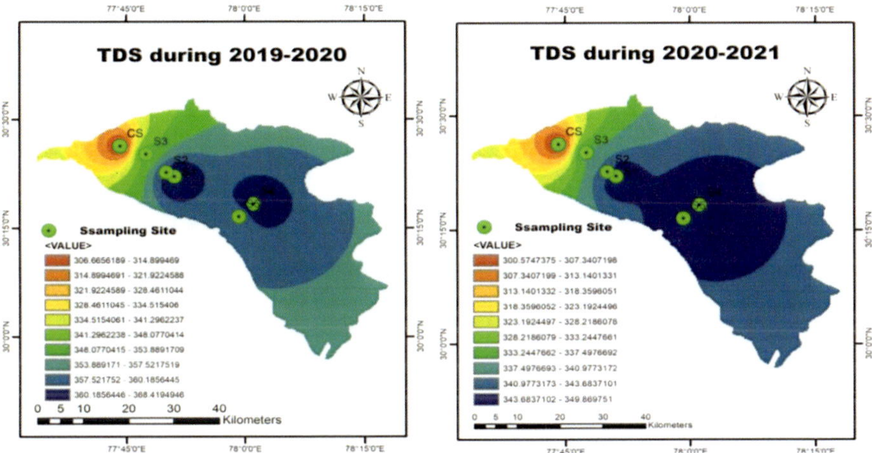

Figure 14.19. Spatio-temporal variations in TDS during both the years of the study in groundwater

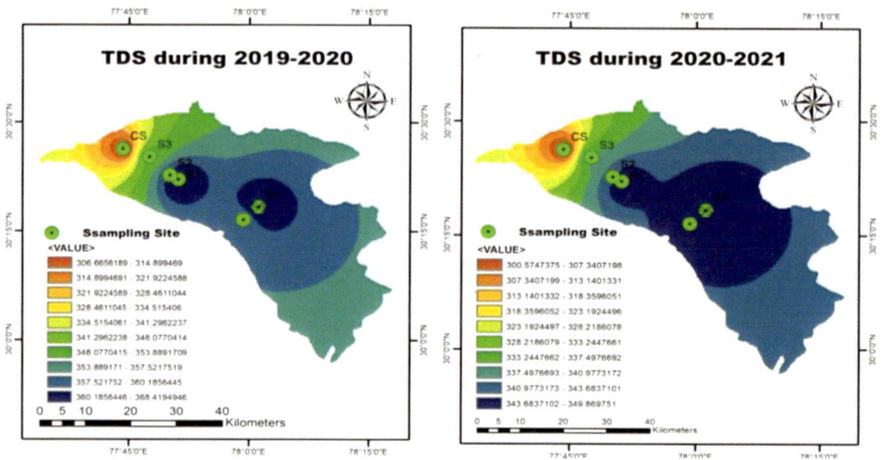

Figure 14.20. Spatio-temporal variations in TDS during both the years of the study in groundwater

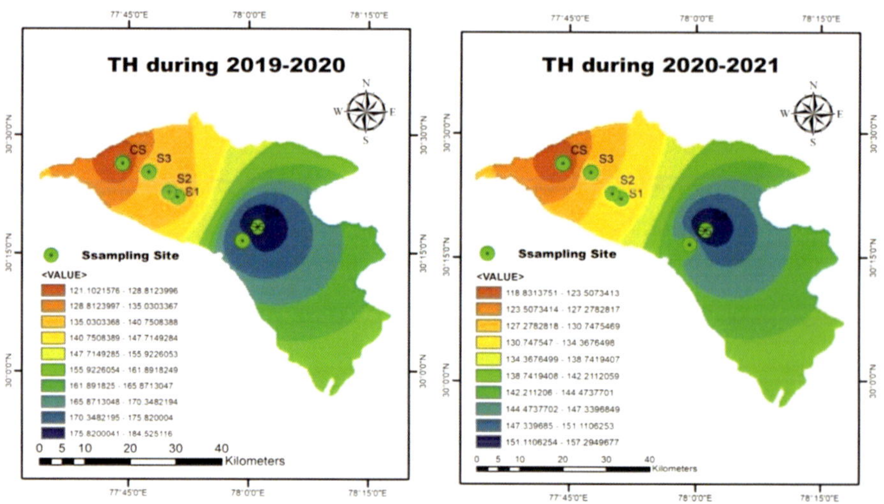

Figure 14.21. Spatio-temporal variations in TH during both the years of the study in groundwater

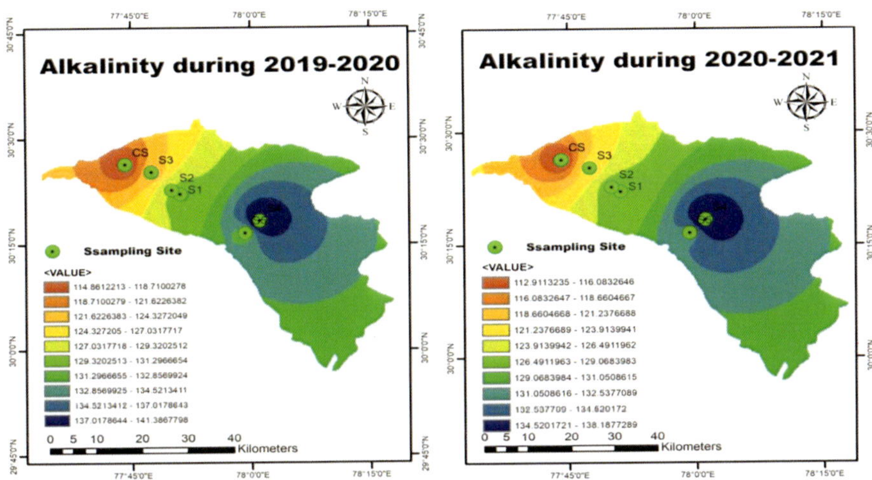

Figure 14.22. Spatio-temporal variations in Alkalinity during both the years of the study in groundwater

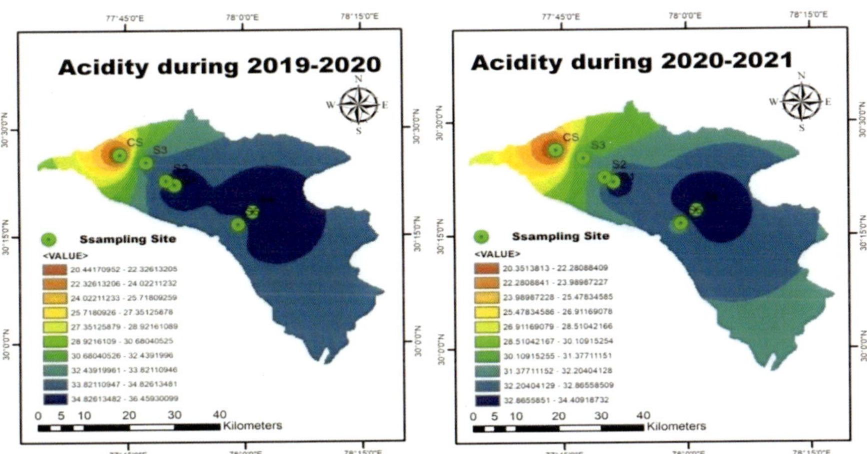

Figure 14.22. Spatio-temporal variations in Acidity during both the years of the study in groundwater

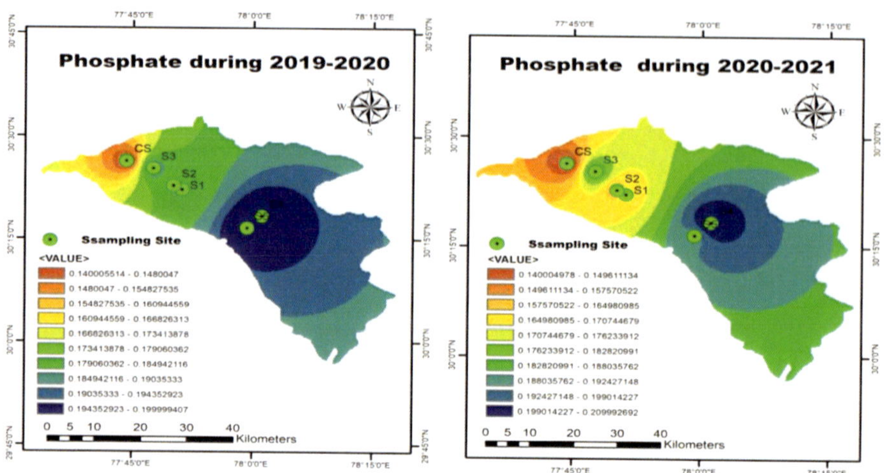

Figure 14.23. Spatio-temporal variations in Phosphate during both the years of the study in groundwater

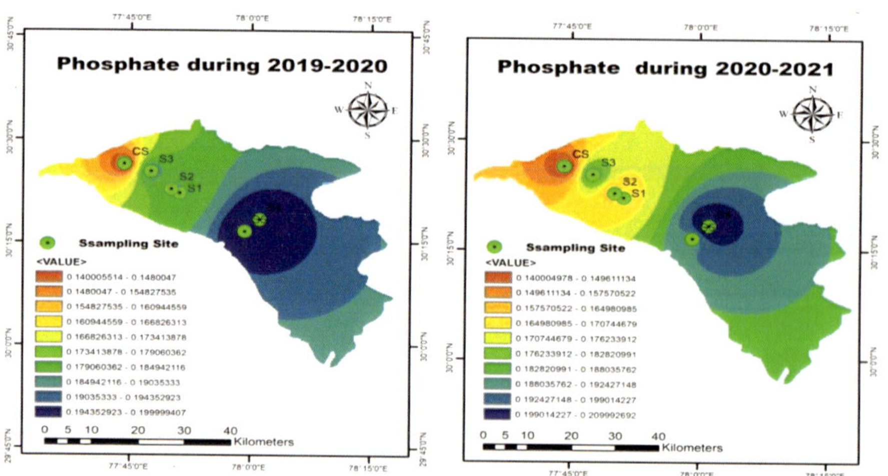

Figure 14.24. Spatio-temporal variations in Bicarbonate during both the years of the study in groundwater

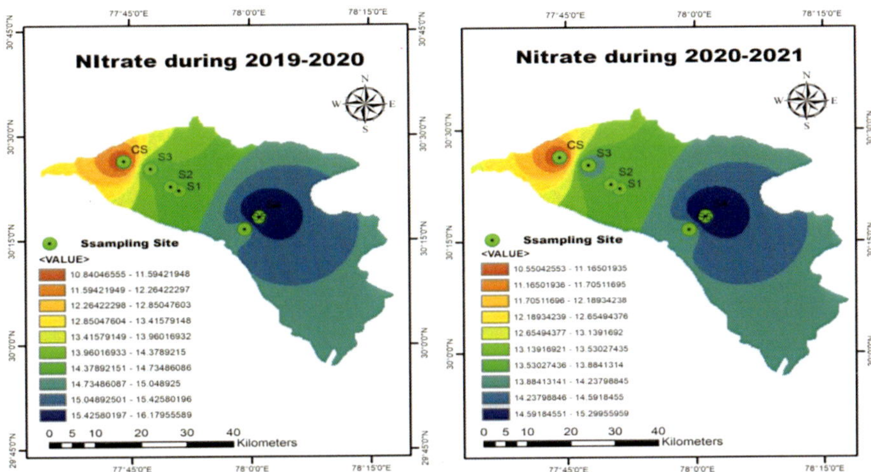

Figure 14.25. Spatio-temporal variations in Nitrate during both the years of the study in groundwater

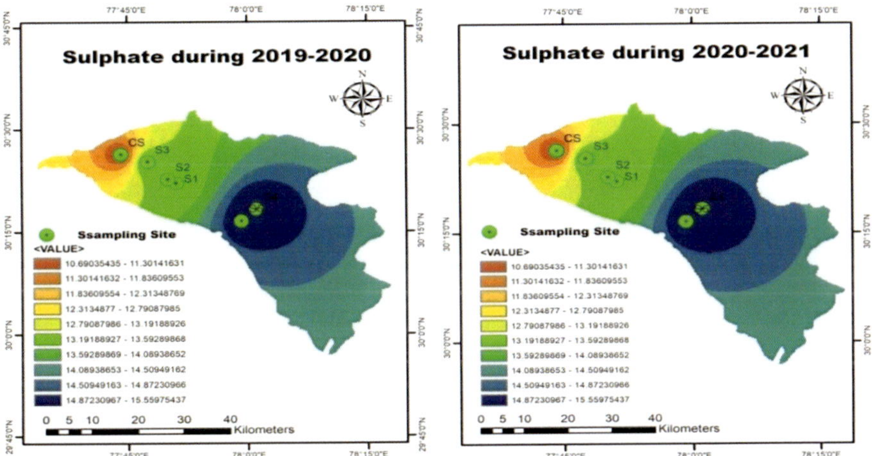

Figure 14.26. Spatio-temporal variations in Sulphate during both the years of the study in groundwater

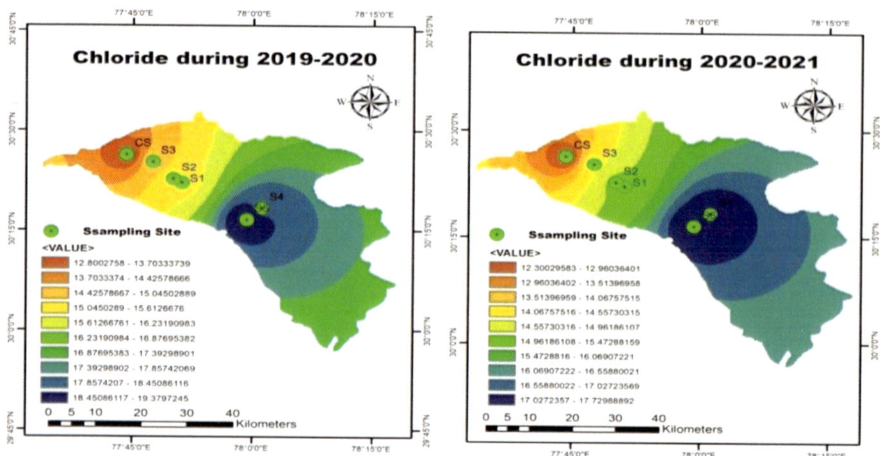

Figure 14.27. Spatio-temporal variations in Chloride during both the years of the study in groundwater

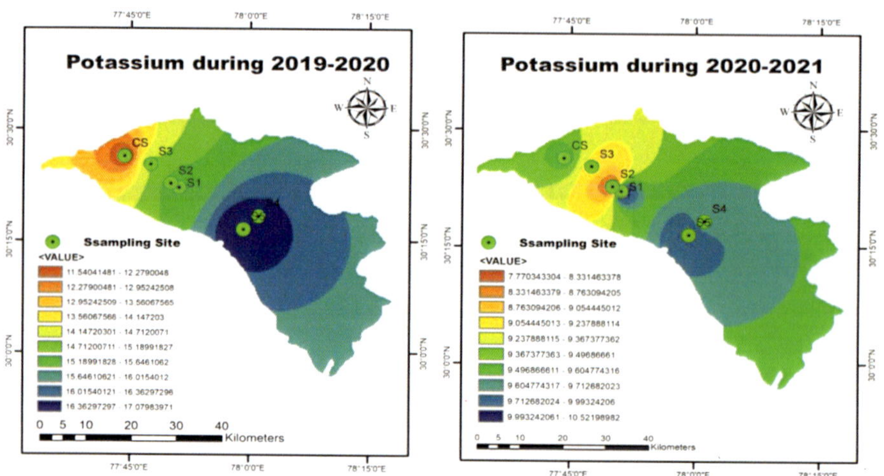

Figure 14.28. Spatio-temporal variations in Potassium during both the years of the study in groundwater

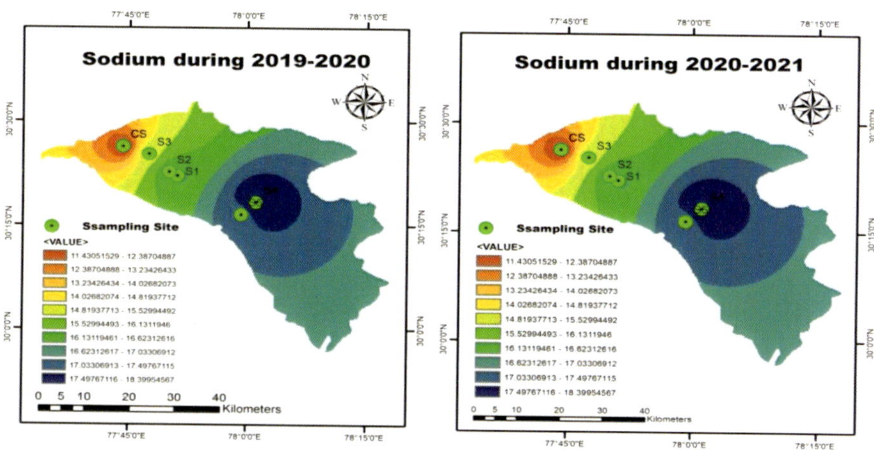

Figure 14.29. Spatio-temporal variations in Sodium during both the years of the study in groundwater

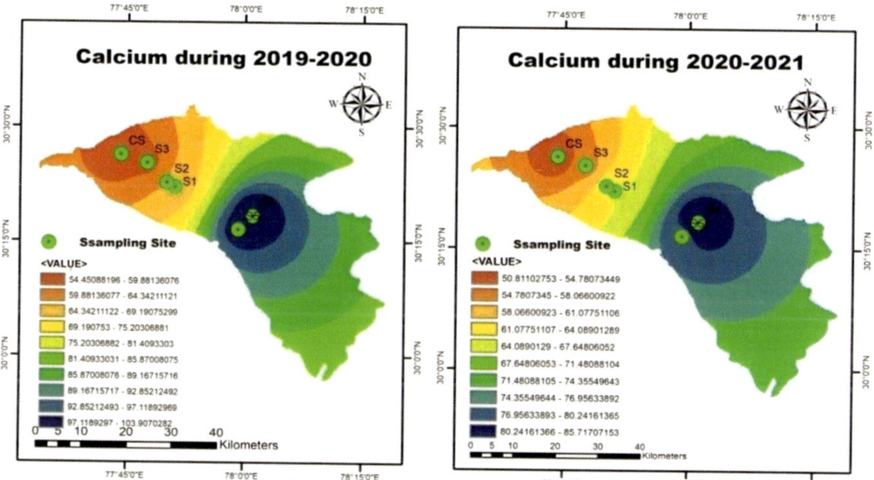

Figure 14.30. Spatio-temporal variations in Calcium during both the years of the study in groundwater

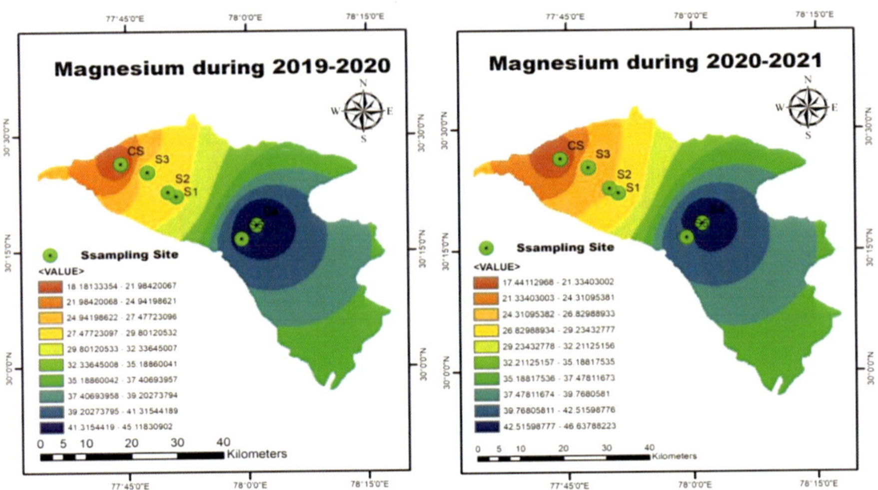

Figure 14.31. Spatio-temporal variations in Magnesium during both the years of the study in groundwater

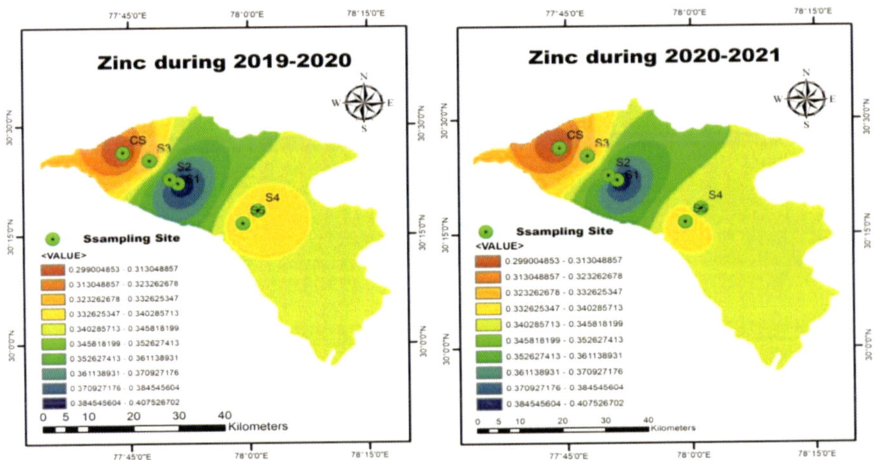

Figure 14.32. Spatio-temporal variations in Zinc during both the years of the study in groundwater

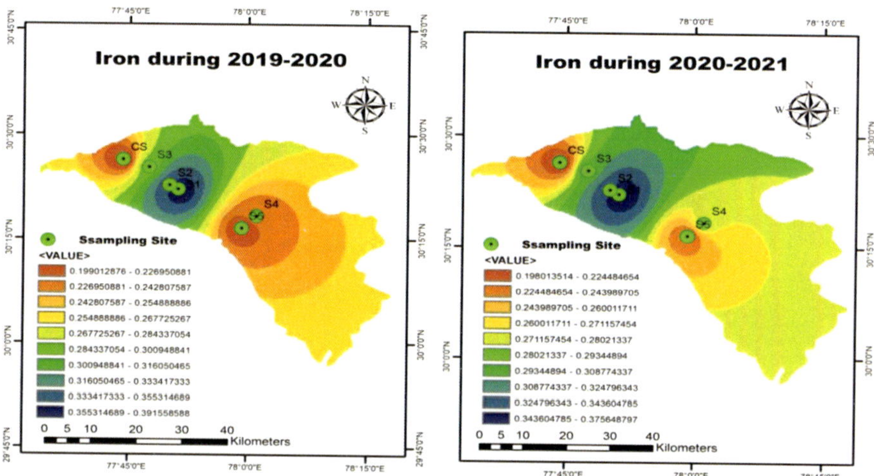

Figure 14.33. Spatio-temporal variations in Iron during both the years of the study in groundwater

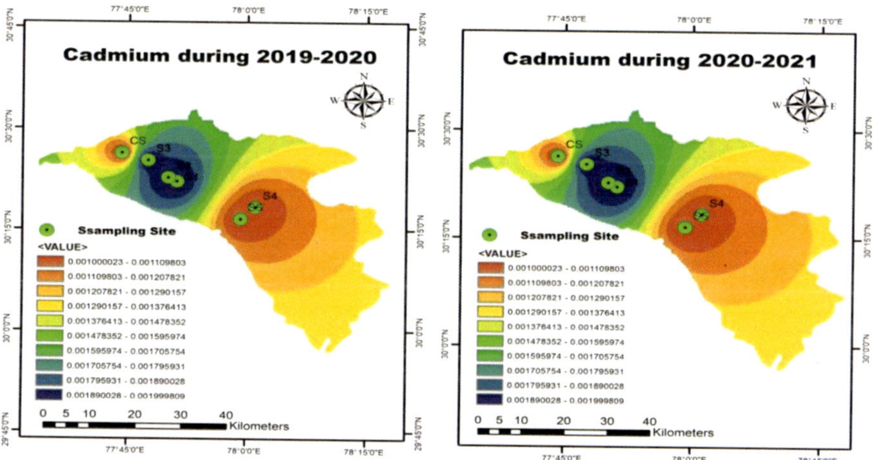

Figure 14.34. Spatio-temporal variations in Cadmium during both the years of the study in groundwater

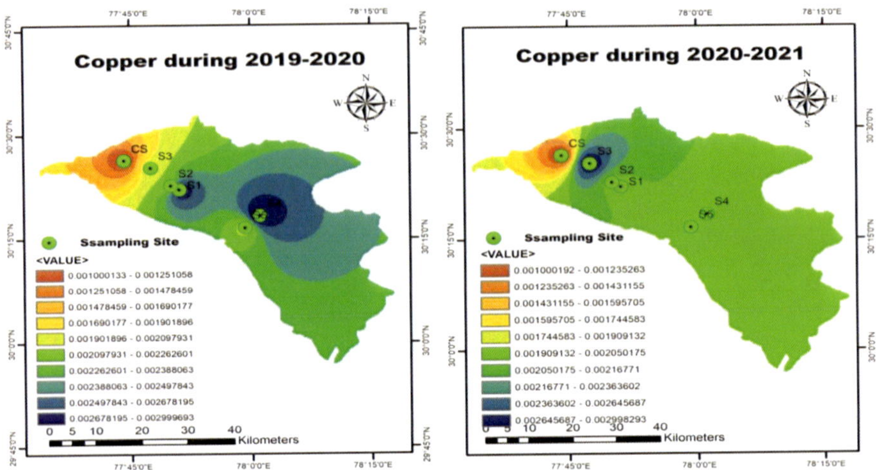

Figure 14.35. Spatio-temporal variations in Copper during both the years of the study in groundwater

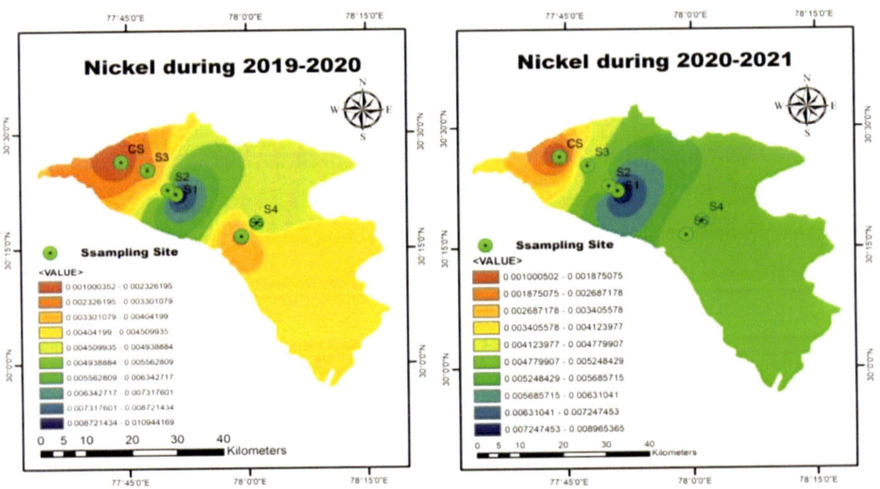

Figure 14.36. Spatio-temporal variations in Nickel during both the years of the study in groundwater

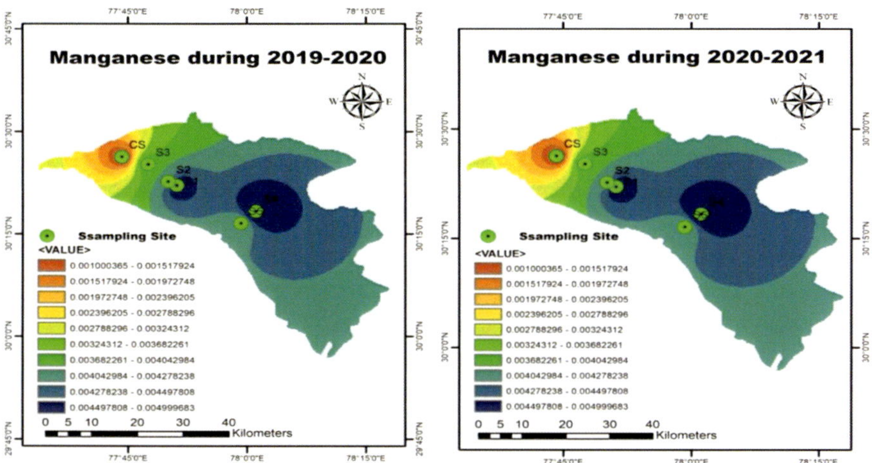

Figure 14.37. Spatio-temporal variations in Manganese during both the years of the study in groundwater

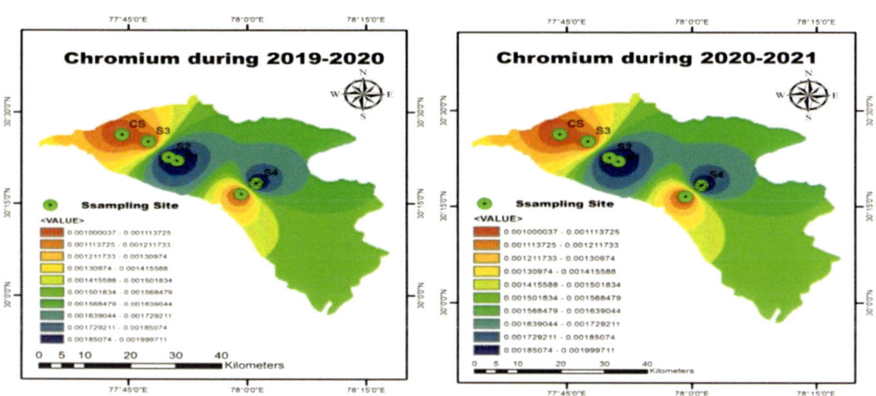

Figure 14.38. Spatio-temporal variations in Chromium during both the years of the study in groundwater

Interpretation of Multivariate Statistical Models

Pollution sources apportionment by PCA

The variability of the original mean dataset was represented by a screen plot and 3D plot of 3 PCs (Figure 4.40a and b), which determines the 3 components, contributes to 95.39%. The KMO test value 0.64 near to 1 indicated the suitability of dataset for PC and the test values of KS (0.155, p=0.200), SW (0.976, p=0.931). All of the parameters showed the communalities values greater than 0.5. Based on the eigen value criteria (Pathak and Limaye, 2011), 3 principal components (PCs) are considered necessary as their eigen values are more significantthan one.

The factor weights values of observed factors related to each PC's represented in Table 4.31. The factor loading values observed by the rotated varimax matrix are classified as weak, moderate, and strong (0.30-0.50, 0.50-0.75, and >0.75, respectively), according to Liu *et al.,* (2003). After selecting the 3 PCs, the factors contributing >0.6 were considered significant (positively or negatively) (Field, 2009). It was determined that PC1 included a substantial part of the variables with 65.52% of the variance, had the load on EC, Temperature, TDS, TU, TH, alkalinity, acidity, phosphate, bicarbonate, nitrate, sulphate, chloride, potassium, sodium, calcium, magnesium and manganese. All these variables had positive factor weight in the development of PC1. The weight of TH, phosphate, bicarbonate, sulphate, chloride, calcium and magnesium showed high positive loadings for the formation of PC1 and represented their strong correlation with each other and accounts for 65.52% of the variance. High TDS values represent the input from domestic sewage (Singh *et al.,* 2005). The PC2 accounts for 22.87% of variance and is associated with a positive pH, zinc, iron, nickel and chromium weight. The weight of acidity, iron, cadmium, and copper showed high positive loadings for the formation of PC3, representing their strong correlation with each other and accounting for 6.99% of variance.

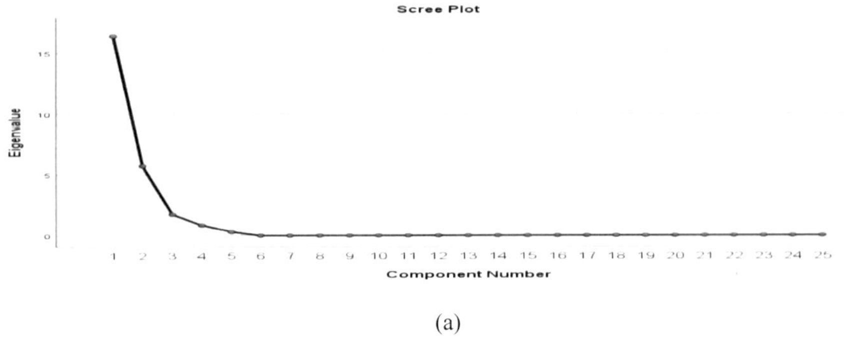

(a)

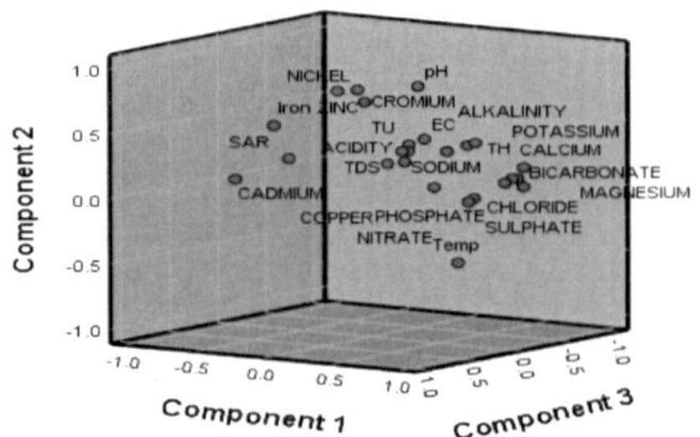

(b)

(a) Screen plot of principal component analysis (PCA); (b) 3D plot of 3 PCs

Pearson Correlation Matrix of Water Quality Parameters of Groundwater Sources at Different Sampling Locations

The correlation between the physico-chemical parameters in groundwater samples of each monitoring location is given in Table 4.32 to Table 4.37 and Table 4.38 to Table 4.43 for the year 2019-2020 and 2020-2021, respectively. A value of r is greator than 0.7 shows a strong correlation, whereas 0.5 to 0.7 r value shows a moderate correlation among the parameters. In 2019-2020, EC showed ahighly positive correlation with TDS (r=0.91), SO_4^{2-} (r=0.90).TH showed the highly positive correlation with Ca^{2+} (r= 0.99), Mg^{2+} (r= 0.98), SO_4^{2-} (r=0.97), HCO^-_3 (r=0.90) at sampling site GW1, alkalinity showed positive correlation with HCO^-_3(r=0.98), SO_4^{2-} (r=0.95), Cl^- (r=0.91), K^+ (r=0.91). HCO^-_3 showed the highly positive correlation withNO_{3-} (r=0.83), SO_4^{2-} (r=0.93), Cl^- (r=0.93).Cl^-showed a highly positive correlation with Na^+ (r=0.95) at sampling site GW2. At sampling location GW3, pH showed a highly positive correlation with TDS (r=0.75), TU (r=0.87), acidity (r=0.78). EC showed a highly positive correlation with TH (r= 0.93), and alkalinity (r=0.91). TDS showed the highly positive correlation with TU (r=0.96), acidity (r=0.81). At sampling location, GW4, TDS showed a highly positive correlation with TU (r=0.90). TH showed the highly positive correlation with Ca^{2+} (r= 0.90), SO_4^{2-} (r=0.92), HCO^-_3 (r=0.98), Fe (r=0.84), Cu (r=0.86) Ni (r=0.95). At sampling location GW5, EC showed the highly positive correlation with TH (r= 0.95),NO_{3-} (r=0.81), SO_4^{2-} (r=0.90), Ca^{2+} (r= 0.88), Na^+ (r=0.89), Zn (r=0.97). TH showed a highly positive correlation with TU (r= 0.99).TH showed the highly positive correlation with Ca^{2+} (r= 0.86), Mg^{2+} (r= 0.82), SO_4^{2-} (r=0.92), HCO^-_3 (r=0.99). Na^+ showed a highly positive correlation with Ca^{2+} (r= 0.97), Mg^{2+} (r= 0.76), and Zn (r=0.94).

In the year 2020-2021, EC showed a highly positive correlation with TDS (r=0.81), TU (r=0.72), TH (r= 0.81), SO_4^{2-}(r=0.77). TDS showed a highly positive correlation with TH (r=0.96), alkalinity (r=0.84), Ca^{2+} (r= 0.91), Mg^{2+} (r= 0.95), SO_4^{2-} (r=0.97), HCO^-_3 (r=0.91). EC showed a highly positive correlation with TDS (r=0.76), acidity (r= 0.79), Ca^{2+} (r= 0.79), Fe (r=0.89), Cu (r=0.84) at sampling location GW1. TDS showed a highly positive correlation with TU (r=0.87), acidity (r=0.79), Ca^{2+} (r= 0.69), Na^+ (r=0.91), Fe (r=0.87), Cu (r=0.80). At sampling location GW2, EC showed a highly positive correlation with TH (r=0.86), alkalinity (r=0.93), acidity (r=0.95), PO_4^{3-} (r=0.92), SO_4^{2-} (r=0.94), Ca^{2+} (r= 0.86), Mg^{2+} (r= 0.88), Zn (r=0.90) and Mn (r=0.83). TDS showed a highly positive correlation with TU (r=0.75), acidity (r=0.78). TDS showed the highly positive correlation with TU (r=0.93), acidity (r=0.91), TH (r=0.87), HCO^-_3(r=0.91), Ca^{2+} (r= 0.86), Mg^{2+} (r= 0.76) at sampling location GW3. TU showed the highly positive correlation with TH (r=0.78), acidity (r=0.82), HCO^-_3 (r=0.84), PO_4^{3-} (r=0.78), K^+ (r=0.75), Ca^{2+} (r= 0.76). At sampling location GW4, pH showed a highly positive correlation with NO_3^- (r=0.87), Na^+ (r=0.78). EC showed a highly positive correlation with TDS (r=0.79), TH (r=0.94), HCO^-_3(r=0.96), alkalinity (r=0.94), Ca^{2+} (r= 0.83), Mg^{2+} (r= 0.81), Zn (r=0.91), Fe (r= 0.91), Mn (r=0.98). At sampling site GW5, pH showed the highly positive correlation with TH (r=0.86), HCO^-_3 (r=0.74), SO_4^{2-} (r=0.88), Na^+ (r=0.77), K^+ (r= 0.77) and Zn (r=0.81). EC showed a highly positive correlation with TDS (r=0.75), TH (r=0.83), HCO^-_3 (r=0.91), alkalinity (r=0.91), Mg^{2+} (r= 0.80), Zn (r=0.91), Ni (r= 0.75), Mn (r=0.77). TDS showed a highly positive correlation with TU (r=0.92), alkalinity (r=0.83), acidity (r=0.92), PO_4^{3-} (r=0.80), HCO^-_3 (r=0.86), Ca^{2+} (r= 0.87), Zn (r=0.75), Ni (r=0.84).

Table 14.31. Factor loading matrix of PCs of selected parameters of groundwater sources

Parameters for GW	Component			Communalities Extraction
	PC1	PC2	PC3	
pH		0.873		.974
EC	.708			.985
Temp.	.840			.960
TDS	.677			.987
TU	.648			.975
TH	.963			.985
Alkalinity	.847			.953
Acidity	.687		.616	.999
Phosphate	.934			.963
Bicarbonate	.959			.984
Nitrate	.828			.976
Sulphate	.955			.992
Chloride	.963			.953
Potassium	.793			.844
Sodium	.812			.972
Calcium	.926			.995
Magnesium	.978			.999
Zinc		.853		.952
Iron		.608	.774	.994
Cadmium			.919	.998
Copper			.659	.916
Nickel		.857		.910
Manganese	.666			.998

Chromium		.742		.638
SAR			.892	.945
% of total variance	65.52	22.87	6.99	
Cumulative %	65.52	88.40	95.39	

Table 4.32 Pearson Correlation Matrix of Water Quality Parameters of groundwater sources at site GWCS during 2019-2020

Values in bold letters show significant correlations

Vari-ables	pH	EC	Temp	TDS	TU	TH	Alky	Acidity	PO_4^{3-}	HCO_3^-	NO_3^-	SO_4^{2-}	Cl^-	K^+	Na^+	Ca^{2+}	Mg^{2+}	Zn	Fe	Cd	Cu	Ni	Mn	Cr
pH	1.00																							
EC	0.52	1.00																						
Temp	-0.21	0.33	1.00																					
TDS	0.21	**0.91**	0.60	1.00																				
TU	-0.57	0.19	0.62	0.46	1.00																			
TH	0.13	**0.89**	0.60	**0.97**	0.58	1.00																		
Alk	0.13	0.71	0.72	**0.84**	0.66	**0.85**	1.00																	
Acidity	-0.61	-0.07	**0.84**	0.31	0.63	0.32	0.40	1.00																
PO_4^{3-}	0.15	**0.82**	0.50	**0.88**	0.60	**0.89**	**0.87**	0.14	1.00															
HCO_3^-	-0.06	**0.80**	0.61	**0.90**	0.69	**0.97**	**0.81**	0.36	**0.88**	1.00														
NO_3^-	0.04	0.73	**0.83**	**0.85**	0.64	**0.90**	**0.86**	0.50	**0.81**	**0.90**	1.00													
SO_4^{2-}	0.27	**0.90**	0.67	**0.97**	0.47	**0.95**	**0.87**	0.30	**0.91**	**0.89**	**0.92**	1.00												
Cl^-	0.05	**0.83**	0.65	**0.98**	0.57	**0.96**	**0.85**	0.43	**0.88**	**0.91**	**0.85**	**0.95**	1.00											
K^+	0.23	0.67	0.76	**0.79**	0.43	**0.80**	0.70	0.51	0.60	0.69	**0.83**	**0.83**	**0.77**	1.00										
Na^+	-0.21	0.67	0.64	**0.89**	**0.75**	**0.92**	**0.79**	0.56	**0.79**	**0.91**	**0.79**	**0.82**	**0.94**	0.71	1.00									
Ca^{2+}	0.27	**0.93**	0.62	**0.99**	0.48	**0.98**	**0.87**	0.29	**0.88**	**0.91**	**0.89**	**0.98**	**0.96**	**0.84**	**0.87**	1.00								
Mg^{2+}	0.21	**0.90**	0.60	**0.98**	0.57	**0.99**	**0.90**	0.28	**0.93**	**0.93**	**0.89**	**0.97**	**0.96**	**0.80**	**0.89**	**0.99**	1.00							
Zn	-0.83	-0.29	0.40	0.07	**0.79**	0.13	0.28	0.70	0.16	0.23	0.15	0.01	0.24	0.04	0.48	0.03	0.12	1.00						
Fe	-0.53	0.04	0.54	0.39	**0.83**	0.40	0.58	0.67	0.47	0.42	0.37	0.34	0.53	0.34	0.68	0.35	0.43	**0.90**	1.00					
Cd	0.26	0.71	0.38	**0.75**	0.58	**0.77**	**0.90**	0.06	**0.90**	0.71	0.65	**0.77**	**0.75**	0.53	0.69	**0.77**	**0.85**	0.21	0.55	1.00				
Cu	-0.22	0.61	0.65	**0.75**	0.83	**0.85**	**0.77**	0.42	**0.88**	**0.92**	**0.88**	**0.80**	**0.80**	0.63	**0.83**	**0.77**	**0.83**	0.38	0.56	0.70	1.00			
Ni	-0.44	0.37	0.73	0.70	0.67	0.66	0.64	0.73	0.64	0.69	0.65	0.66	**0.81**	0.52	**0.84**	0.63	0.65	0.63	**0.76**	0.47	0.70	1.00		
Mn	-0.07	0.68	0.21	0.65	0.39	0.69	0.49	-0.03	0.73	**0.80**	0.61	0.63	0.64	0.19	0.59	0.62	0.65	0.00	0.07	0.47	0.70	0.47	1.00	
Cr	-0.36	0.48	0.44	0.66	**0.83**	0.71	0.75	0.34	**0.83**	**0.82**	0.64	0.63	0.74	0.29	**0.81**	0.63	0.72	0.58	0.69	0.75	**0.84**	**0.75**	**0.75**	1.00

Table 4.33 Pearson Correlation Matrix of Water Quality Parameters of groundwater sources at site GW1 during 2019-2020

Values in bold letters show significant correlations

Variables	pH	EC	Temp	TDS	TU	TH	Alky	Acidity	PO_4^{3-}	HCO_3^-	NO_3^-	SO_4^{2-}	Cl^-	K^+	Na^+	Ca^{2+}	Mg^{2+}	Zn	Fe	Cd	Cu	Ni	Mn	Cr
pH	1.00																							
EC	0.62	1.00																						
Temp	**0.89**	0.59	1.00																					
TDS	0.60	0.62	0.59	1.00																				
TU	0.68	**0.77**	0.59	**0.83**	1.00																			
TH	0.50	**0.82**	0.41	0.25	0.66	1.00																		
Alk	0.58	**0.84**	0.45	0.33	0.75	**0.98**	1.00																	
Acidity	**0.77**	**0.82**	**0.75**	0.65	**0.83**	0.74	0.77	1.00																
PO_4^{3-}	**0.82**	0.61	0.64	0.23	0.42	0.52	0.59	0.58	1.00															
HCO_3^-	0.61	**0.88**	0.55	0.35	0.72	**0.98**	**0.98**	**0.78**	0.62	1.00														
NO_3^-	0.35	**0.78**	0.52	0.25	0.46	**0.82**	0.74	0.67	0.29	**0.83**	1.00													
SO_4^{2-}	0.60	**0.85**	0.46	0.55	**0.88**	**0.92**	**0.95**	**0.79**	0.48	**0.93**	0.68	1.00												
Cl^-	0.53	0.72	0.45	0.18	0.55	**0.93**	**0.91**	0.58	0.53	**0.93**	**0.78**	**0.85**	1.00											
K^+	0.49	**0.87**	0.47	0.57	**0.87**	**0.90**	**0.91**	**0.83**	0.37	**0.91**	**0.77**	**0.95**	0.76	1.00										
Na^+	0.73	**0.77**	**0.79**	**0.77**	**0.86**	0.61	0.67	0.71	0.46	0.73	0.61	**0.77**	0.66	0.77	1.00									
Ca^{2+}	0.60	**0.90**	0.59	0.45	**0.75**	**0.95**	**0.94**	**0.78**	0.52	**0.98**	**0.88**	**0.93**	**0.93**	**0.92**	**0.82**	1.00								
Mg^{2+}	0.56	**0.85**	0.48	0.30	0.68	**0.99**	**0.98**	0.74	0.56	**0.99**	**0.83**	**0.93**	**0.96**	**0.90**	0.70	**0.97**	1.00							
Zn	0.58	**0.76**	0.55	0.20	0.53	**0.89**	**0.88**	0.59	0.61	**0.93**	**0.80**	**0.80**	**0.98**	0.74	0.71	**0.93**	**0.94**	1.00						
Fe	0.33	0.28	0.54	**0.78**	0.40	-0.10	-0.09	0.30	-0.14	0.01	0.23	0.13	-0.04	0.20	0.59	0.19	-0.03	0.01	1.00					
Cd	0.29	-0.07	0.15	0.12	0.23	0.18	0.22	-0.04	0.01	0.19	-0.01	0.32	0.45	0.10	0.35	0.25	0.25	0.37	0.16	1.00				
Cu	0.41	**0.91**	0.35	0.41	0.71	**0.95**	**0.94**	0.75	0.42	**0.94**	**0.83**	**0.93**	**0.84**	**0.94**	0.64	**0.94**	**0.94**	**0.81**	0.05	0.06	1.00			
Ni	0.51	**0.76**	0.50	**0.80**	0.68	0.50	0.50	0.72	0.20	0.52	0.58	0.65	0.42	0.66	0.63	0.63	0.52	0.39	0.65	0.08	0.66	1.00		
Mn	0.55	**0.80**	0.51	0.33	0.69	**0.94**	**0.94**	0.66	0.50	**0.96**	**0.81**	**0.91**	**0.97**	**0.87**	**0.78**	**0.98**	**0.97**	**0.96**	0.08	0.39	**0.89**	0.48	1.00	
Cr	0.00	0.17	0.34	0.17	0.10	0.18	0.09	0.10	-0.32	0.22	0.60	0.14	0.33	0.25	0.46	0.38	0.24	0.36	0.59	0.29	0.20	0.28	0.38	1.00

Table 4.34 Pearson Correlation Matrix of Water Quality Parameters of groundwater sources at site GW2 during 2019-2020

Values in bold letters show significant correlations

Variables	pH	EC	Temp	TDS	TU	TH	Alky	Acidity	PO_4^{3-}	HCO_3^-	NO_3^-	SO_4^{2-}	Cl^-	K^+	Na^+	Ca^{2+}	Mg^{2+}	Zn	Fe	Cd	Cu	Ni	Mn	Cr
pH	1.00																							
EC	0.09	1.00																						
Temp	0.55	0.17	1.00																					
TDS	0.18	0.34	0.10	1.00																				
TU	0.19	0.66	0.60	0.63	1.00																			
TH	0.19	**0.85**	0.52	-0.01	0.65	1.00																		
Alk	0.41	**0.89**	0.46	0.12	0.61	**0.93**	1.00																	
Acidity	0.35	0.50	0.56	0.63	0.67	0.49	0.47	1.00																
PO_4^{3-}	0.02	0.61	0.66	0.08	**0.76**	**0.86**	0.65	0.60	1.00															
HCO_3^-	0.26	**0.82**	0.55	-0.06	0.61	**0.99**	**0.94**	0.42	**0.82**	1.00														
NO_3^-	0.44	**0.84**	0.32	0.02	0.37	**0.84**	**0.95**	0.42	0.48	**0.86**	1.00													
SO_4^{2-}	0.54	0.68	0.62	-0.12	0.51	**0.91**	**0.92**	0.39	0.70	**0.94**	**0.86**	1.00												
Cl^-	0.29	**0.81**	0.48	-0.02	0.58	**0.95**	**0.92**	0.51	**0.77**	**0.93**	**0.85**	**0.91**	1.00											
K^+	0.09	0.56	0.24	-0.54	0.09	**0.79**	0.73	-0.01	0.53	**0.83**	**0.77**	**0.79**	0.73	1.00										
Na^+	0.34	**0.77**	0.62	0.03	0.70	**0.97**	**0.91**	0.52	**0.87**	**0.96**	**0.78**	**0.94**	**0.95**	0.71	1.00									
Ca^{2+}	0.34	**0.80**	0.67	0.11	**0.76**	**0.96**	**0.93**	0.51	**0.85**	**0.97**	**0.80**	**0.92**	**0.90**	0.69	**0.97**	1.00								
Mg^{2+}	0.07	**0.78**	0.31	-0.27	0.39	**0.95**	**0.87**	0.25	0.72	**0.95**	**0.84**	**0.87**	**0.92**	**0.92**	**0.89**	**0.85**	1.00							
Zn	0.62	**0.76**	0.36	0.52	0.61	0.65	**0.79**	**0.75**	0.45	0.63	**0.77**	0.68	0.72	0.27	0.69	0.67	0.50	1.00						
Fe	-0.03	**0.94**	0.18	0.02	0.54	**0.92**	**0.90**	0.29	0.66	**0.90**	**0.83**	0.75	**0.86**	**0.77**	**0.82**	**0.84**	**0.92**	0.56	1.00					
Cd	-0.27	0.31	-0.51	0.09	-0.09	-0.03	0.15	-0.47	-0.37	0.00	0.17	-0.13	-0.11	0.11	-0.17	-0.03	0.06	-0.10	0.33	1.00				
Cu	0.38	**0.87**	0.52	0.42	**0.86**	**0.86**	**0.89**	0.63	0.74	**0.84**	0.74	**0.80**	**0.86**	0.43	**0.90**	**0.90**	0.70	**0.84**	**0.78**	0.00	1.00			
Ni	0.39	**0.89**	0.39	0.45	**0.79**	**0.78**	**0.90**	0.46	0.56	**0.78**	**0.76**	0.74	0.73	0.43	**0.79**	**0.85**	0.64	**0.79**	**0.80**	0.30	**0.94**	1.00		
Mn	0.31	**0.86**	0.21	-0.06	0.37	**0.87**	**0.94**	0.30	0.51	0.88	**0.96**	**0.86**	**0.90**	**0.80**	**0.82**	**0.80**	**0.91**	0.73	**0.90**	0.22	**0.77**	**0.77**	1.00	
Cr	-0.01	0.16	-0.18	0.48	0.17	-0.15	0.01	-0.11	-0.25	-0.11	0.00	-0.22	-0.36	-0.19	-0.23	-0.04	-0.26	0.01	0.03	0.65	0.00	0.31	-0.11	1.00

Table 4.35 Pearson Correlation Matrix of Water Quality Parameters of groundwater sources at site GW3 during 2019-2020

Values in bold letters show significant correlations

Variables	pH	EC	Temp	TDS	TU	TH	Alky	Acidity	PO_4^{3-}	HCO_3^-	NO_3^-	SO_4^{2-}	Cl^-	K^+	Na^+	Ca^{2+}	Mg^{2+}	Zn	Fe	Cd	Cu	Ni	Mn	Cr
pH	1.00																							
EC	0.32	1.00																						
Temp	0.18	0.21	1.00																					
TDS	**0.75**	0.49	0.60	1.00																				
TU	**0.87**	0.51	0.47	**0.96**	1.00																			
TH	0.50	**0.93**	0.45	0.67	0.72	1.00																		
Alk	0.43	**0.91**	0.27	0.44	0.54	**0.93**	1.00																	
Acidity	**0.78**	0.67	0.59	**0.81**	**0.88**	**0.87**	**0.79**	1.00																
PO_4^{3-}	0.18	-0.03	**0.83**	0.36	0.25	0.20	0.13	0.44	1.00															
HCO_3^-	0.39	**0.89**	0.54	0.62	0.65	**0.98**	**0.93**	**0.85**	0.31	1.00														
NO_3^-	0.22	**0.87**	0.12	0.17	0.26	**0.80**	**0.93**	0.59	0.04	**0.79**	1.00													
SO_4^{2-}	0.12	**0.87**	0.61	0.53	0.47	**0.89**	**0.80**	0.67	0.33	**0.93**	0.73	1.00												
Cl^-	0.47	0.68	0.38	0.37	0.47	**0.78**	**0.89**	**0.78**	0.46	**0.82**	**0.83**	0.66	1.00											
K^+	0.20	**0.89**	0.27	0.33	0.42	**0.90**	**0.95**	0.67	0.07	**0.93**	**0.89**	**0.86**	**0.80**	1.00										
Na^+	0.20	**0.94**	0.32	0.44	0.48	**0.93**	**0.93**	0.67	0.05	**0.95**	**0.85**	**0.92**	0.73	**0.98**	1.00									
Ca^{2+}	0.24	**0.76**	0.71	0.70	0.61	**0.84**	0.71	0.71	0.47	**0.89**	0.55	**0.94**	0.63	0.74	**0.83**	1.00								
Mg^{2+}	0.43	**0.91**	0.51	0.69	0.66	**0.94**	**0.88**	**0.80**	0.34	**0.94**	**0.76**	**0.92**	**0.78**	**0.83**	**0.90**	**0.93**	1.00							
Zn	0.23	**0.87**	0.48	0.67	0.61	**0.88**	0.74	0.64	0.09	**0.88**	0.60	**0.92**	0.49	**0.79**	**0.90**	**0.92**	**0.89**	1.00						
Fe	0.28	**0.77**	0.48	0.69	0.55	0.71	0.55	0.54	0.24	0.67	0.47	**0.79**	0.38	0.49	0.65	**0.83**	**0.84**	**0.84**	1.00					
Cd	-0.12	0.34	-0.56	-0.10	-0.07	0.10	0.09	-0.21	-0.89	-0.04	0.19	0.03	-0.33	0.13	0.21	-0.13	-0.01	0.26	0.18	1.00				
Cu	0.61	0.58	0.43	0.61	0.69	0.73	0.62	**0.81**	0.11	0.64	0.54	0.52	0.48	0.51	0.53	0.43	0.54	0.52	0.40	0.17	1.00			
Ni	0.41	**0.81**	0.58	0.74	0.70	**0.88**	0.75	**0.77**	0.17	**0.84**	0.63	**0.85**	0.53	0.71	**0.81**	**0.84**	**0.85**	**0.91**	**0.80**	0.22	**0.76**	1.00		
Mn	0.14	**0.87**	0.37	0.36	0.31	**0.79**	**0.83**	0.55	0.31	**0.81**	**0.84**	**0.86**	**0.77**	**0.80**	**0.84**	**0.81**	**0.91**	0.74	**0.78**	0.00	0.31	0.68	1.00	
Cr	-0.32	-0.61	-0.16	-0.16	-0.19	-0.56	-0.70	-0.50	-0.38	-0.56	**-0.78**	-0.49	**-0.83**	-0.55	-0.52	-0.43	-0.66	-0.29	-0.46	0.22	-0.25	-0.33	**-0.82**	1.00

Table 4.36 Pearson Correlation Matrix of Water Quality Parameters of groundwater sources at site GW4 during 2019-2020

Values in bold letters show significant correlations

Variables	pH	EC	Temp	TDS	TU	TH	Alky	Acidity	PO_4^{3-}	HCO_3^-	NO_3^-	SO_4^{2-}	Cl^-	K^+	Na^+	Ca^{2+}	Mg^{2+}	Zn	Fe	Cd	Cu	Ni	Mn	Cr
pH	1.00																							
EC	-0.04	1.00																						
Temp	-0.10	0.16	1.00																					
TDS	-0.37	0.61	0.23	1.00																				
TU	-0.36	0.68	0.55	**0.90**	1.00																			
TH	-0.05	**0.95**	0.32	0.48	0.64	1.00																		
Alk	-0.37	0.76	0.58	0.51	0.71	**0.79**	1.00																	
Acidity	-0.02	-0.12	0.49	0.38	0.49	-0.17	0.03	1.00																
PO_4^{3-}	-0.15	0.48	0.72	0.59	0.71	0.60	0.57	0.15	1.00															
HCO_3^-	0.08	**0.94**	0.17	0.37	0.51	**0.98**	0.68	-0.25	0.48	1.00														
NO_3^-	0.14	**0.78**	0.21	0.14	0.34	0.75	0.73	-0.06	0.10	**0.78**	1.00													
SO_4^{2-}	0.20	**0.87**	0.35	0.30	0.53	**0.92**	0.67	0.01	0.44	**0.94**	**0.85**	1.00												
Cl^-	0.50	0.22	-0.18	0.29	0.26	0.08	-0.22	0.54	-0.15	0.16	0.18	0.31	1.00											
K^+	0.03	-0.04	0.06	0.41	0.34	-0.24	-0.04	**0.83**	-0.20	-0.26	0.07	-0.07	0.66	1.00										
Na^+	0.50	0.66	-0.23	0.26	0.24	0.53	0.16	0.07	-0.06	0.65	0.66	0.70	**0.75**	0.32	1.00									
Ca^{2+}	-0.11	**0.89**	0.32	0.36	0.58	**0.90**	**0.88**	-0.09	0.35	**0.87**	**0.91**	**0.87**	0.05	-0.09	0.50	1.00								
Mg^{2+}	-0.12	0.31	0.12	0.34	0.39	0.12	0.47	0.47	-0.23	0.09	0.57	0.23	0.35	0.73	0.37	0.45	1.00							
Zn	-0.18	0.78	0.13	0.40	0.56	0.72	0.70	0.16	0.07	0.72	**0.87**	**0.78**	0.34	0.29	0.65	**0.87**	0.66	1.00						
Fe	-0.22	**0.83**	0.44	0.40	0.62	**0.84**	**0.96**	-0.04	0.40	**0.78**	**0.87**	**0.78**	-0.07	-0.05	0.36	**0.97**	0.52	**0.82**	1.00					
Cd	0.35	0.43	-0.03	-0.15	-0.14	0.44	0.21	-0.33	0.09	0.56	0.57	0.56	0.01	-0.19	0.54	0.34	-0.06	0.33	0.27	1.00				
Cu	-0.31	**0.89**	0.40	0.59	0.75	**0.86**	**0.96**	0.02	0.48	**0.78**	**0.80**	**0.76**	-0.01	0.04	0.38	**0.94**	0.53	**0.83**	**0.97**	0.23	1.00			
Ni	-0.06	**0.95**	0.20	0.39	0.56	**0.95**	**0.82**	-0.20	0.36	**0.94**	**0.88**	**0.90**	0.10	-0.16	0.59	**0.98**	0.35	**0.84**	**0.92**	0.42	**0.91**	1.00		
Mn	-0.22	0.74	0.40	0.20	0.49	**0.80**	**0.88**	-0.07	0.28	**0.75**	**0.88**	**0.79**	-0.10	-0.13	0.32	**0.96**	0.46	**0.84**	**0.96**	0.30	**0.89**	**0.90**	1.00	
Cr	0.10	0.13	-0.22	0.37	0.13	-0.14	0.03	0.35	-0.26	-0.10	0.21	-0.06	0.42	**0.77**	0.44	-0.03	0.68	0.26	0.02	0.22	0.12	-0.02	-0.13	1.00

Table 4.37 Pearson Correlation Matrix of Water Quality Parameters of groundwater sources at site GW5 during 2019-2020

Values in bold letters show significant correlations

Vari-ables	pH	EC	Temp	TDS	TU	TH	Alky	Acidity	PO_4^{3-}	HCO_3^-	NO_3^-	SO_4^{2-}	Cl^-	K^+	Na^+	Ca^{2+}	Mg^{2+}	Zn	Fe	Cd	Cu	Ni	Mn	Cr
pH	1.00																							
EC	0.49	1.00																						
Temp	-0.27	0.10	1.00																					
TDS	-0.39	0.60	0.43	1.00																				
TU	-0.36	0.61	0.51	**0.99**	1.00																			
TH	0.58	**0.95**	0.14	0.50	0.53	1.00																		
Alk	0.12	0.59	0.75	0.53	0.59	0.58	1.00																	
Acidity	-0.15	0.43	**0.91**	0.63	0.70	0.41	**0.91**	1.00																
PO_4^{3-}	0.18	0.19	0.45	0.18	0.22	0.43	0.41	0.35	1.00															
HCO_3^-	0.61	**0.95**	0.15	0.47	0.50	**0.99**	0.61	0.43	0.41	1.00														
NO_3^-	0.45	**0.81**	0.53	0.45	0.52	**0.82**	**0.91**	**0.77**	0.41	**0.86**	1.00													
SO_4^{2-}	0.59	**0.90**	0.21	0.42	0.47	**0.92**	0.52	0.46	0.25	**0.93**	**0.80**	1.00												
Cl^-	0.14	0.53	0.17	0.48	0.51	0.52	0.08	0.24	-0.02	0.45	0.27	0.67	1.00											
K^+	-0.26	0.24	-0.04	0.41	0.41	0.06	-0.11	0.13	-0.63	0.06	-0.03	0.31	0.59	1.00										
Na^+	0.47	**0.89**	0.34	0.49	0.53	**0.82**	**0.79**	0.64	0.07	**0.84**	**0.92**	**0.83**	0.44	0.22	1.00									
Ca^{2+}	0.51	**0.88**	0.36	0.45	0.50	**0.86**	**0.85**	0.65	0.24	**0.89**	**0.97**	**0.81**	0.30	0.04	**0.97**	1.00								
Mg^{2+}	0.08	0.56	0.18	0.42	0.41	0.32	0.55	0.48	-0.48	0.35	0.53	0.37	0.22	0.44	**0.76**	0.65	1.00							
Zn	0.41	**0.97**	0.20	0.62	0.65	**0.91**	0.67	0.54	0.11	**0.92**	**0.85**	**0.90**	0.55	0.33	**0.94**	**0.92**	0.65	1.00						
Fe	0.23	0.43	0.61	0.24	0.28	0.38	**0.91**	0.75	0.25	0.43	**0.80**	0.32	-0.15	-0.28	0.71	**0.77**	0.62	0.49	1.00					
Cd	0.21	-0.16	0.21	-0.27	-0.22	0.10	0.19	0.02	0.63	0.06	0.15	-0.08	-0.23	-0.74	-0.10	0.07	-0.49	-0.14	0.17	1.00				
Cu	0.34	0.32	0.47	0.14	0.16	0.37	0.46	0.41	0.48	0.34	0.46	0.29	0.28	-0.42	0.41	0.41	0.19	0.25	0.56	0.22	1.00			
Ni	-0.13	-0.04	**0.92**	0.15	0.24	0.00	0.64	**0.78**	0.27	0.01	0.44	0.12	0.16	-0.10	0.31	0.30	0.23	0.09	0.63	0.28	0.54	1.00		
Mn	0.26	0.56	0.64	0.35	0.42	0.58	**0.96**	**0.80**	0.38	0.63	**0.92**	0.53	0.00	-0.18	**0.78**	**0.86**	0.50	0.66	**0.90**	0.33	0.36	0.59	1.00	
Cr	-0.53	0.02	-0.02	0.48	0.48	-0.07	-0.27	0.02	-0.41	-0.12	-0.27	0.09	0.59	**0.85**	-0.09	-0.23	0.06	0.11	-0.54	-0.45	-0.49	-0.12	-0.35	1.00

Table 4.38 Pearson Correlation Matrix of Water Quality Parameters of groundwater sources at site GWCS during 2020-2021

Values in bold letters show significant correlations

Vari-ables	pH	EC	Temp	TDS	TU	TH	Alky	Acidity	PO_4^{3-}	HCO_3^-	NO_3^-	SO_4^{2-}	Cl^-	K^+	Na^+	Ca^{2+}	Mg^{2+}	Zn	Fe	Cd	Cu	Ni	Mn	Cr
pH	1.00																							
EC	0.41	1.00																						
Temp	0.32	0.08	1.00																					
TDS	0.69	**0.81**	0.46	1.00																				
TU	0.49	**0.72**	0.06	0.58	1.00																			
TH	0.69	**0.81**	0.37	**0.89**	**0.80**	1.00																		
Alk	0.51	**0.75**	0.42	**0.84**	**0.84**	**0.93**	1.00																	
Acidity	0.36	0.28	**0.92**	0.62	0.34	0.62	0.71	1.00																
PO_4^{3-}	0.46	0.71	-0.21	0.62	**0.82**	**0.78**	**0.76**	0.15	1.00															
HCO_3^-	0.66	0.67	0.65	**0.91**	0.69	**0.88**	**0.92**	**0.82**	0.56	1.00														
NO_3^-	0.50	0.47	**0.86**	**0.75**	0.45	0.63	0.71	**0.87**	0.15	**0.89**	1.00													
SO_4^{2-}	0.74	**0.77**	0.51	**0.97**	0.69	**0.89**	**0.87**	0.66	0.64	**0.96**	**0.81**	1.00												
Cl^-	0.59	**0.78**	0.58	**0.87**	**0.79**	**0.94**	**0.95**	**0.78**	0.61	**0.95**	**0.83**	**0.91**	1.00											
K^+	0.59	0.56	0.68	0.70	0.71	0.72	**0.76**	**0.75**	0.38	**0.86**	**0.88**	**0.83**	**0.88**	1.00										
Na^+	0.61	**0.72**	0.66	**0.86**	0.71	**0.93**	**0.92**	**0.83**	0.53	**0.95**	**0.85**	**0.89**	**0.99**	**0.88**	1.00									
Ca^{2+}	0.66	**0.85**	0.27	**0.91**	**0.85**	**0.92**	**0.93**	0.51	**0.81**	**0.90**	0.67	**0.94**	**0.90**	0.73	**0.84**	1.00								
Mg^{2+}	0.64	**0.85**	0.45	**0.95**	**0.78**	**0.95**	**0.95**	0.67	0.72	**0.95**	**0.77**	**0.97**	**0.96**	**0.82**	**0.93**	**0.97**	1.00							
Zn	0.64	**0.80**	0.19	**0.85**	**0.87**	**0.90**	**0.90**	0.46	**0.89**	**0.84**	0.56	**0.90**	**0.85**	0.74	**0.79**	**0.95**	**0.93**	1.00						
Fe	0.16	0.50	0.28	0.30	0.61	0.64	0.61	0.48	0.39	0.43	0.32	0.31	0.66	0.47	0.67	0.41	0.49	0.39	1.00					
Cd	0.40	0.64	0.37	0.53	**0.75**	**0.77**	**0.77**	0.55	0.45	0.67	0.57	0.56	**0.82**	0.64	**0.81**	0.66	0.69	0.56	**0.89**	1.00				
Cu	0.49	**0.74**	0.58	0.70	0.71	**0.76**	**0.75**	0.65	0.35	**0.79**	**0.81**	**0.76**	**0.90**	**0.89**	**0.90**	0.73	**0.81**	0.66	0.68	**0.84**	1.00			
Ni	0.44	0.46	0.08	0.66	0.26	0.65	0.59	0.33	0.58	0.53	0.22	0.53	0.47	0.05	0.46	0.60	0.56	0.49	0.20	0.33	0.12	1.00		
Mn	**0.78**	0.69	0.13	0.65	0.65	**0.75**	0.58	0.22	0.49	0.60	0.44	0.65	0.67	0.52	0.65	0.72	0.67	0.60	0.49	0.75	0.70	0.45	1.00	
Cr	0.40	0.64	0.37	0.53	**0.75**	**0.77**	**0.77**	0.55	0.45	0.67	0.57	0.56	**0.82**	0.64	**0.81**	0.66	0.69	0.56	**0.89**	0.42	**0.84**	0.33	**0.75**	1.00

Table 4.39 Pearson Correlation Matrix of Water Quality Parameters of groundwater sources at site GW1 during 2020-2021

Values in bold letters show significant correlations

Vari-ables	pH	EC	Temp	TDS	TU	TH	Alky	Acidity	PO_4^{3-}	HCO_3^-	NO_3^-	SO_4^{2-}	Cl^-	K^+	Na^+	Ca^{2+}	Mg^{2+}	Zn	Fe	Cd	Cu	Ni	Mn	Cr
pH	1.00																							
EC	0.15	1.00																						
Temp	**0.89**	0.50	1.00																					
TDS	0.62	**0.76**	**0.85**	1.00																				
TU	0.56	0.70	0.74	**0.87**	1.00																			
TH	0.02	0.62	0.20	0.60	0.64	1.00																		
Alk	0.14	0.65	0.26	0.58	0.69	**0.95**	1.00																	
Acid-ity	0.24	**0.79**	0.42	**0.79**	**0.79**	**0.91**	**0.91**	1.00																
PO_4^{3-}	0.30	0.55	0.39	0.71	0.72	**0.90**	**0.90**	**0.90**	1.00															
HCO_3^-	0.05	0.66	0.27	0.68	0.59	**0.96**	**0.86**	**0.89**	**0.84**	1.00														
NO_3^-	0.25	0.57	0.54	0.69	0.68	0.39	0.37	0.50	0.58	0.36	1.00													
SO_4^{2-}	0.18	0.68	0.31	0.71	0.71	**0.93**	**0.92**	**0.98**	**0.96**	**0.89**	0.50	1.00												
Cl^-	-0.01	0.61	0.15	0.52	0.59	**0.95**	**0.96**	**0.85**	**0.91**	**0.86**	0.47	**0.91**	1.00											
K^+	0.10	0.29	0.33	0.59	0.32	0.44	0.18	0.41	0.39	0.63	0.36	0.39	0.22	1.00										
Na^+	0.52	0.63	0.72	**0.91**	**0.91**	0.72	0.70	**0.80**	**0.86**	0.70	**0.81**	**0.79**	0.69	0.52	1.00									
Ca^{2+}	0.16	**0.79**	0.40	**0.75**	**0.69**	**0.89**	**0.89**	**0.90**	**0.91**	**0.87**	0.66	**0.92**	**0.93**	0.38	**0.82**	1.00								
Mg^{2+}	0.10	0.66	0.27	0.61	0.65	**0.95**	**0.97**	**0.88**	**0.93**	**0.88**	0.50	**0.92**	**0.99**	0.28	**0.75**	**0.95**	1.00							
Zn	0.43	0.52	0.49	0.68	**0.74**	**0.86**	**0.92**	**0.82**	**0.93**	**0.79**	0.46	**0.85**	**0.89**	0.25	**0.82**	**0.86**	**0.92**	1.00						
Fe	0.49	**0.89**	**0.77**	**0.87**	**0.90**	0.54	0.62	0.74	0.59	0.53	0.70	0.62	0.54	0.21	**0.80**	0.73	0.61	0.63	1.00					
Cd	-0.12	0.34	-0.56	-0.10	-0.07	0.10	0.09	-0.21	-0.89	-0.04	0.19	0.03	-0.33	0.13	0.21	-0.13	-0.01	0.26	0.18	1.00				
Cu	0.21	**0.84**	0.45	**0.80**	**0.83**	**0.85**	**0.87**	**0.95**	**0.89**	**0.80**	0.71	**0.93**	**0.86**	0.35	**0.86**	**0.94**	**0.88**	**0.79**	**0.82**	0.47	1.00			
Ni	0.46	0.65	0.59	0.72	**0.87**	**0.80**	**0.89**	**0.81**	**0.83**	0.70	0.52	**0.78**	**0.81**	0.14	**0.83**	**0.81**	**0.86**	**0.94**	**0.80**	0.36	**0.82**	1.00		
Mn	0.31	0.46	0.42	0.63	0.63	**0.80**	**0.81**	0.72	**0.92**	0.72	0.64	**0.80**	**0.88**	0.29	**0.83**	**0.88**	**0.91**	**0.93**	0.55	0.49	**0.77**	**0.83**	1.00	
Cr	-0.01	0.16	-0.18	0.48	0.17	-0.15	0.01	-0.11	-0.25	-0.11	0.00	-0.22	-0.36	-0.19	-0.23	-0.04	-0.26	0.01	0.03	0.65	0.00	0.31	-0.11	1.00

Table 4.40 Pearson Correlation Matrix of Water Quality Parameters of groundwater sources at site GW2 during 2020-2021
Values in bold letters show significant correlations

Variables	pH	EC	Temp	TDS	TU	TH	Alky	Acidity	PO_4^{3-}	HCO_3^-	NO_3^-	SO_4^{2-}	Cl^-	K^+	Na^+	Ca^{2+}	Mg^{2+}	Zn	Fe	Cd	Cu	Ni	Mn	Cr
pH	1.00																							
EC	0.35	1.00																						
Temp	0.23	0.10	1.00																					
TDS	0.21	0.57	**0.79**	1.00																				
TU	0.36	0.59	**0.84**	**0.95**	1.00																			
TH	0.52	**0.86**	0.44	0.72	**0.75**	1.00																		
Alk	0.54	**0.93**	0.15	0.55	0.57	**0.90**	1.00																	
Acidity	0.45	**0.95**	0.39	**0.75**	**0.78**	**0.96**	**0.93**	1.00																
PO_4^{3-}	0.41	**0.92**	0.37	0.65	0.71	**0.96**	**0.90**	**0.97**	1.00															
HCO_3^-	0.34	0.74	0.39	0.69	0.67	**0.90**	0.69	**0.84**	**0.82**	1.00														
NO_3^-	0.02	0.69	-0.39	0.01	-0.04	0.56	0.69	0.56	0.63	0.48	1.00													
SO_4^{2-}	0.43	**0.94**	0.37	0.70	0.74	**0.95**	**0.92**	**0.99**	**0.98**	**0.79**	0.58	1.00												
Cl^-	0.62	**0.83**	0.29	0.56	0.60	**0.97**	**0.91**	**0.91**	**0.92**	**0.86**	0.63	**0.90**	1.00											
K^+	0.55	0.55	**0.79**	**0.83**	**0.87**	**0.85**	0.68	**0.78**	**0.77**	0.69	0.15	**0.77**	**0.79**	1.00										
Na^+	0.52	**0.94**	0.30	0.65	0.69	**0.93**	**0.97**	**0.97**	**0.96**	0.71	0.60	**0.98**	**0.90**	**0.76**	1.00									
Ca^{2+}	0.43	**0.86**	0.45	0.71	**0.75**	**0.96**	**0.92**	**0.95**	**0.97**	**0.78**	0.59	**0.95**	**0.91**	**0.85**	**0.95**	1.00								
Mg^{2+}	0.57	**0.88**	0.26	0.59	0.60	**0.96**	**0.96**	**0.94**	**0.94**	**0.78**	0.67	**0.94**	**0.98**	**0.78**	**0.96**	**0.94**	1.00							
Zn	0.60	**0.90**	0.39	0.72	**0.76**	**0.95**	**0.96**	**0.97**	**0.93**	**0.77**	0.51	**0.95**	**0.92**	**0.83**	**0.98**	**0.95**	**0.95**	1.00						
Fe	0.50	0.74	0.23	0.57	0.51	**0.86**	**0.90**	**0.81**	**0.79**	0.70	0.65	**0.79**	**0.89**	**0.76**	**0.84**	**0.86**	**0.93**	**0.87**	1.00					
Cd	0.08	-0.04	-0.36	-0.03	-0.15	-0.28	-0.04	-0.16	-0.37	-0.19	-0.21	-0.25	-0.26	-0.32	-0.16	-0.33	-0.20	-0.08	-0.06	1.00				
Cu	0.66	0.72	0.45	0.55	0.65	**0.94**	**0.80**	**0.85**	**0.90**	**0.80**	0.47	**0.87**	**0.96**	**0.84**	**0.85**	**0.87**	**0.91**	**0.86**	**0.77**	-0.45	1.00			
Ni	0.38	0.68	0.65	**0.85**	**0.83**	**0.90**	**0.78**	**0.86**	**0.83**	**0.78**	0.38	**0.82**	**0.82**	**0.94**	**0.81**	**0.92**	**0.84**	**0.87**	**0.86**	-0.24	**0.79**	1.00		
Mn	0.52	**0.83**	0.18	0.49	0.53	**0.90**	**0.92**	**0.87**	**0.88**	**0.78**	0.72	**0.83**	**0.92**	0.69	**0.87**	**0.92**	**0.91**	**0.89**	**0.87**	-0.18	**0.82**	**0.80**	1.00	
Cr	0.00	0.17	0.34	0.17	0.10	0.18	0.09	0.10	-0.32	0.22	0.60	0.14	0.33	0.25	0.46	0.38	0.24	0.36	0.59	0.29	0.20	0.28	0.38	1.00

Table 4.41 Pearson Correlation Matrix of Water Quality Parameters of groundwater sources at site GW3 during 2020-2021

Values in bold letters show significant correlations

Vari-ables	pH	EC	Temp	TDS	TU	TH	Alky	Acidity	PO_4^{3-}	HCO_3^-	NO_3^-	SO_4^{2-}	Cl^-	K^+	Na^+	Ca^{2+}	Mg^{2+}	Zn	Fe	Cd	Cu	Ni	Mn	Cr
pH	1.00																							
EC	0.61	1.00																						
Temp	0.18	0.17	1.00																					
TDS	0.68	0.70	**0.76**	1.00																				
TU	0.55	0.48	**0.86**	**0.93**	1.00																			
TH	0.68	**0.91**	0.45	**0.87**	**0.78**	1.00																		
Alk	0.38	**0.82**	-0.01	0.49	0.33	**0.76**	1.00																	
Acidity	0.62	0.74	0.58	**0.91**	**0.82**	**0.89**	**0.75**	1.00																
PO_4^{3-}	0.29	0.64	0.67	0.73	**0.78**	**0.80**	0.40	0.60	1.00															
HCO_3^-	0.55	**0.80**	0.63	**0.91**	**0.84**	**0.91**	0.59	**0.87**	**0.77**	1.00														
NO_3^-	0.29	0.73	-0.23	0.27	0.10	0.63	**0.95**	0.58	0.26	0.41	1.00													
SO_4^{2-}	0.42	**0.86**	0.26	0.69	0.44	**0.76**	**0.86**	**0.83**	0.43	0.73	0.74	1.00												
Cl^-	0.41	**0.87**	0.14	0.61	0.48	**0.86**	**0.98**	**0.81**	0.56	0.71	**0.89**	**0.87**	1.00											
K^+	0.40	0.37	**0.78**	0.74	**0.75**	0.59	0.17	0.60	0.72	0.56	0.09	0.30	0.28	1.00										
Na^+	0.34	**0.86**	0.15	0.58	0.44	**0.81**	**0.89**	**0.76**	0.52	**0.80**	**0.81**	**0.87**	**0.92**	0.17	1.00									
Ca^{2+}	0.41	**0.85**	0.59	**0.86**	**0.76**	**0.90**	0.68	**0.84**	**0.81**	**0.94**	0.49	**0.83**	**0.79**	0.51	**0.84**	1.00								
Mg^{2+}	0.58	**0.92**	0.31	**0.76**	0.67	**0.97**	**0.81**	**0.82**	**0.76**	**0.90**	0.70	**0.76**	**0.89**	0.42	**0.90**	**0.89**	1.00							
Zn	0.69	**0.93**	0.30	**0.80**	0.64	**0.93**	**0.77**	**0.86**	0.59	**0.91**	0.65	**0.84**	**0.83**	0.37	**0.89**	**0.87**	**0.94**	1.00						
Fe	0.18	0.72	-0.09	0.34	0.16	0.59	**0.89**	0.63	0.22	0.56	**0.89**	**0.83**	**0.86**	-0.03	**0.93**	0.62	0.71	0.74	1.00					
Cd	0.50	-0.09	-0.38	-0.14	-0.16	-0.11	-0.30	-0.21	-0.38	-0.10	-0.20	-0.34	-0.34	-0.26	-0.21	-0.36	-0.09	0.06	-0.22	1.00				
Cu	0.11	-0.09	-0.38	-0.24	-0.51	-0.34	-0.12	-0.23	-0.58	-0.46	-0.02	0.06	-0.25	-0.12	-0.33	-0.38	-0.45	-0.23	-0.13	0.15	1.00			
Ni	0.31	**0.83**	0.43	0.72	0.63	**0.87**	**0.79**	**0.83**	0.72	**0.90**	0.69	**0.83**	**0.87**	0.44	**0.95**	**0.92**	**0.91**	**0.88**	**0.81**	-0.33	-0.44	1.00		
Mn	0.39	**0.88**	0.35	0.67	0.61	**0.91**	0.73	0.70	**0.87**	**0.83**	0.66	0.68	**0.83**	0.54	**0.82**	**0.86**	**0.94**	**0.82**	0.62	-0.26	-0.45	**0.90**	1.00	
Cr	-0.14	-0.55	-0.08	-0.29	-0.34	-0.64	-0.66	-0.44	-0.66	-0.40	-0.71	-0.34	-0.70	-0.44	-0.55	-0.45	-0.67	-0.43	-0.47	0.33	0.45	-0.61	-0.79	1.00

Table 4.42 Pearson Correlation Matrix of Water Quality Parameters of groundwater sources at site GW4 during 2020-2021

Values in bold letters show significant correlations

Variables	pH	EC	Temp	TDS	TU	TH	Alky	Acidity	PO_4^{3-}	HCO_3^-	NO_3^-	SO_4^{2-}	Cl^-	K^+	Na^+	Ca^{2+}	Mg^{2+}	Zn	Fe	Cd	Cu	Ni	Mn	Cr
pH	1.00																							
EC	0.46	1.00																						
Temp	-0.21	0.24	1.00																					
TDS	0.23	**0.79**	0.67	1.00																				
TU	-0.13	0.63	**0.84**	**0.89**	1.00																			
TH	0.37	**0.94**	0.44	**0.82**	0.74	1.00																		
Alk	0.19	**0.94**	0.37	**0.80**	0.72	**0.93**	1.00																	
Acidity	0.01	0.35	0.70	**0.76**	0.70	0.33	0.36	1.00																
PO_4^{3-}	-0.04	0.71	**0.76**	**0.82**	**0.88**	**0.82**	**0.86**	0.58	1.00															
HCO_3^-	0.51	**0.96**	0.33	**0.83**	0.65	**0.97**	**0.91**	0.33	0.74	1.00														
NO_3^-	**0.87**	0.74	-0.18	0.51	0.14	0.59	0.50	0.19	0.16	0.73	1.00													
SO_4^{2-}	0.72	0.91	0.15	0.73	0.48	**0.84**	**0.75**	0.29	0.50	**0.92**	**0.91**	1.00												
Cl^-	0.65	0.09	-0.40	0.11	-0.28	-0.12	-0.15	0.16	-0.40	0.09	0.69	0.40	1.00											
K^+	0.33	**0.96**	0.48	**0.90**	**0.79**	**0.93**	**0.95**	0.58	**0.86**	**0.93**	0.61	**0.83**	0.01	1.00										
Na^+	**0.78**	0.60	0.09	0.63	0.30	0.51	0.37	0.38	0.17	0.66	**0.90**	**0.86**	**0.75**	0.55	1.00									
Ca^{2+}	0.19	**0.83**	0.63	**0.95**	**0.86**	**0.84**	**0.89**	0.67	**0.90**	**0.85**	0.47	0.73	0.01	**0.93**	0.53	1.00								
Mg^{2+}	0.36	**0.81**	0.09	0.67	0.45	0.59	0.74	0.56	0.52	0.66	0.69	0.71	0.35	**0.81**	0.55	0.73	1.00							
Zn	0.46	**0.91**	0.39	**0.77**	0.70	**0.91**	**0.83**	0.32	0.69	**0.90**	0.68	**0.92**	0.07	**0.88**	0.66	**0.80**	0.64	1.00						
Fe	0.49	**0.91**	-0.08	0.56	0.38	0.74	**0.79**	0.18	0.43	**0.78**	**0.80**	**0.86**	0.28	**0.79**	0.60	0.63	**0.86**	**0.82**	1.00					
Cd	0.56	0.26	0.30	0.59	0.26	0.22	0.14	0.65	0.19	0.38	0.57	0.49	0.71	0.36	**0.79**	0.48	0.39	0.27	0.15	1.00				
Cu	-0.25	0.54	0.62	**0.81**	**0.81**	0.52	0.69	**0.81**	**0.79**	0.49	0.11	0.32	-0.11	0.72	0.18	**0.84**	0.68	0.43	0.38	0.34	1.00			
Ni	0.18	0.59	0.59	**0.80**	0.72	0.50	0.54	**0.93**	0.63	0.50	0.41	0.52	0.21	0.74	0.50	**0.75**	**0.77**	0.56	0.49	0.56	**0.80**	1.00		
Mn	0.42	**0.98**	0.34	**0.81**	0.70	**0.95**	**0.91**	0.43	0.74	**0.93**	0.69	**0.88**	0.02	**0.96**	0.57	**0.82**	**0.77**	**0.92**	**0.87**	0.24	0.56	0.65	1.00	
Cr	-0.53	0.02	-0.02	0.48	0.48	-0.07	-0.27	0.02	-0.41	-0.12	-0.27	0.09	0.59	**0.85**	-0.09	-0.23	0.06	0.11	-0.54	-0.45	-0.49	-0.12	-0.35	1.00

Table 4.43 Pearson Correlation Matrix of Water Quality Parameters of groundwater sources at site GW5 during 2020-2021

Values in bold letters show significant correlations

Variables	pH	EC	Temp	TDS	TU	TH	Alky	Acidity	PO_4^{3-}	HCO_3^-	NO_3^-	SO_4^{2-}	Cl^-	K^+	Na^+	Ca^{2+}	Mg^{2+}	Zn	Fe	Cd	Cu	Ni	Mn	Cr
pH	1.00																							
EC	0.70	1.00																						
Temp	-0.15	0.18	1.00																					
TDS	0.36	**0.75**	0.68	1.00																				
TU	0.11	0.60	**0.86**	**0.92**	1.00																			
TH	**0.86**	**0.83**	0.19	0.67	0.51	1.00																		
Alk	0.59	**0.91**	0.43	**0.83**	**0.75**	**0.85**	1.00																	
Acidity	0.25	**0.81**	0.54	**0.92**	**0.87**	0.55	**0.80**	1.00																
PO_4^{3-}	0.31	0.73	0.67	**0.80**	**0.86**	0.68	**0.93**	**0.76**	1.00															
HCO_3^-	**0.74**	**0.91**	0.36	**0.86**	0.69	**0.93**	**0.95**	**0.76**	**0.80**	1.00														
NO_3^-	0.73	**0.88**	0.23	0.63	0.56	**0.89**	**0.88**	0.59	**0.75**	**0.84**	1.00													
SO_4^{2-}	**0.88**	0.71	0.03	0.61	0.29	**0.78**	0.62	0.44	0.33	**0.81**	0.61	1.00												
Cl^-	0.61	**0.81**	-0.14	0.62	0.34	0.71	0.68	0.67	0.40	**0.78**	0.59	0.74	1.00											
K^+	**0.77**	**0.90**	0.31	**0.85**	0.64	**0.89**	**0.86**	0.74	0.66	**0.95**	**0.85**	**0.88**	**0.78**	1.00										
Na^+	**0.77**	**0.90**	0.17	**0.76**	0.57	**0.88**	**0.79**	0.73	0.58	**0.88**	**0.87**	**0.81**	**0.82**	**0.96**	1.00									
Ca^{2-}	0.04	0.57	**0.86**	**0.87**	**0.95**	0.38	0.74	**0.85**	**0.85**	0.64	0.45	0.25	0.28	0.56	0.43	1.00								
Mg^{2+}	0.29	**0.80**	0.21	0.68	0.58	0.41	0.61	**0.89**	0.49	0.60	0.50	0.43	0.68	0.63	0.69	0.61	1.00							
Zn	**0.81**	**0.91**	0.11	**0.75**	0.51	**0.88**	**0.80**	0.70	0.54	**0.90**	**0.84**	**0.88**	**0.87**	**0.97**	**0.99**	0.41	0.68	1.00						
Fe	0.40	0.52	-0.37	0.03	-0.08	0.35	0.44	0.10	0.27	0.32	0.61	0.26	0.40	0.36	0.38	-0.05	0.19	0.41	1.00					
Cd	0.02	0.43	0.28	0.18	0.38	0.12	0.36	0.43	0.48	0.16	0.41	-0.20	-0.06	0.12	0.20	0.43	0.55	0.12	0.19	1.00				
Cu	0.34	0.32	0.47	0.14	0.16	0.37	0.46	0.41	0.48	0.34	0.46	0.29	0.28	-0.42	0.41	0.41	0.19	0.25	0.56	0.22	1.00			
Ni	0.45	**0.75**	0.63	**0.84**	**0.78**	0.56	0.74	**0.83**	0.70	0.76	0.58	0.60	0.43	**0.76**	0.67	**0.83**	**0.76**	0.67	-0.02	0.47	0.24	1.00		
Mn	0.47	**0.77**	0.56	0.74	**0.80**	**0.81**	**0.89**	0.71	**0.92**	**0.80**	**0.89**	0.39	0.43	0.74	0.74	0.69	0.48	0.67	0.31	0.51	0.39	0.64	1.00	
Cr	-0.24	-0.05	0.04	0.19	0.05	-0.33	-0.19	0.17	-0.29	-0.12	-0.23	0.15	0.13	0.11	0.06	0.10	0.25	0.09	0.00	-0.29	0.48	0.11	-0.35	1.00

Water Quality Indices of Groundwater Quality for both the Years

Assessment of water quality of analysed water samples for the suitability for drinking purposes have been determined on the basis of Groundwater contamination index (C_d), comprehensive pollution index (CPI) and Heavy metal pollution index (HPI). Computation of C_d, CPI and HPI included water quality parameters having significant values and variations during the 2019-2020 and 2020-2021. The suitability and safety for drinking purposes in accordance to C_d, CPI and HPI has been determined and summarised in Table 4.44 to 4.46.

Contamination index (C_d)

For 2019-2020

In the year 2019-20 for sampling site GWCS, the minimum C_d value 0.184 was observed in January, while the maximum C_d value 0.678 was observed in August. The annual mean of C_d value was recorded 0.484 under low contamination category for the year 2019-20. For sampling site GW1, the minimum C_d value 0.725 was observed in December, while the maximum C_d value 1.418 was observed in August. The annual mean of C_d value was recorded 1.108 under medium contamination category for the year 2019-2020.

For sampling site GW2, the minimum C_d value 0.310 was observed in January, while the maximum C_d value 1.087 was observed in August. The annual mean of C_d value was recorded 0.529 under slightly polluted category for the year 2019- 2020. For sampling site GW3, the minimum C_d value 0.269 was observed in January, while the maximum C_d value 1.018 was observed in August. The annual mean of C_d value was recorded 0.588 under slightly polluted category for the year 2019- 2020. For sampling site GW4, the minimum C_d value 0.797 was observed in January, while the maximum C_d value 1.417 was observed in August. The annual mean of C_d value was recorded 1.129 under medium contamination category for the year 2019- 2020. For sampling site GW5, the minimum C_d value 0.835 was observed in January, while the maximum C_d value 1.342 was observed in August. The annual mean of C_d value was recorded 1.075 under slightly polluted category for the year 2019- 2020

For 2020-2021

In the year 2020-21 for sampling site GWCS, the minimum C_d value 0.174 was observed in January, while the maximum C_d value 0.557 was observed in August. The annual mean of C_d value was recorded 0.313 under low contamination category for 2020-21. For sampling site GW1, the minimum C_d value 0.615 was observed in January, while the maximum C_d value 1.434 was observed in January, while the maximum C_d value 1.434 was observed in August. The annual mean of C_d value was recorded 1.074 under medium contamination category for the year 2020-2021. For sampling site GW2, the minimum C_d value 0.207 was observed in January, while the maximum C_d value 1.037 was observed in August. The annual mean of C_d value was recorded 0.508 under medium contamination category for 2020-2021. For sampling site GW3, the minimum C_d value 0.359 was observed in January, while the maximum C_d value 1.048 was observed in August. The annual mean of C_d value was recorded 0.513 under low contamination category for the year 2020-2021. For sampling site GW4, the minimum C_d value 0.876 was observed in January, while the maximum C_d value 1.416 was observed in August. The annual mean of C_d value was recorded 1.114 under medium contamination category for 2020-2021. For sampling site GW5, the minimum C_d value 0.581 was observed in January, while the maximum C_d value 1.089 was observed in August. The annual mean of C_d value was recorded 1.013 under medium contamination category for 2020-2021.

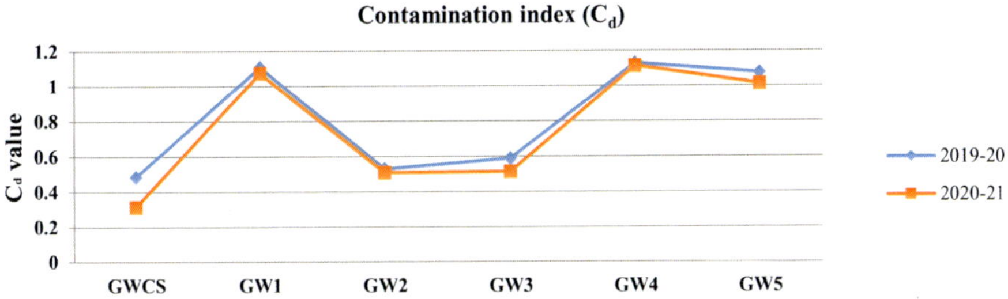

Figure 4.41 Site wise variation of C_d values of groundwater quality during the year of 2019-20 and 2020-21

Comprehensive pollution index (CPI)

For 2019-2020: In the year 2019-20 for sampling site GWCS, the minimum CPI value 0.136 was observed in the month of December, while the maximum CPI value 0.558 was observed in the month of August. The annual mean of CPI value was recorded 0.348 under sub-clean category for the year 2019-20. For sampling site GW1, the minimum CPI value 0.618 was observed in the month of January, while the maximum CPI value 1.364 was observed in the month of August. The annual mean of CPI values was recorded 1.056 under medium contamination category for the year 2019-20. For sampling site GW2, The minimum CPI value 0.311 was observed in the month of January, while the maximum CPI value 1.089 was observed in the month of August. The annual mean of CPI value was recorded 0.562 under slightly polluted category for the year 2019-2020. For sampling site GW3, minimum CPI value 0.297 was observed in the month of January, while the maximum CPI value 0.749 was observed in the month of August. The annual mean of CPI value was recorded 0.471 under slightly polluted category for the year 2019- 2020. For sampling site GW4, The minimum CPI value 0.828 was observed in the month of January, while the maximum CPI value 1.192 was observed in the month of August. The annual mean of CPI value was recorded 1.062 under medium contamination category for the year 2019- 2020. For sampling site GW5, the minimum CPI value 0.612 was observed in the month of January, while the maximum CPI value 1.212 was observed in the month of August. The annual mean of C_d values was recorded 1.039 under slightly polluted category for the year 2019- 2020.

For 2020-2021: In the year 2020-21 for sampling site GWCS, the minimum CPI value 0.148 was observed in the month of January, while the maximum CPI value 0.364 was observed in the month of August. The annual mean of CPI values was recorded 0.249 under sub-clean category for 2020-21. For sampling site GW1, the minimum CPI value 0.421 was observed in the month of January, while the maximum CPI value 1.272 was observed in the month of August. The annual mean of CPI value was recorded 1.011 under moderately polluted category for the year 2020-21. For sampling site GW2, the minimum CPI value 0.318 was observed in the month of January, while the maximum CPI value 1.012 was observed in the month of August. The annual mean of CPI value was recorded 0.479 under medium contamination category for 2020-2021. For sampling site GW3, the minimum CPI value 0.326 was observed in the month of January, while the maximum CPI value 0.786 was observed in the month of August. The annual mean of CPI value was recorded 0.472 under low contamination category for the year 2020-2021. For sampling site GW4, the minimum CPI value 0.838 was observed in the month of January, while the maximum CPI value 1.187 was observed in the month of August. The annual mean of CPI value was recorded 1.032 under medium contamination category for 2020-2021. For sampling site GW5, the minimum CPI value 0.662 was observed in the month of January, while the maximum CPI value 1.142 was observed in the month of August. The annual mean of CPI value was recorded 1.009 under medium contamination category for 2020-2021.

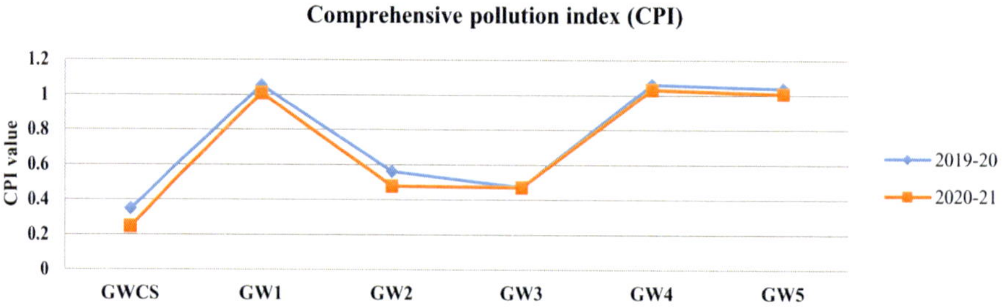

Figure1 4.42. Site wise variation of CPI values of groundwater quality during the year of 2019-20 and 2020-21

Heavy Metal Pollution Index (HPI)

For 2019-2020: For sampling site GWCS, the HPI monthly mean variation ranged from 7.48 to 12.78 during the year 2019-2020. The annual mean of HPI values was recorded 10.25 under low contamination category for 2019-2020 respectively. For sampling site GW1, The HPI monthly variation ranged from 20.08 to 29.68 during the year 2019-2020. The annual mean of HPI values was recorded 25.72 under medium contamination category for the year 2019-2020 respectively. For sampling site GW2, The HPI monthly variation ranged from 19.84 to 28.21 during 2019-2020 respectively. The annual mean of HPI values was recorded 23.55 under medium contamination category for 2019-2020 respectively. For sampling site GW3, the HPI monthly variation ranged from 10.28 to 18.37 during 2019-2020 respectively. The annual mean of HPI values was recorded 13.87 under low contamination category for 2019-2020 respectively. For sampling site GW4, The HPI monthly variation ranged from 18.19 to 26.34 during 2019-2020 respectively. The annual mean of HPI values was recorded 21.26 under medium contamination category for 2019-2020 respectively. For sampling site GW5, The HPI monthly variation ranged from 16.46 to 25.08 during 2019-2020 respectively. The annual mean of HPI values was recorded 20.93 under medium contamination category for 2019-2020.

For 2020-2021: For sampling site GWCS, the HPI monthly mean variation ranged from 6.17 to 11.21 during the year 2020-2021. The annual mean of HPI values was recorded 8.54 under low contamination category for 2020-2021 respectively.For sampling site GW1, The HPI monthly variation ranged from 15.48 to 28.37 during the year 2020-2021 respectively. The annual mean of HPI values was recorded 22.42 under medium contamination category for the year 2020-2021 respectively. For sampling site GW2, The HPI monthly variation ranged from 17.29 to 26.44 during 2020-2021 respectively. The annual mean of HPI values was recorded 21.74 under medium contamination category for 2020-2021 respectively. For sampling site GW3, the HPI monthly variation ranged from 8.43 to 15.48 during 2020-2021 respectively. The annual mean of HPI values was recorded 11.73 under low contamination category for 2020-2021 respectively. For sampling site GW4, The HPI monthly variation ranged from 17.72 to 22.11 during 2020-2021 respectively. The annual mean of HPI values was recorded 17.72 under medium contamination category for 2020-2021 respectively. For sampling site GW5, The HPI monthly variation ranged from 15.87 to 24.89 during 2020-2021 respectively. The annual mean of HPI values was recorded 18.67 under medium contamination category for 2020-2021.

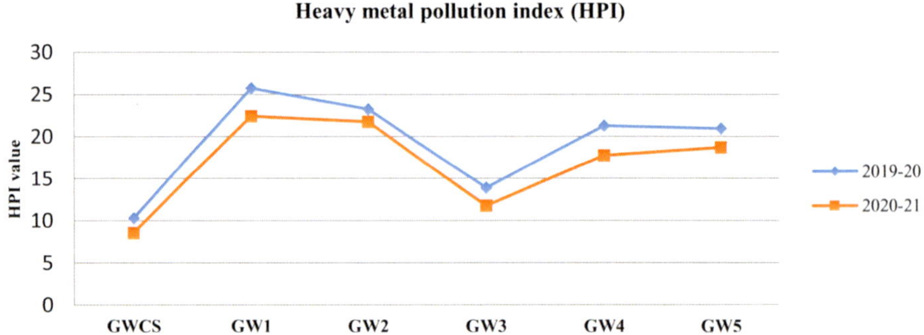

Figure 14.43. Site wise variation of HPI values of groundwater quality during the year of 2019-20 and 2020-21

Groundwater Suitability for Irrigation Purposes

Groundwater suitability for irrigation purpose depends on the concentration of dissolved ions present in it. Excessive ion concentration can affect the soil, plant growth, and agriculture productivity (Wilcox, 1955; Jasrotia et al., 2019; Mandal et al., 2019; Xu et al., 2019; Snousy et al., 2021). Table 4.47 and 4.48 shows the water sample quality status for all irrigation water quality parameters, which have been grouped into numerous categories. SAR is an important criterion for determining if groundwater is suitable for irrigation. SAR values vary between 1.66 to 6.89 for 2019-2020 and 2020 2021 the SAR value ranged from 0.59 to 1.78. SAR values in the research area were less than 10 in both the years, indicating that the water samples collected in both years were of excellent quality. Sodium (Na^+) combines with carbonate and chloride to form alkaline and saline soil. The Base Exchange process further dissociates Na^+, replaces Ca^{2+} in soil, and reduces permeability. Sodium percentage is found ranged from 24.89 to 38.98 for 2019-2020, and for 2020-2021 the Na% value ranged from 14.84 to 29.45. According to the Wilcox diagram and USSL diagram (Figure 4.44a, b), GWCS water sample fall under the excellent category, and GW1 to GW5 fall under the good category, respectively. Electrical conductivity (EC), measures the concentration of dissolved salts in groundwater and signifies salinity hazard to crops. Groundwater with high salinity is unsuitable for plants and poses a salinity hazard. Water with a low EC has a significant impact on crop productivity. In the study area, all samples fall under the in good category, and hence, groundwater is suitable for the growth of crops. Magnesium hazard (MH), is used to evaluate groundwater suitability for irrigation by determining the concentration of Ca^{2+} over Mg^{2+} (Ragunath, 1987). Excess quantity of Mg^{2+} ions in irrigation water adversely impacts soil quality (makes soil alkaline) and reduces crop production (Snousy et al., 2021).

Table 14.44 Monthly and Site wise variation of Contamination index (C_d) of Ground water quality during the year of 2019-20 and 2020-21

Sampling location	**Contamination index (C_d)**	**Monthly variation of C_d at sampling locations**								**Site wise mean variation of C_d at sampling location**
		May	**June**	**July**	**Aug**	**Oct**	**Nov**	**Dec**	**Jan**	
GWCS	2019-20	0.365	0.521	0.513	0.678	0.487	0.376	0.245	0.184	0.484
	QS	Low Contamination	Low Contamination	Low Contamination	Low Contamination	Low Contamination	Low Contamination	Low Contamination	Low Contamination	Low Contamination
	2020-21	0.216	0.214	0.393	0.557	0.417	0.348	0.185	0.174	0.313
	QS	Low Contamination	Low Contamination	Low Contamination	Low Contamination	Low Contamination	Low Contamination	Low Contamination	Low Contamination	Low Contamination
GW1	2019-20	1.076	1.011	1.186	1.418	1.156	1.042	0.865	0.725	1.108
	QS	Medium Contamination	Medium Contamination	Medium Contamination	Medium Contamination	Medium Contamination	Medium Contamination	Medium Contamination	Medium Contamination	Medium Contamination
	2020-21	1.091	1.134	1.163	1.434	1.256	1.173	0.724	0.615	1.074
	QS	Medium Contamination	Medium Contamination	Medium Contamination	Medium Contamination	Medium Contamination	Medium Contamination	Low Contamination	Low Contamination	Medium Contamination

GW2	2019-20	0.412	0.429	0.736	1.087	0.627	0.413	0.217	0.310	0.529
	QS	Low Contamination	Low Contamination	Low Contamination	Medium Contamination	Low Contamination	Low Contamination	Low Contamination	Low Contamination	Low Contamination
	2020-21	0.367	0.389	0.629	1.037	0.427	0.387	0.321	0.207	0.508
	QS	Low Contamination	Low Contamination	Low Contamination	Medium Contamination	Low Contamination	Low Contamination	Low Contamination	Low Contamination	Low Contamination
GW3	2019-20	0.277	0.289	0.512	1.018	0.476	0.312	0.287	0.269	0.588
	QS	Low Contamination	Low Contamination	Low Contamination	Medium Contamination	Low Contamination	Low Contamination	Low Contamination	Low Contamination	Low Contamination
	2020-21	0.367	0.321	0.639	1.048	0.528	0.432	0.407	0.359	0.513
	QS	Low Contamination	Low Contamination	Low Contamination	Medium Contamination	Low Contamination	Low Contamination	Low Contamination	Low Contamination	Low Contamination
GW4	2019-20	1.042	1.102	1.402	1.417	1.322	1.062	0.886	0.797	1.129
	QS	Medium Contamination	Medium Contamination	Medium Contamination	Medium Contamination	Medium Contamination	Medium Contamination	Low Contamination	Low Contamination	Medium Contamination
	2020-21	1.059	1.067	1.113	1.416	1.098	1.023	1.019	0.876	1.114
	QS	Medium Contamination	Medium Contamination	Medium Contamination	Medium Contamination	Medium Contamination	Medium Contamination	Medium Contamination	Medium Contamination	Medium Contamination

GW5	2019-20	1.012	1.109	1.135	1.342	1.217	1.052	0.901	0.835	1.075
	QS	Medium Contami-nation	Medium Contami-nation	Medium Contami-nation	Medium Contamina-tion	Medium Contamina-tion	Medium Contamina-tion	Low Contamina-tion	Low Contami-nation	Medium Contamina-tion
	2020-21	1.122	1.069	1.025	1.089	1.022	1.012	0.749	0.581	1.013
	QS	Medium Contami-nation	Medium Contami-nation	Medium Contami-nation	Medium Contamina-tion	Medium Contamina-tion	Medium Contamina-tion	Medium Contamina-tion	Medium Contami-nation	Medium Contamina-tion

Table 14.45 Monthly and Site wise variation of Comprehensive pollution index (CPI) of Ground water quality during the year of 2019-20 and 2020-21

Sampling location	Compre-hensive pollution index (CPI)	Monthly variation of CPI at sampling locations								Site wise mean variation of CPI at sam-pling location
		May	June	July	Aug	Oct	Nov	Dec	Jan	
GWCS	2019-20	0.244	0.318	0.407	0.558	0.364	0.248	0.136	0.211	0.348
	QS	Clean	Sub-clean	Sub-clean	Sub-clean	Sub-clean	Clean	Clean	Clean	Sub-Clean
	2020-21	0.189	0.236	0.313	0.347	0.364	0.238	0.156	0.148	0.249
	QS	Clean	Clean	Sub-clean	Sub-clean	Sub-clean	Sub-clean	Clean	Clean	Sub-clean
GW1	2019-20	1.013	0.876	1.088	1.364	1.197	1.089	0.768	0.618	1.056
	QS	Moderately Polluted	Moderately Polluted	Moderately Polluted	Moderately Polluted	Moderately Polluted	Moderately Polluted	Slightly Polluted	Slightly Polluted	Moderately Polluted
	2020-21	1.112	1.089	1.144	1.272	1.197	1.185	0.665	0.421	1.011
	QS	Moderately Polluted	Moderately Polluted	Moderately Polluted	Moderately Polluted	Moderately Polluted	Moderately Polluted	Slightly Polluted	Slightly Polluted	Moderately Polluted
GW2	2019-20	0.424	0.418	0.775	1.089	0.454	0.418	0.356	0.311	0.562
	QS	Slightly Polluted	Slightly Polluted	Slightly Polluted	Moderately Polluted	Slightly Polluted	Slightly Polluted	Sub-clean	Sub-clean	Slightly Polluted
	2020-21	0.237	0.268	0.683	1.012	0.534	0.425	0.351	0.318	0.479
	QS	Sub-clean	Sub-clean	Slightly Polluted	Moderately Polluted	Slightly Polluted	Slightly Polluted	Sub-clean	Sub-clean	Slightly Polluted

GW3	2019-20	0.401	0.347	0.583	0.749	0.487	0.419	0.308	0.297	0.471
	QS	Sub-clean	Sub-clean	Slightly Polluted	Slightly Polluted	Slightly Polluted	Slightly Polluted	Sub-clean	Sub-clean	Slightly Polluted
	2020-21	0.342	0.367	0.713	0.786	0.467	0.423	0.348	0.326	0.472
	QS	Sub-clean	Sub-clean	Slightly Polluted	Slightly Polluted	Slightly Polluted	Slightly Polluted	Sub-clean	Sub-clean	Slightly Polluted
GW4	2019-20	1.014	1.021	1.094	1.192	1.046	1.088	0.979	0.828	1.062
	QS	Moderately Polluted	Moderately Polluted	Moderately Polluted	Moderately Polluted	Moderately Polluted	Moderately Polluted	Slightly Polluted	Slightly Polluted	Moderately Polluted
	2020-21	1.083	1.046	1.114	1.187	1.118	1.079	0.787	0.838	1.032
	QS	Moderately Polluted	Moderately Polluted	Moderately Polluted	Moderately Polluted	Moderately Polluted	Moderately Polluted	Slightly Polluted	Slightly Polluted	Moderately Polluted
GW5	2019-20	1.048	1.032	1.119	1.212	1.124	1.086	1.077	0.612	1.039
	QS	Moderately Polluted	Moderately Polluted	Moderately Polluted	Moderately Polluted	Moderately Polluted	Moderately Polluted	Moder-ately Polluted	Slightly Polluted	Moderately Polluted
	2020-21	1.019	1.029	1.079	1.142	1.109	0.836	0.848	0.662	1.009
	QS	Moderately Polluted	Moderately Polluted	Moderately Polluted	Moderately Polluted	Moderately Polluted	Slightly Polluted	Slightly Polluted	Slightly Polluted	Moderately Polluted

Table 14.46 Heavy metal pollution index (HPI)) of Ground water quality at sampling sites during the year of 2019-20 and 2020-21

Sampling location	HPI	Monthly variation of HPI at sampling locations								Site wise variation of HPI at sampling location
		May	June	July	Aug	Oct	Nov	Dec	Jan	
GWCS	2019-20	10.13	11.24	11.44	12.78	10.89	9.77	7.48	8.29	10.25
	QS	Low Contamination	Low Contamination	Low Contamination	Low Contamination	Low Contamination	Low Contamination	Low Contamination	Low Contamination	Low Contamination
	2020-21	9.97	8.09	10.18	11.21	7.33	8.21	6.17	7.19	8.54
	QS	Low Contamination	Low Contamination	Low Contamination	Low Contamination	Low Contamination	Low Contamination	Low Contamination	Low Contamination	Low Contamination
GW1	2019-20	24.33	26.86	29.68	28.49	27.48	26.48	22.37	20.08	25.72
	QS	Medium Contamination	Medium Contamination	Medium Contamination	Medium Contamination	Medium Contamination	Medium Contamination	Medium Contamination	Medium Contamination	Medium Contamination
	2020-21	21.58	23.48	28.37	26.36	24.47	21.49	18.11	15.48	22.42
	QS	Medium Contamination	Medium Contamination	Medium Contamination	Medium Contamination	Medium Contamination	Medium Contamination	Medium Contamination	Medium Contamination	Medium Contamination
GW2	2019-20	21.13	22.48	26.48	28.21	24.28	23.58	20.03	19.84	23.25
	QS	Medium Contamination	Medium Contamination	Medium Contamination	Medium Contamination	Medium Contamination	Medium Contamination	Medium Contamination	Medium Contamination	Medium Contamination
	2020-21	20.38	21.34	24.58	26.44	22.46	22.03	19.37	17.29	21.74
	QS	Medium Contamination	Medium Contamination	Medium Contamination	Medium Contamination	Medium Contamination	Medium Contamination	Medium Contamination	Medium Contamination	Medium Contamination

GW3	2019-20	12.28	14.28	17.49	18.37	14.48	12.28	11.48	10.28	13.87
	QS	Low Contamination	Low Contamination	Medium Contamination	Medium Contamination	Low Contamination	Low Contamination	Low Contamination	Low Contamination	Low Contamination
	2020-21	10.48	12.59	14.37	15.48	12.58	10.52	9.37	8.43	11.73
	QS	Low Contamination	Low Contamination	Medium Contamination	Medium Contamination	Low Contamination	Low Contamination	Low Contamination	Low Contamination	Low Contamination
GW4	2019-20	18.37	19.77	23.86	26.34	22.89	21.35	19.32	18.19	21.26
	QS	Medium Contamination	Medium Contamination	Medium Contamination	Medium Contamination	Medium Contamination	Medium Contamination	Medium Contamination	Medium Contamination	Medium Contamination
	2020-21	15.87	17.54	20.59	22.11	18.68	16.73	15.23	15.03	17.72
	QS	Medium Contamination	Medium Contamination	Medium Contamination	Medium Contamination	Medium Contamination	Medium Contamination	Medium Contamination	Medium Contamination	Medium Contamination
GW5	2019-20	17.35	20.43	24.65	25.08	23.67	21.55	18.23	16.46	20.93
	QS	Medium Contamination	Medium Contamination	Medium Contamination	Medium Contamination	Medium Contamination	Medium Contamination	Medium Contamination	Medium Contamination	Medium Contamination
	2020-21	15.68	17.78	22.45	24.89	17.45	18.78	16.43	15.87	18.67
	QS	Medium Contamination	Medium Contamination	Medium Contamination	Medium Contamination	Medium Contamination	Medium Contamination	Medium Contamination	Medium Contamination	Medium Contamination

In the present study, GWCS, GW1, GW2, GW3 had a MAR value less than 50 and were suitable for irrigation. The remaining two samples, GW4 and GW5, yielded MAR values more than 50 and were considered unsuitable for irrigation. Ion chemistry process in groundwater samples is interpreted using Gibbs diagrams (Gibbs, 1977). Precipitation dominance, evaporation dominance, and rock dominance are depicted in the Gibbs diagram. The graph of the ratio of cations [(Na + K)/(Na + K + Ca)] and anions [Cl/(Cl + HCO_3)] against TDS is known as a Gibbs plot. The Gibbs plot indicates that the research region's whole sample is made up of rock dominance (Figs. 4.45c, d). The rock–water interface is the fundamental mechanism affecting alluvial plains' groundwater chemistry (Alam, 2013; Raju *et al.*, 2011).

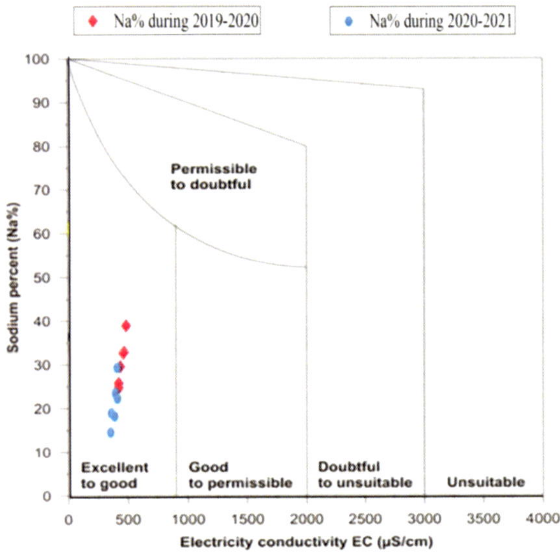

(a)

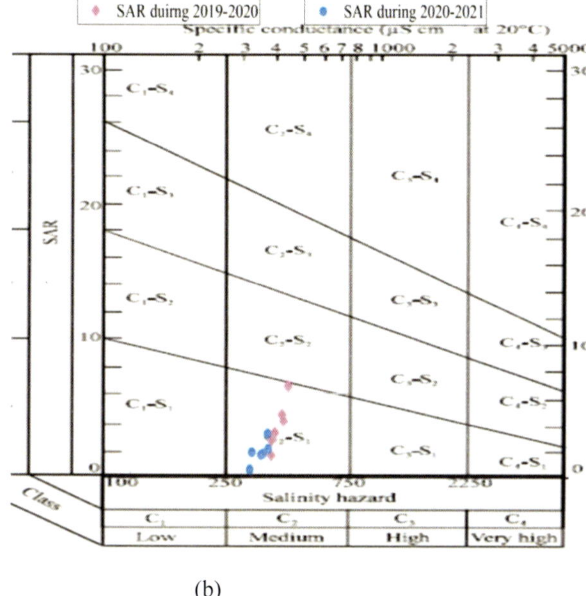

(b)

Figure 14.44. Assessment of irrigation water suitability by (a) Wilcox diagram (b) USSL diagram

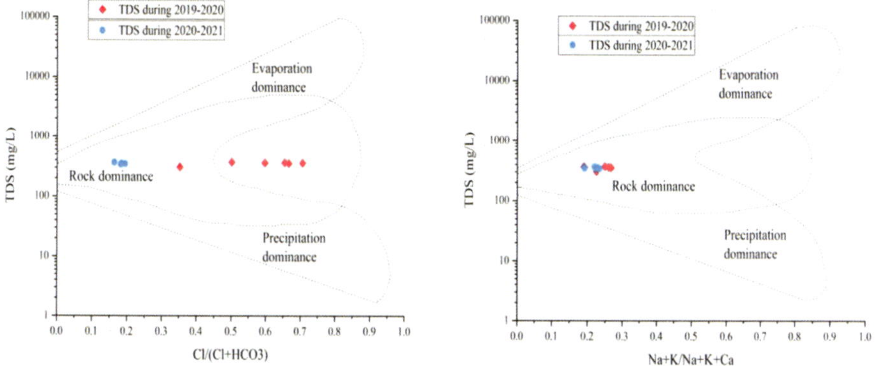

Figure 14.45. Assessment of irrigation water suitability by Gibbs diagram for cation and anion

Table 14.47. Classification of Groundwater for Irrigation Purposes

Index	Equation	Suitability classification	Water condition	Water samples
Sodium adsorption ratio (SAR)	$SAR = \frac{Na^{+}}{\sqrt{(Ca^{2+} + Mg^{+})/2}}$	$SAR < 10$ $10 < SAR < 18$ $18 < SAR < 26$ $SAR > 26$	Excellent Good Doubtful Unsuitable	GWCS,GW1,GW2,GW3,GW4,GW5 - - -
Salinity and alkalinity hazard (SAH)	Total salt concentration	$EC < 250$ $250 < EC < 750$ $750 < EC < 2250$ $EC > 2250$	Excellent Good Permissible Unsuitable	- GWCS,GW1,GW2,GW3,GW4,GW5 - -
Sodium percentage (Na %)	$Na\% = \frac{(Na^{+} + K^{+})}{Ca^{2+} + Mg^{2+} + Na^{+} + K^{+}} \times 100$	< 20 20–40 40–60 60–80 > 80	Excellent Good Acceptable Doubtful Unsuitable	GWCS GW1,GW2,GW3,GW4,GW5 - -
Magnesium adsorption ratio (MAR)	$MAR = \frac{Mg^{2+}}{(Ca^{2+} + Mg^{+})} \times 100$	$MH < 50\%$ $MH > 50\%$	Suitable Harmful and Unsuitable	GWCS,GW1,GW2,GW3 GW4,GW5 -
Gibbs Ratio I (For Anion)	$Cl^{-}/(Cl^{-}+HCO_3)$	-		-
Gibbs Ratio II (ForCation)	$(Na^{+} + K^{+})/(Na^{+}+K^{+}+Ca^{2+})$	-		-

Table 14.48. Suitability of Groundwater for Irrigation Purposes in the year 2019-2020 and 2020-2021

Sample ID	2019-2020					2020-2021				
	SAR	Na%	SAH	TDS	MH	SAR	SAH	Na%	TDS	MH
GWCS	1.66	24.89	356.48	306.66	31.21	0.59	343.11	14.84	364.57	24.43
GW1	4.74	32.74	430.48	368.5	40.26	2.14	400.34	22.47	349.94	35.28
GW2	3.36	29.67	404.80	358.49	35.37	1.89	379.69	19.15	340.64	31.37
GW3	2.78	25.95	400.22	351.37	32.45	1.69	386.93	18.46	339.61	29.47
GW4	6.89	38.98	423.67	363.38	57.27	3.29	398.80	29.45	348.12	53.27
GW5	4.25	32.98	416.27	358.12	54.73	1.78	393.08	23.69	345.16	50.08

Risk assessment on human health

Non-carcinogenic health risk assessment

According to the USEPA (2005), there are three well-known ways that people might be exposed to or put at risk from the possible risks associated with certain toxic metals (such as

Cd, Cr, Zn and Fe) in groundwater: ingestion and skin contact. Calculating the chemical daily intake (CDI), dermal absorbed dose (DAD), hazard quotient (HQ), and hazard index (HI) was the goal of this investigation. The estimated result of hazard quotient ($HQ_{ingestion}$ and HQ_{dermal}) and hazard index (HI) for non-carcinogenic concern related to adult and child by ingestion and dermal contact of groundwater is presented in Table 4.49 to 4.52. The HQ and HI values of children for each metal were generally observed higher than the HQ and HI values for adults at each location. However, the HQ value for each metal was observed below the health risk level (less than 1) at every monitoring station for both adults and children. The HI_{CDI} (2019-2020) value for adults obtained by this study ranged from 3.34E-02 to 6.89E-02. For children the maximum HI_{CDI} value were found at sampling site GW1 (3.75E-01) and the minimum HI value were at sampling site GWCS (1.82E-01). The HI_{DAD} (2019-2020) value calculated for adults at all the sampling locations were ranged from 1.23E-02 to 8.25E-03. For children the maximum HI_{DAD} value were found at GW4 (9.11E-02) and the minimum HI_{DAD} value were found at sampling site GWCS (1.23E-03).

The HI_{CDI} (2020-2021) value for an adults obtained by this study ranged from 3.38E-02 to 6.64E-02. For children the maximum HI_{CDI} value were found at sampling site GW1 (3.61E-01) and the minimum HI value were at sampling site GWCS (1.84E-01). The HI_{DAD} (2019-2020) value calculated for adults at all the sampling locations were ranged from 1.23E-02 to 8.26E-03. For children the maximum HI_{DAD} value were found at GW4 (9.10E-02) and the minimum

HI_{DAD} value at sampling site GWCS (1.37E-02). The $HQ_{ingestion}$ is multiple times higher than the HQ_{dermal}, indicating that ingestion of ground water poses a higher risk than dermal exposure. The overall result of HI suggests no potential health risk to both adult and child on consuming groundwater for drinking purposes. The health risk value for children was higher than the values for adults, indicating that children were more sensitive to groundwater pollution. The contribution of pollutants to health risks varies. The highest contribution for non-carcinogenic risk to adults and children was caused by Fe, Cr, Mn, Cd to the value of HI, while rest of the metals contributed least to total HI.

Table 14.49. Summary of Health risk assessment for heavy metals in water samples through chemical daily intake (CDI) at different sampling locations during the year of 2019- 2020

Sampling location	Adults								Child							
	HQ for Cu	HQ for Zn	HQ for Cr	HQ for Fe	HQ for Mn	HQ for Ni	HQ for Cd	HI	HQ for Cu	HQ for Zn	HQ for Cr	HQ for Fe	HQ for Mn	HQ for Ni	HQ for Cd	HI
GWCS	3.30E-04	1.18E-02	4.08E-03	3.48E-03	8.75E-04	6.12E-04	1.22E-02	3.34E-02	1.67E-03	6.42E-02	2.22E-02	1.90E-02	4.76E-03	3.33E-03	6.67E-02	1.82E-01
GW1	9.18E-04	1.73E-02	8.16E-03	6.86E-03	4.37E-03	6.73E-03	2.45E-02	6.89E-02	5.00E-03	9.44E-02	4.44E-02	3.73E-02	2.38E-02	3.67E-02	1.33E-01	3.75E-01
GW2	6.12E-04	1.52E-02	8.16E-03	5.91E-03	3.50E-03	1.84E-03	2.45E-02	5.97E-02	3.33E-03	8.27E-02	4.44E-02	3.22E-02	1.90E-02	1.00E-02	1.33E-01	3.25E-01
GW3	6.12E-04	1.33E-02	4.08E-03	5.14E-03	3.50E-03	1.84E-03	2.45E-02	5.29E-02	3.33E-03	7.22E-02	2.22E-02	2.80E-02	1.90E-02	1.00E-02	1.33E-01	2.88E-01
GW4	9.18E-04	1.38E-02	8.16E-03	4.08E-03	4.37E-03	3.06E-03	1.22E-02	4.66E-02	5.00E-03	7.51E-02	4.44E-02	2.22E-02	2.38E-02	1.67E-02	6.67E-02	2.54E-01
GW5	6.12E-04	1.39E-02	4.08E-03	3.67E-03	3.50E-03	1.84E-03	1.22E-02	3.98E-02	3.33E-03	7.56E-02	2.22E-02	2.00E-02	1.90E-02	1.00E-02	6.67E-02	2.17E-01

Table 14.50. Summary of Health risk assessment for heavy metals in water samples through dermal absorbed dose (DAD) at different sampling locations during the year of 2019- 2020

Sampling location	Adults								Child							
	HQ for Cu	HQ for Zn	HQ for Cr	HQ for Fe	HQ for Mn	HQ for Ni	HQ for Cd	HI	HQ for Cu	HQ for Zn	HQ for Cr	HQ for Fe	HQ for Mn	HQ for Ni	HQ for Cd	HI
GWCS	1.10E-05	1.91E-04	1.76E-07	1.10E-04	1.32E-03	2.45E-05	1.32E-02	1.23E-02	1.10E-04	1.27E-03	1.17E-06	1.23E-03	3.52E-02	1.63E-04	4.40E-02	1.23E-03
CW1	5.51E-06	2.81E-04	8.82E-08	9.40E-05	6.61E-03	2.69E-04	6.61E-03	8.25E-03	3.67E-05	1.87E-03	5.87E-07	6.25E-04	8.80E-03	1.79E-03	8.80E-02	5.49E-02
GW2	1.65E-05	2.46E-04	1.76E-07	1.60E-04	5.29E-03	7.35E-05	1.32E-02	1.90E-02	7.33E-05	1.64E-03	1.17E-06	1.06E-03	3.52E-02	4.89E-04	8.80E-02	1.37E-01
GW3	1.10E-05	2.15E-04	8.82E-08	1.39E-04	5.29E-03	7.35E-05	1.32E-02	1.90E-02	7.33E-05	1.43E-03	5.87E-07	9.24E-04	4.40E-02	4.89E-04	8.80E-02	1.26E-01
GW4	1.65E-05	2.23E-04	1.76E-07	1.85E-04	6.61E-03	1.22E-04	6.61E-03	1.37E-02	1.10E-04	1.49E-03	1.17E-06	7.32E-04	4.40E-02	8.15E-04	4.40E-02	9.11E-02
GW5	1.10E-05	2.25E-04	8.82E-08	9.92E-05	5.29E-03	7.35E-05	6.61E-03	2.06E-02	7.33E-05	1.50E-03	5.87E-07	6.60E-04	3.52E-02	4.89E-04	4.40E-02	8.19E-02

Table 14.51. Summary of Health risk assessment for heavy meta s in water samples through chemical daily intake (CDI) at different sampling locations during the year of 2020- 2021

Sampling location	Adults								Child							
	HQ for Cu	HQ for Zn	HQ for Cr	HQ for Fe	HQ for Mn	HQ for Ni	HQ for Cd	HI	HQ for Cu	HQ for Zn	HQ for Cr	HQ for Fe	HQ for Mn	HQ for Ni	HQ for Cd	HI
GWCS	3.06E-04	1.22E-02	4.08E-03	3.46E-03	8.75E-04	6.12E-04	1.22E-02	3.38E-02	1.67E-03	6.64E-02	2.22E-02	1.89E-02	4.76E-03	3.33E-03	6.67E-02	1.84E-01
GW1	6.12E-04	1.67E-02	8.16E-03	6.58E-03	4.37E-03	5.51E-03	2.45E-02	6.64E-02	3.33E-03	9.07E-02	4.44E-02	3.58E-02	2.38E-02	3.00E-02	1.33E-01	3.61E-01
GW2	6.12E-04	1.40E-02	8.16E-03	5.83E-03	3.50E-03	2.45E-03	2.45E-02	5.90E-02	3.33E-03	7.60E-02	4.44E-02	3.17E-02	1.90E-02	1.33E-02	1.33E-01	3.21E-01
GW3	9.18E-04	1.34E-02	4.08E-03	5.30E-03	3.50E-03	3.06E-03	2.45E-02	5.47E-02	5.00E-03	7.29E-02	2.22E-02	2.89E-02	1.90E-02	1.67E-02	1.33E-01	2.98E-01
GW4	6.12E-04	1.40E-02	8.16E-03	4.92E-03	4.37E-03	3.06E-03	1.22E-02	4.74E-02	3.33E-03	7.62E-02	4.44E-02	2.68E-02	2.38E-02	1.67E-02	6.67E-02	2.58E-01
GW5	6.12E-04	1.36E-02	4.08E-03	3.87E-03	3.50E-03	3.06E-03	1.22E-02	4.10E-02	3.33E-03	7.42E-02	2.22E-02	2.10E-02	1.90E-02	1.67E-02	6.67E-02	2.23E-01

Table 14.52. Summary of Health risk assessment for heavy metals in water samples through dermal absorbed dose (DAD) at different sampling locations during the year of 2020- 2021

Sampling location	Adults								Child							
	HQ for Cu	HQ for Zn	HQ for Cr	HQ for Fe	HQ for Mn	HQ for Ni	HQ for Cd	HI	HQ for Cu	HQ for Zn	HQ for Cr	HQ for Fe	HQ for Mn	HQ for Ni	HQ for Cd	HI
GWCS	1.10E-05	1.98E-04	1.76E-07	1.04E-04	1.32E-03	1.22E-04	1.32E-02	1.23E-02	1.10E-04	1.24E-03	1.17E-07	1.05E-04	3.52E-03	1.63E-04	4.40E-02	1.37E-02
GW1	5.51E-06	2.70E-04	8.82E-08	9.35E-05	6.61E-03	2.45E-05	6.61E-03	8.26E-03	7.33E-05	1.80E-03	5.87E-06	6.22E-03	8.80E-02	1.47E-03	8.80E-02	5.49E-01
GW2	1.10E-05	2.26E-04	1.76E-07	1.57E-04	5.29E-03	9.80E-05	1.32E-02	1.90E-02	7.33E-05	1.50E-03	1.17E-06	1.18E-03	4.40E-02	6.52E-04	8.80E-02	1.26E-01
GW3	1.65E-05	2.17E-04	8.82E-08	1.43E-04	5.29E-03	2.20E-04	1.32E-02	1.90E-02	3.67E-05	1.44E-03	5.87E-07	9.52E-04	3.52E-02	8.15E-04	8.80E-02	1.27E-01
GW4	1.10E-05	1.86E-04	1.76E-07	1.33E-04	6.61E-03	1.22E-04	6.61E-03	1.37E-02	7.33E-05	1.32E-03	1.17E-06	8.83E-04	4.40E-02	8.15E-04	4.40E-02	9.10E-02
GW5	1.10E-05	1.46E-04	8.82E-08	1.78E-04	5.29E-03	1.22E-04	6.61E-03	2.05E-02	7.33E-05	9.72E-04	5.87E-07	6.95E-04	3.52E-02	8.15E-04	4.40E-02	8.18E-02

Questionnaire Survey Analysis during in 2019-2020

An offline, composed of 15 questions was conducted to spread the awareness, assess the pollution status and impact of industrial wastewater quality on groundwater quality and human health during the year 2019-2020. 480 people had participated in survey in studied region in 2019-2020. A short survey to assess the personal information includes the person's name, age and occupation.

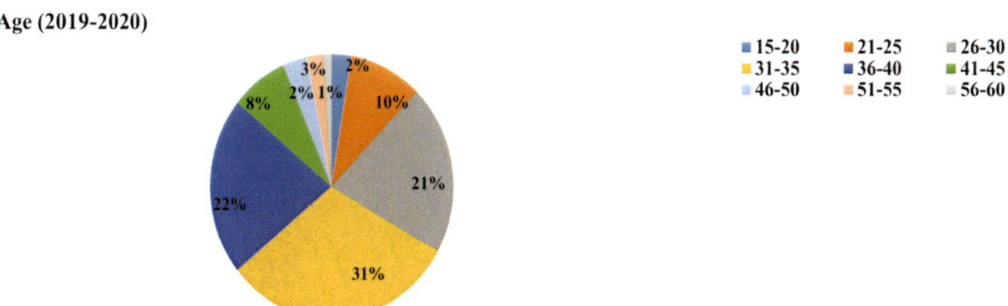

Figure 14.46. Percent distribution of Age in the study area

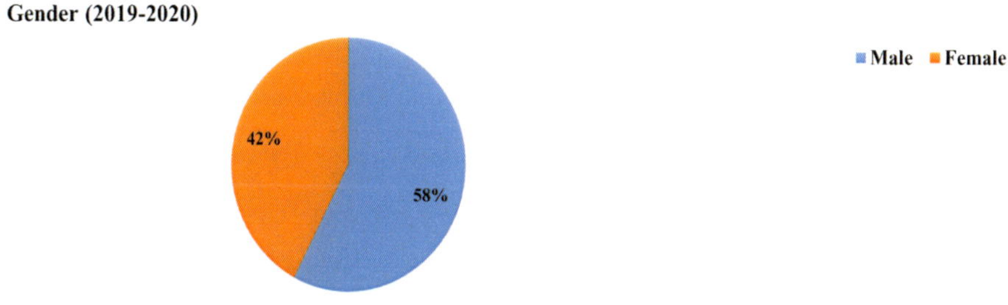

Figure 14.47. Percent distribution of Gender in the study area

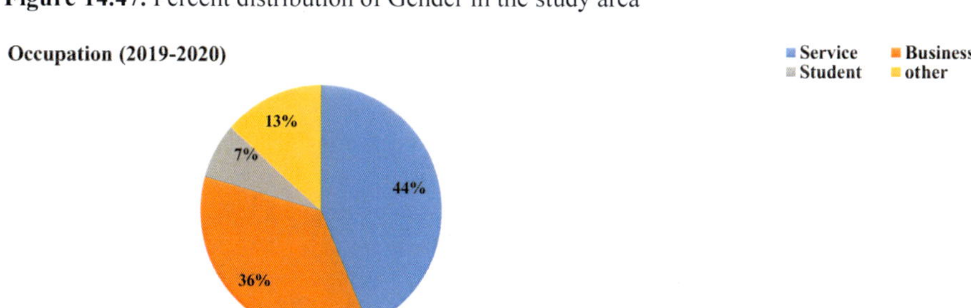

Percent distribution of occupation of people in the study area

Figure 14.48. Percent distribution of drinking water resources for household use in the study area

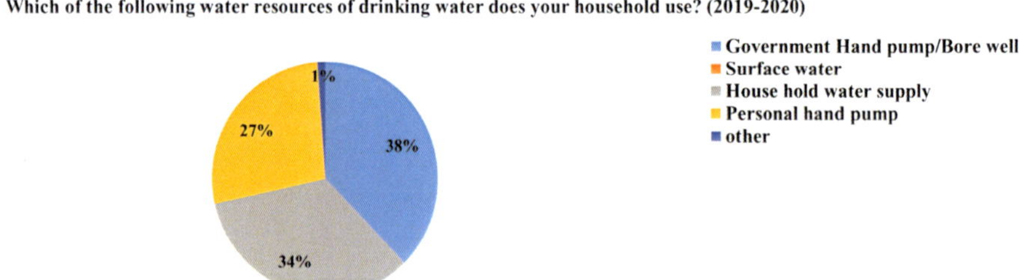

Figure 14.49. Percent distribution of drinking water source depth use in the study area

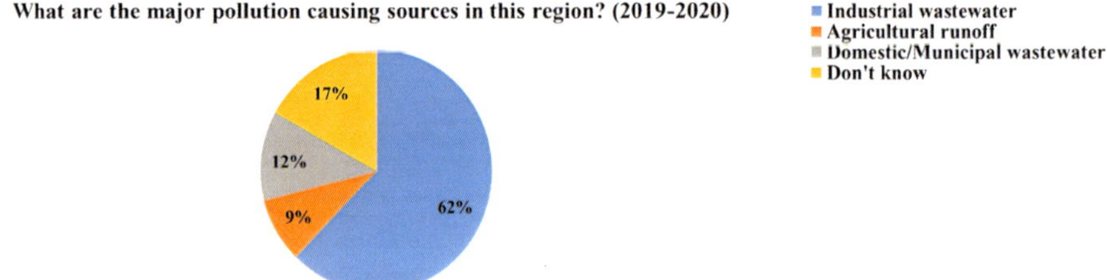

Figure 14.50. Percent distribution of major pollution causing sources in the study area

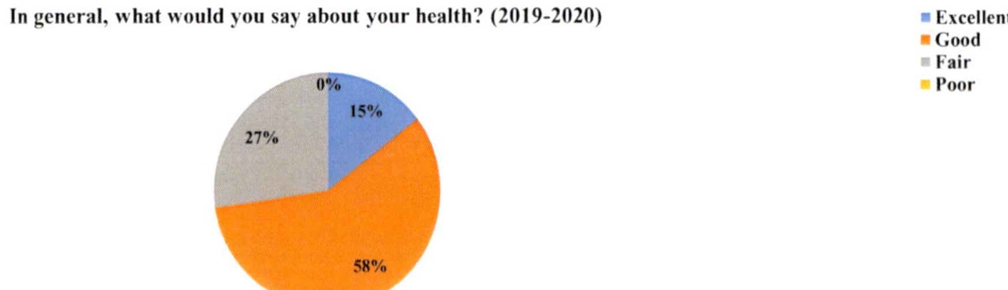

Figure 14.51. Percent distribution of health status of people in the study area

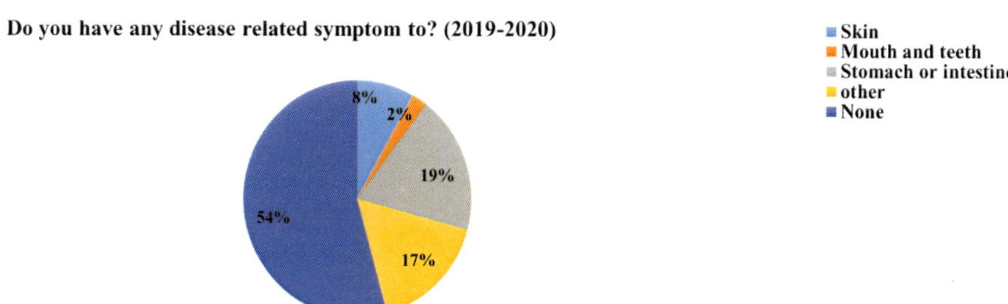

Figure 14.52. Percent distribution of disease in people in the study area

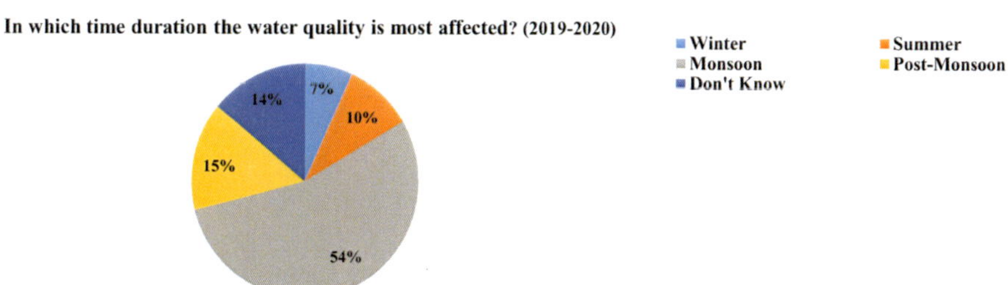

Figure 14.53. Percent distribution of seasonal variation of water quality in the study area

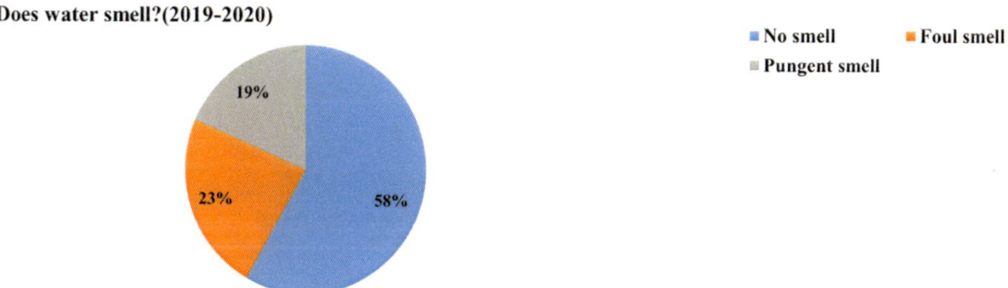

Figure 14.54. Percent distribution of water smell in the study area

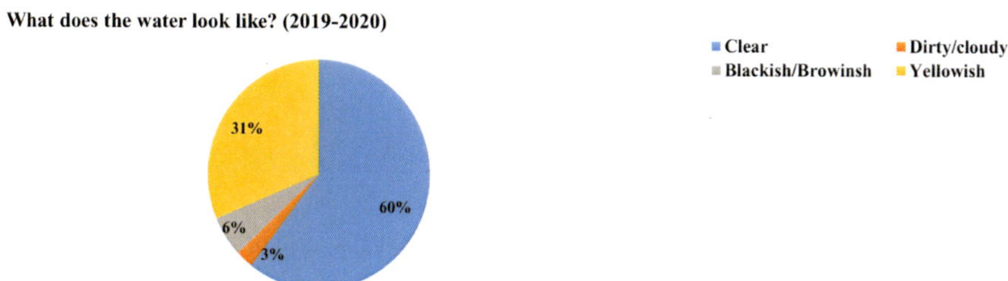

Figure 14.55. Percent distribution of water appearance in the study area

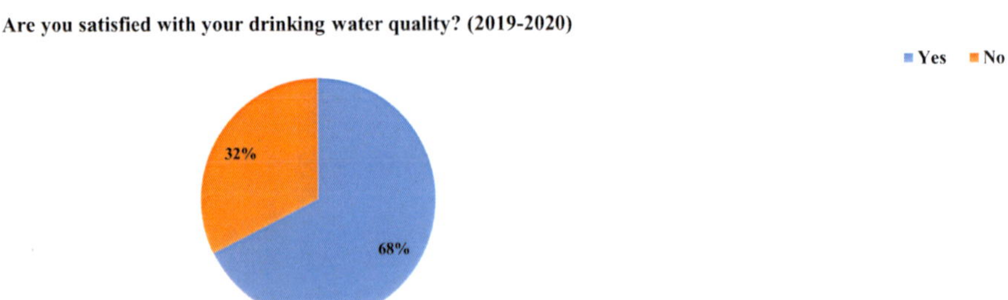

Figure 14.56. Percent distribution of water quality in the study area

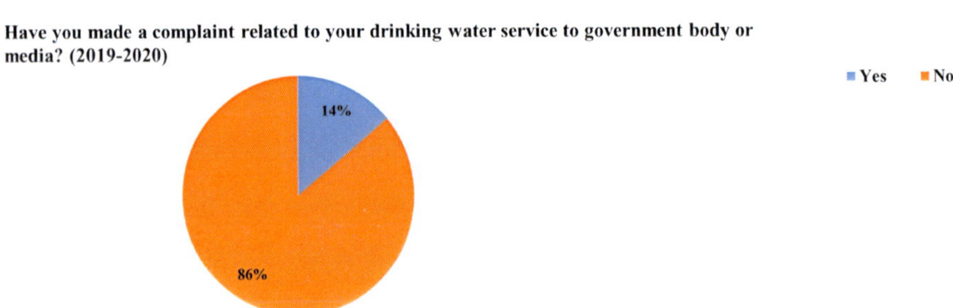

Figure 14.57. Percent distribution of complaint of drinking water quality in the study area

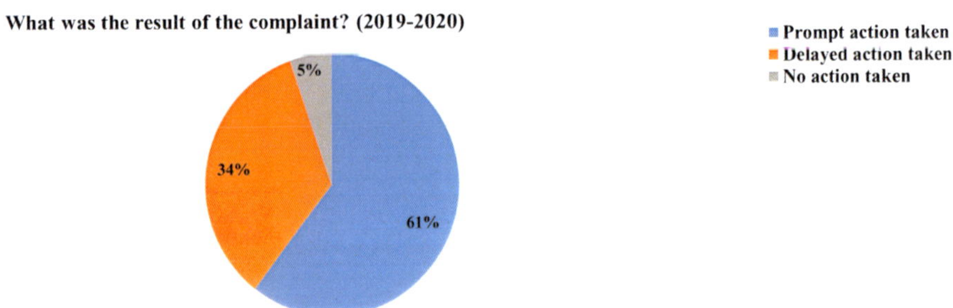

Figure 14.58. Percent distribution of complaint results of drinking water quality in the study area

Questionnaire Survey Analysis in Year 2020-2021

An offline questionnaire survey, composed of 15 questions was conducted to spread the awareness, assess the pollution status and impact of industrial wastewater quality on groundwater quality and human health during the year 2020-2021. 480 people had participated in survey in studied region in 2020-2021. A short survey to assess the personal information includes the person's name, age and occupation.

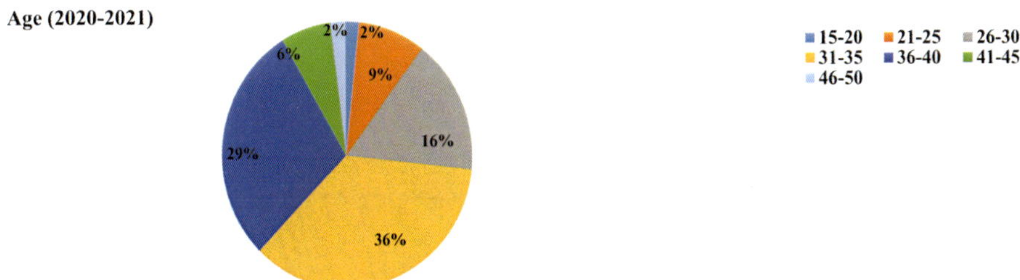

Figure 14.59. Percent distribution of Age in the study area

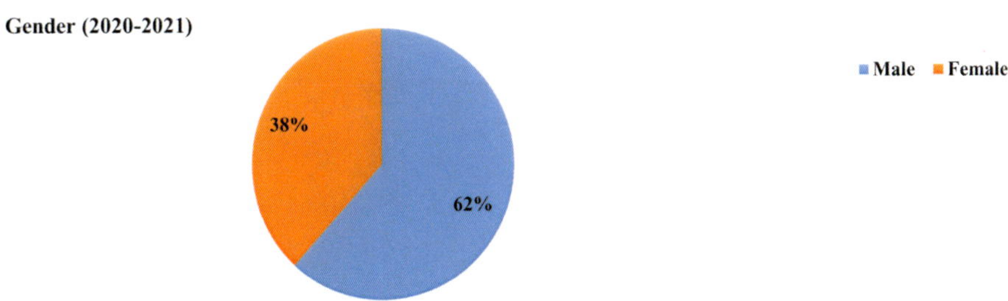

Figure 14.60. Percent distribution of Gender in the study area

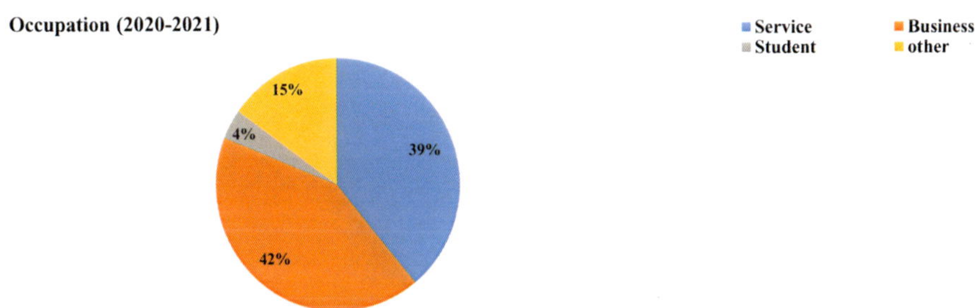

Figure 14.60. Percent distribution of occupation of people in the study area

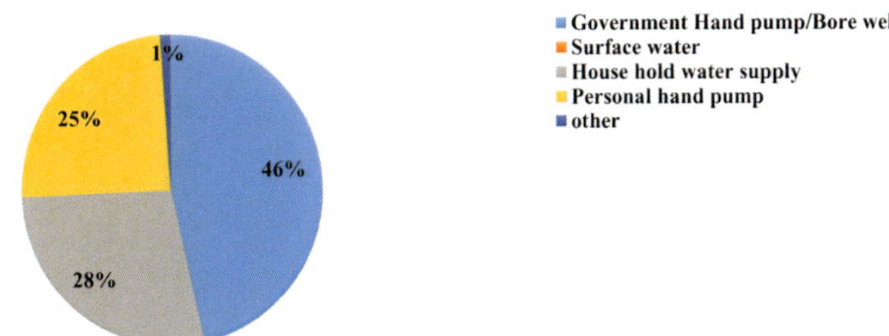

Figure 14.61. Percent distribution of drinking water resources for household use in the study area

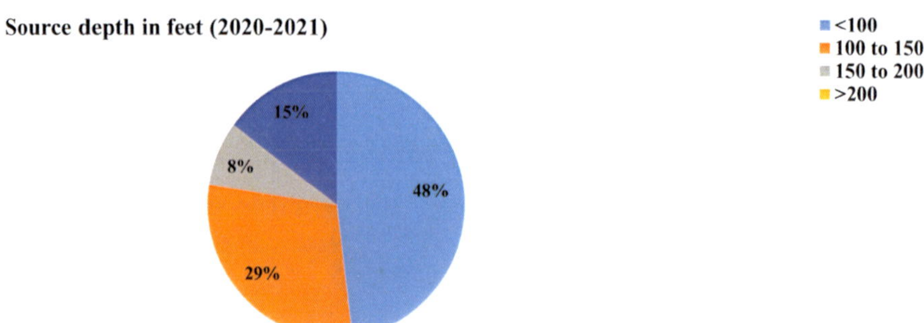

Figure 14.62. Percent distribution of drinking water source depth use in the study area

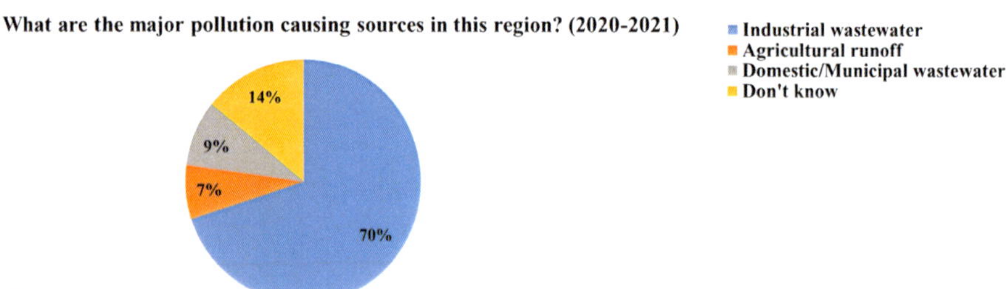

Figure 14.63. Percent distribution of major pollution causing sources in the study area

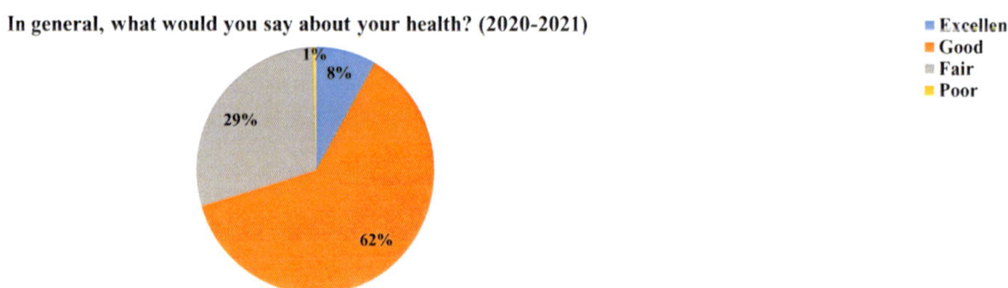

Figure 14.64. Percent distribution of health status of people in the study area

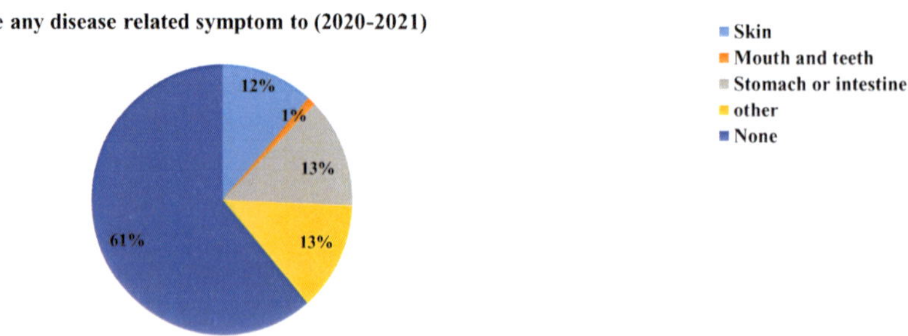

Figure 14.65. Percent distribution of disease in people in the study area

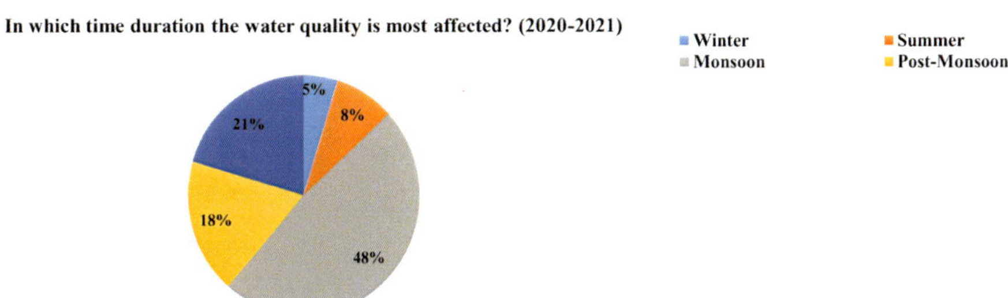

Figure 14.66. Percent distribution of seasonal variation of water quality in the study area

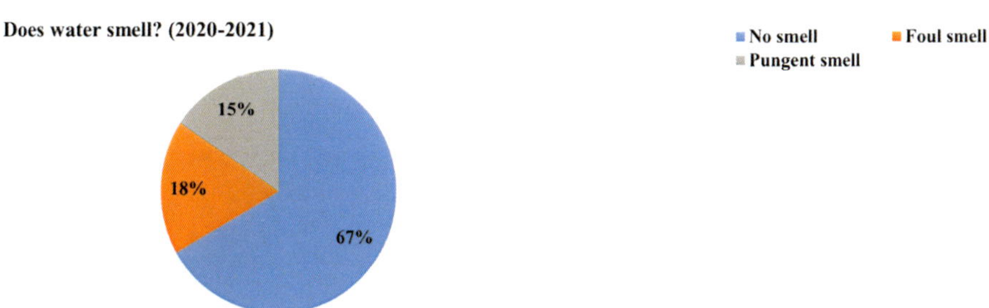

Figure 14.67. Percent distribution of water smell in the study area

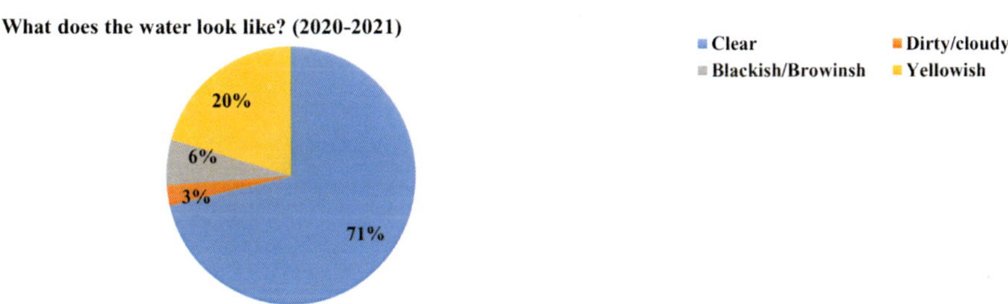

Figure 14.68. Percent distribution of water appearance in the study area

Figure 14.69. Percent distribution of water quality in the study area

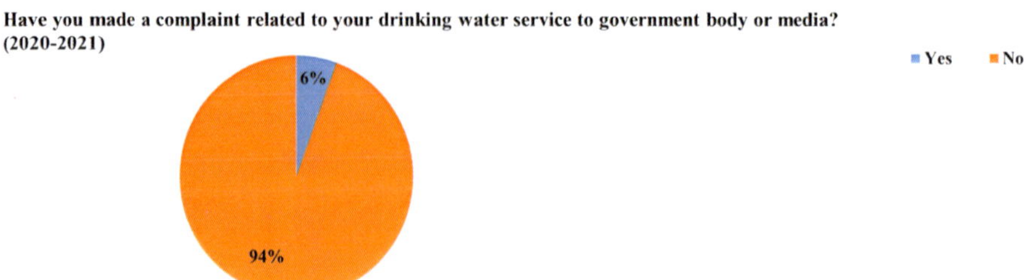

Figure 14.70. Percent distribution of complaint of drinking water quality in the study area

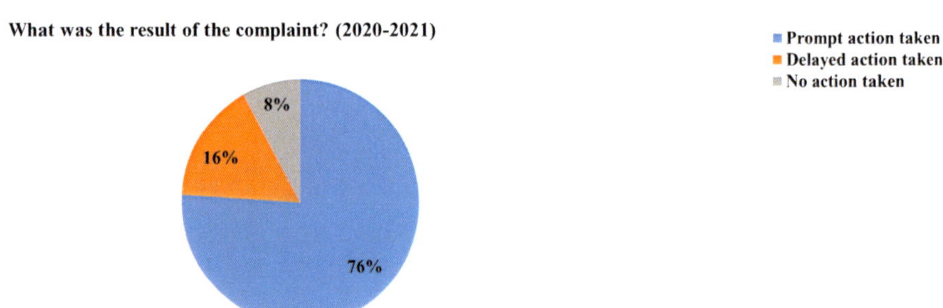

Figure 14.71. Percent distribution of complaint results of drinking water quality in the study area

People of age group (15-60 years) in 2019-2020 and (15-50 years) in 2020-2021 had participated in this survey. In this survey, 58% male and 42% female in the year 2019-2020 and 62% male and 38% female in the year 2020-2021 had participated in this survey. The people who answered these questions were 44 % servicemen's, labour, 7% student, 13% others such as maid, housewivesetc and 36% businessmen small vendors who runs their restaurants, and shops in the studied region in 2019-2020. In the year 2020-2021, 39% servicemen's, labour, 4% student, 15% others such as house helper, housewifes etc and 42% shop keepers, small vendors in the study area. In other questions, they were asked which water resources they use for their drinking and household purposes, water resources depth and what is the major pollution causing sources in their region etc. What problems they faced regarding their health issues such as skin, oral, gastrointestinal related or any other issues, and some other problems.

In this survey, 62 % in 2019-2020 and 70% in 2020-2021 people have found industrial sector is one of the biggest sources of pollution in context of water because of their wastewater discharge. Most of the participants have found excellent and good health conditions, but due to development of industrial sector and urbanization it might be chances of increasing cases of health risk in the studied region. Therefore,government agencies and nongovernmental organizations (NGOs) should strengthen theawareness and health concerns about drinking water quality among the general public. According to this survey, in the year 2019-2020, 54% of people were unaffected with any kind of disease but 19% of people have found gastrointestinal related issue and 8% people have skin-related problems. In monsoon season, 54% of people find water polluted in terms of visualisation and smell. They found water become more turbid like yellowish and cloudy in that period. 23% people are found foul smell in water. They found it unable to use for drinking purposes. In 2020-2021, 12% of people have found skin-related problems such as dryness, itching, acne, rashes and skin darkening. In 12%, most of them are female participants facing these issues. Some people also suffering from other issues like hair-related issues such as hair fall, itchy scalp, dryness. 13% people have also found some gastrointestinal-related issues.

In this survey, 68% (2019-2020) and 71% (2020-2021) people have found satisfied with the drinking water quality and 32% (2019-2020) and 29% (2020-2021) people said they are not much satisfied with the drinking water quality because of their health related issues and seasonal variations. In other questions, have you ever made a complaint related to your drinking water service to government body or other water related agencies such Jal sansthan or media. In the year, 2019-2020 and 2020-2021, 86% and 94% people never made a complaint about it but who people which are not satisfied with the drinking water services they made a complaint about this issues and they found some actions were promptly taken some were taken delayed. In this survey, some people were aware of the pollution regarding industrial wastewater, drinking water, waterborne impacts and most importantly the wastewater discharge openly in open surface, nalla or drainage. They were also asked if any pollution source opened in open land or agricultural areas. Most of the people answered yes and some of them even specify the location of them.

Many of the population had frequent waterborne and skin diseases, specifically in monsoon season followed by post-monsoon summer and winter season high in people living near the industrial areas or using the water sources for their drinking and household purposes. Thus it can be concluded that pollution is constantly increasing due to industrialisation and urbanization, deteriorating water quality, and a great risk to human health. If this situation is continuously going on in the future, there may be chances of higher risk and water borne diseases in human health. A better understanding of thefactors influencing public awareness of drinking water can contribute to improvements in water management, consumer services and water pollution accident prevention and control. Many factorsare involved in the public awareness of drinking water quality. Water sources, water treatment processes, and water supply networks can easily affect water quality and safety. There are limited studies on publicawareness about drinking water safety and the relationship between public awareness and the preventionof drinking water accidents in rural areas. The knowledge provided by this study will contribute to preventing drinking water contamination and improving water management; especially from the perspective of public participation. The study provides baseline information in policy framing to and enhance the public awareness level about water quality and subsequent minimization of water pollution accidents in this region. Moreover, it can also act as a baseline for similar worldwide studies regarding assessment and the improvement of public awareness levels about drinking water and the minimization of the occurrence of pollution accidents.

15

Case Study – Discussion

This chapter discusses results obtained during the study period through physico-chemical and heavy metal analysis of water, water quality indices, interpolation, Principal Component Analysis (PCA) and Correlation matrix (CM), irrigational parameters and health risk assessment. The study made it possible to identify more major factors affecting the water quality by comparing the data with those from national and international agencies. The objective of the research was then achieved by statistically enumerating the acquired values. The findings that are reported here will assist the controlling agencies in their future efforts to reduce water pollution at the designated site.

Physico-chemical variables

To assess the impact of growing industries on groundwater quality, heavy metal and physico-chemical characteristics have been investigated. Following is a discussion of the analysis of the results.

pH generally shows the alkaline and acidic nature of the water. It is vital because it significantly affects water quality by altering the alkalinity and solubility of hardness and metals (Sener *et al.,* 2017; Mitra *et al.,* 2017). According to Müller and Cornel (2017), Souza *et al.,* (2000), and Goswami *et al.,* (2000), irrigation water samples should have a pH range of 6.5 to 8.4. The pH of the sewage treatment plants and drinking water sources was found to be within permissible ranges as per BIS (2012) and the CPCB surface water discharge limit at the sampling sites. The contour map was constructed using the pH values at the precise sampling locations, and the contours were generated by interpolating data points. Low-pH water can dissolve or corrode metals and other materials, whereas high pH water tastes bitter (USGS, 2019). High pH levels are known to be caused by various contaminants and substances. Carbonate, bicarbonate, and hydroxide compounds present in the soil or bedrock sources dissolve and move with the water.

Electrical conductivity (EC; μS/cm) represents the current carrying capacity and presence of dissolved ions, i.e., carbonates, bicarbonates, and other minerals, in a water sample. Gupta *et al.,* (2017) also reported that anthropogenic activities enhanced the nutrient level, increasing the electrical conductivity. According to Ehya and Marbouti (2016) and Ravikumar and Somashekar (2017), a higher EC value suggests high salt and mineral content in groundwater with limited drainage, rapid intrusion, and discharges. The reverse is additionally accurate: low EC values are frequently associated with terrain, high runoff, low infiltration and recharge water types, and low salt enrichment. According to Kouser *et al.,* (2022), the salt enrichment of groundwater can be classified as type I if it is low (less than 1500 μS/cm), type II if it is between 1500 and 3000 μS/cm, and type III if it is higher than 3000 μS/cm. The majority of the groundwater samples in the research region are classified as type I, or water with minimal salt enrichment.

Temperature (T, °C) play an essential role in regulating nutrients as well as biological activity such as help to producers for food production through photosynthesis (Ali *et al.,* 2016). The annual mean of temperature values was recorded 23.87±0.11°C and 23.19±0.06 °C for 2019-20 and 2020-21, respectively. A similar temperature value was observed by several researchers in Uttarakhand (Anoop and Renu, 2014; Deoli *et al.,* 2017). Variations in groundwater temperature may be due to climatic variation and heat received from the Sun and the geothermal gradient due to conductive and convective motion received from the earth's interior.

Dissolved oxygen (DO, mg/L) is a vital parameter that regulates an aquatic system's physico-chemical and biological processes. It is used as an indicator in water quality assessment by showing the pollution load. The DO of the aquatic system depends on various environmental factors and hydrological characteristics. In 2019–20, wastewater minimum DO (0.24 mg/l) and highest DO (0.58 mg/l) were recorded in May and January, respectively. In the second year 2020–21, site IW1 recorded the lowest DO (0.25 mg/l) in June, whereas site IW2 recorded the highest DO (0.46 mg/l) in September. According to several studies, anthropogenic activity reduces the amount of dissolved oxygen in the water (Soo *et al.,* 2016; Bhatia *et al.,* 2018; Ogungbe *et al.,* 2020).

Total dissolved solids (TDS, mg/L) refer to the presence of ions soluble in water. The different kinds of activities that materialize in a specific location impact the values of TDS. In industrial wastewater, the annual mean of TDS values was recorded as 1059.46±18.63 mg/L and 1030.97±16.75 mg/L for 2019-20 and 2020-21, respectively. In groundwater, the annual mean of TDS values was recorded as 351.09±22.51 mg/L and 337.34±18.46 mg/L for 2019-20 and 2020-21, respectively. It has been noted that at all selected sampling sites, the TDS was recorded as high in the monsoon months compared to the summer, autumn, and winter months. The correlation between TDS and water turbidity is strong; if TDS concentrations rise, turbidity will follow (Raheja *et al.*, 2022; Panneerselvam *et al.*, 2023). Nearly all of the samples are below the permissible level (500 mg/L), according to the groundwater analysis results.

COD refers to the chemical oxidization of organic and inorganic matter. COD also reflects the industrial pollution load. In the monsoon, COD will be lesser in comparison to other seasons. The annual mean of COD values was recorded as 1084.18±19.62 mg/L and 1057.21±17.46 mg/L for 2019-20 and 2020-21, respectively, in the industrial wastewater. The annual mean of BOD values was recorded as 370.99±12.87mg/L and 353.45±19.25 mg/L for 2019-20 and 2020-21, respectively. When considerable concentrations of biologically resistant organic matter are present, COD readings are always greater than BOD values and may even be much higher. BOD is a sign that the wastewater contains organic contaminants. The monsoon's effluents have COD values as high as 900 mg/L, which may be caused by the high mineral content in their surface runoff. Bhatia *et al.*, (2018) also reported higher COD and BOD values in industrial wastewater.

Turbidity (TU, NTU) refers to the higher sediment and debris load that significantly affects the food chain, abundance, and diversity of aquatic organisms because of the lower sunlight penetration in aquatic systems (Verma and Saksena, 2010; Matta *et al.*, 2022). It is an ecological indicator parameter for assessing the water quality of an aquatic ecosystem. The annual mean of TU values was recorded at 1.09±0.15 NTU and 1.00±0.09 NTU for 2019-20 and 2020-21, respectively. The turbidity of all drinking water sources during the monsoon season was higher than the permissible limit of BIS (2012). Besides, the level of TU was higher in monsoon months at all selected sampling sites because of the heavy rainfall compared to winter and summer months. Studies conducted by Bhadula *et al.*, 2013; Rawat *et al.*, 2017 also reported a higher turbidity level in monsoon seasons at different sampling locations.

Total hardness (TH, mg/L) describes the type of geological strata that have come into contact with it. This is caused by the presence of anions such as carbonate, bicarbonate, chloride, and sulphate, as well as cations like calcium and Magnesium (Saha *et al.*, 2018; Raheja *et al.*, 2021; Panneerselvam *et al.*, 2023). Although hard water consumption may offer modest health advantages, scaling in pipes, boilers, cooling towers, and home use can result in major issues in domestic and industrial environments. The inhibition of the detergent reaction in water caused by the precipitation of magnesium and calcium salts such as chlorides, sulphates, and carbonates is known as water hardness. Both transient and persistent hardness are possible (Saha *et al.*, 2018). According to Singh *et al.*, (2020), the total hardness of groundwater may be divided into four categories: gentle (0–75), mild (75–150), hard (150–300), and very hard (> 300). The yearly mean of TH values for 2019-20 and 2020-21, were148.77±24.77 mg/L and 135.36±13.30 mg/L. Untreated sewage and industrial waste disposal may be blamed for excessive hardness (Anwar *et al.*, 2014).

Alkalinity (A, mg/L) is defined as the capability of an alkaline solution to neutralize an acid, as well as the combination of bicarbonates and carbonates. Alkalinity is an essential component for evaluating water quality since it includes essential elements like bicarbonates and carbonates (Panghal *et al.*, 2020). The annual mean of alkalinity values was recorded as 128.59±9.12 mg/L and 126.85±8.67 mg/L for 2019-20 and 2020-21, respectively. According to Kumar *et al.*, (2017), a high alkalinity range results from the interaction between water and minerals and various human activities such as urban drainage, erosion of soil, and wastewater from industries.

Bicarbonate (HCO_3^-, mg/l) is the most essential chemical component in water. Bicarbonate often results from the weathering of silicate rocks, but it may also arise from primary carbonate and calcareous rocks (Chandan, 2017), and it aids in giving groundwater an alkaline character (Ram *et al.*, 2021). The annual mean of HCO^-_3 values was recorded as 131.79±19.64 mg/L and 118.21±9.19 mg/L for 2019-20 and 2020-21, respectively. The researchers observed a similar temperature value (Kouser *et al.*, 2022; Goswami *et al.*, 2022). Groundwater containing higher bicarbonates and carbonates exhibits the major characteristic of mineral dissolution (Ghalib, 2017).

Nitrate (NO_3^-, mg/L) The presence of nitrate in groundwater as a soluble anionic form makes determining its concentration in drinking water crucial due to its detrimental effects on human health (Jayarajan and Kuriachan, 2021). Nitrate levels were less than the BIS permissible limits (45 mg/l) in groundwater evaluations (BIS, 2012). The nitrate concentrations in groundwater often result from anthropogenic sources such as agricultural operations (excessive fertilizer usage), septic tank output, incorrect animal waste disposal, and residential waste disposal (Perez *et al.*, 2019; Singh *et al.*, 2020). Several studies on nitrate that were conducted by Singh *et al.*, (2020), Vasanthavigar *et al.*, (2010) in Tamil Nadu, India, and Kaur *et al.*, (2017) in the south-western part of Punjab, India, revealed similar findings.

Sulphate (SO_4^2, mg/L) A range of minerals, including pyrite, gypsum and galena, that are natural sources of sulphate are present in the research area's geology. Ion exchange or oxidation-reduction processes and the location in the groundwater aquifer system increase the concentration of sulphate in groundwater (Gyeltshen *et al.*, 2019; Shawai *et al.*, 2019). The decomposition of sulphide minerals found in granite is one source of sulphate, along with precipitation, fertilizers, and other factors. The annual mean of SO_4^{2-} values was recorded as 13.61±1.72 mg/L and 13.18±1.72 mg/L for 2019-20 and 2020-21, respectively. Sulphate in groundwater may be due to geochemical weathering and the leaching process by which the sulphate in wastewater reaches the depth of the water table. Saleem and Mahmood (2016) and Kanagaraj and Elango (2016) found a similar sulphate finding in their research.

Chloride (Cl^-, mg/L) is the salt of sodium (NaCl), potassium (KCl), and calcium (CaC12), abundantly dispersed in nature. Chloride is a significant anion present in groundwater and can come from various sources, including leaching, the weathering of various minerals, and specific human causes. As a result of its high concentration in groundwater, it imparts a salty flavor and negatively affects human health by causing kidney stones (Mohamed *et al.*, 2019). The annual mean of Cl^- values was recorded as 16.02±2.41 mg/L and 15.37±2.02 mg/L for the year 2019-20 and 2020-21, respectively, indicating that the groundwater is safe to drink and within allowed limits (BIS, 2012; WHO, 2017).

Potassium (K^+, mg/L) is the 7th essential parameter not present in higher amounts and is rarely found free in nature. Due to the solubility of potassium-rich rocks, typically found in a range of rocks, groundwater has the greatest potassium concentration. According to Mallick *et al.*, (2018), the main sources of potassium in groundwater are industrial chemical emissions and excessive fertilizer usage. Plagioclase and potassium feldspars weather naturally and dissolve in the area to produce Na^+ and K^+ (Dudeja *et al.*, 2011).

Sodium (Na^+, mg/L) is the sixth essential element found almost in all types of natural water. Generally, it is found in lower amounts as compared to other ions such as calcium and magnesium, and it is highly soluble in water. The higher concentration of sodium in water is not suitable for irrigation. It leads to soil erosion, and the use of water with high sodium content becomes unfit for human consumption. High salt levels (> 200 mg/L) in people can lead to serious health problems such as hypertension and abnormalities of the kidney and nervous system (WHO, 2017).

Phosphate (PO_4^{3-}, mg/L) is prevalent in groundwater due to soaps, surfactants, manures, and biological activity (Selvaraj *et al.*, 2020). The annual mean of PO_4^{3-} values was recorded at 0.18±0.02 mg/L and 0.18±0.03 mg/L for 2019-20 and 2020-21, respectively. The use of detergents can significantly increase the phosphate content of water. Most of the time, home sewage and overflow from rural areas are the sources of this phosphorus.

Acidity results from chemicals that hydrolyze and release hydrogen ions. Another stated that a material is considered acidic if it contains a hydroxyl anion. The annual mean of acidity values was recorded as 32.88±6.16 mg/L and 30.72±5.24 mg/L for 2019-20 and 2020-21, respectively. Considering the weathering resistance of the sandy soils and sediments making up the watershed surface geology, the water acidity could be caused by a lack of weathering products to neutralize the inputs of acidic atmospheric deposition.

Calcium (Ca, mg/L) and Magnesium (mg/L) are essential elements for the human body's bone growth. In contrast, Calcium is easily available in all types of natural water and is easily soluble, while magnesium is present in lower concentrations than calcium because of the slow formation process. The annual mean of Ca^{2+} values was recorded at 74.42±21.62 mg/L and 66.43±13.41 mg/L for 2019-20 and 2020-21, respectively. The annual mean of Mg^{2+} values was recorded as 32.05±9.80 mg/L and 31.10±11.04 mg/L for 2019-20 and 2020-21, respectively. When calcium levels are at their maximum, it can lead to significant health issues such as kidney stones and heart problems (WHO, 2017). The primary source of carbonate (Ca^{2+}) in groundwater is the disintegration of carbonate through sedimentary rocks and minerals, including calcite, dolomite, and limestone. The hydrolysis of the magnesium-rich minerals $CaCO_3$ and Ca-Mg $(CO_3)^2$, as well as some harsh agricultural practices, can all play a significant role in the development of calcium and magnesium weathering processes in silicate minerals (Alsuhaimi *et al.,* 2019; Sharma and Chhipa, 2016; Sharma *et al.,* 2019).

Potential Toxic Elements (PTEs)

During the study period from 2019-2020 to 2020-2021, Eight heavy metals such as iron (Fe), Zinc (Zn), Lead (Pb), Chromium (Cr), Nickel (Ni), Manganese (Mn), Copper (Cu), and Cadmium (Cd) were analysed for assessing the concentration of heavy metals in industrial waste water and groundwater in the Dehradun industrial region. Heavy metal contamination is not a modern issue emerging from industrialisation. PTEs are pollutants that are created by a variety of anthropogenic activities. When exposed to biological ecosystems and people, ecotoxicity, high biomagnification, and carcinogenicity can be lethal (Shan *et al.,* 2021; Briffa *et al.,* 2020). Widespread industrialization, insufficient wastewater treatment, and a lack of public awareness are the major causes of PTEs in groundwater. The discussion of selected heavy metals parameters is discussed below:

Zinc (Zn) is an active heavy metal due to its many soluble forms in groundwater (Malecki *et al.,* 2017). The annual mean of Zn values was recorded as 0.348±0.05 mg/L and 0.342±0.04 mg/L for 2019-20 and 2020-21, respectively. Anthropogenic activities such as waste incineration, metal manufacturing operations, and some release from the E' 'Earth's crust may be the sources of Zn in specific locations (Malecki *et al.*, 2017; Oyem *et al.*, 2015). The research ' 'region's zinc concentrations were within the advised 5 mg/L level (BIS, 2012). The poor rate of zinc penetration into groundwater may be due to its limited adsorptive capabilities in soil. The monsoon has higher Zn concentrations than other times of the year. Furthermore, several studies (Bharti *et al.*, 2013) revealed that the Zn levels in groundwater in industrial districts and the areas around them were lower than allowed.

Iron (Fe) is the most plentiful element on Earth's crust and is found in all types of natural water. The annual mean Fe values were recorded as 0.278±0.08 mg/L and 0.285±0.07 mg/L for 2019-20 and 2020-21 respectively, in groundwater. In industrial wastewater, the annual mean Fe values were recorded as 5.09±1.15 mg/L and 5.02±1.07 mg/L for 2019-20 and 2020-21, respectively, which is above the prescribed range by CPCB. Winter, summer, and autumn had the lowest average iron levels over the study period, whereas the monsoon had the highest. During the monsoon season, it was most likely caused by toxic materials eroding and dissolving the castings of hand pumps and tube well pipes. According to research by Hu *et al.*, (2018) and Xu *et al.*, (2020), corrosion of the subterranean water pipeline and metal surfaces caused colored water with reddish-brown precipitates to be discovered in the water, indicating high iron content at these sources. The water at these sources has a significant amount of iron because of the corrosion of ancient metal pipes and other metal water supply and lift equipment.

Cadmium (Cd) is considered a highly toxic element that threatens the growth of aquatic flora and fauna as well as human health. Aquatic animals easily absorb and accumulate it (Rzetala, 2016). Cd is regularly discovered in the environment as a result of mineral extraction, utilizing fossil fuels for smelting, smoldering biodegradable waste, producing nickel-cadmium power sources, and other chemical products (Agency for Toxic Substances and Disease Registry, 2012). Acceptable Cd amounts were found in every groundwater sample from our research area. According to Wu *et al.*, (2016), the human carcinogen Cd may cause damage to the kidneys, lungs, skeletal system, and respiratory system even at low exposure levels. Copper (Cu), is prevalent in human body tissues and necessary for forming red blood cells, maintaining nerve cells, and maintaining the immune system. Cu contents in the research area consistently remained below the 2 mg/L suggested limit (BIS, 2012). A poisonous and soluble element is nickel, whose presence in the groundwater column is due to hydro-geochemical and man-made activities. According to Patel *et al.*, (2018) and Anyanwu and Nwachukwu (2020), nickel ferrous limonite and hydrous nickel silicate (granite) are the two most prevalent Ni compounds. Batteries that have been abandoned, power plants, the glass and ceramics sector, color manufacturers, and waste production are all common sources of nickel pollution (Govil *et al.*, 2008). A lower Ni concentration was found in this experiment.

Manganese (Mn) is one of the crucial elements naturally occurring in groundwater due to its importance in promoting healthy human growth. Both geogenic and human-caused contamination of groundwater with Mn are possible. The Mn content was acceptable in all groundwater samples in this research location. The annual mean of Mn values was recorded at 0.004±0.00 mg/L for 2019-20 and 2020-21, respectively. Chromium (Cr) is frequently found in chromite and magnesiochromite minerals, which are abundant in the environment. Because of its hardness and corrosion resistance, Cr is commonly used in the industrial, medicinal, and ceramic sectors (Wugan and Tao, 2012). Water contains typically less than 0.05 mg/L of Cr. Every sample from the present study area had Cr concentrations less than the allowed limit. Lead (Pb) is a vital heavy metal found in a small amount in the natural environment. In the natural environment, the presence of lead is increased due to the anthropogenic activities of mining, manufacturing, and burning fossil fuels (Tiwari *et al.*, 2016). The AAS was unable to identify the levels of lead in groundwater. In industrial wastewater, the annual mean of Pb values was recorded as 0.08±0.02 mg/L and 0.07±0.02 mg/L for the year 2019-20 and 2020-21.the highest levels of Pb and Hg are linked to emissions from industries as well as atmospheric input and runoff from vehicle emissions into the environment. Pregnant women, children under six, and fetuses are the main groups impacted by high Pb concentrations. Lead harms the central nervous system, causing symptoms such as diminished memory, illusions, irritability, diminished attention span, headaches and migraines, and muscle tremor (WHO, 2017).

Interpretation of Multivariate Statistical Models

Pollution Sources Spportionment by PCA and Correlation Matrix of Water Quality Parameters of Groundwater Sources

The PCA calculation helps to find those parameters that are more concentrated and influence the quality by condensing the vast data set into a defined and standardized form. The communalities values were more significant than 0.5 for each parameter. Three principal components (PCs) were determined to be essential based on the eigenvalue criterion since their eigenvalues were larger than one. With 65.52% of the variance, PC1 comprised a considerable portion of the variables and had a load on EC, temperature, TDS, TU, TH, alkalinity, acidity, phosphate, bicarbonate, nitrate, sulphate, chloride, potassium, sodium, calcium, magnesium, and manganese. This demonstrated the heterogeneous source of contamination from anthropogenic and natural processes, such as runoff from ore producers, air deposition, soil erosion or weathering, and bio-fertilizers (Sharma *et al.,* 2015; Singh and Pandey, 2019; Bowes *et al.,* 2020).

According to Singh *et al.,* (2005), the elevated TDS levels reflect the inflow from household sewage. The loading weights of PC2 reflected domestic sewage discharge and industrial wastewater. The PC2 explains 22.87% of the variation and positively correlates with pH, zinc, iron, nickel, and chromium. Acidity, iron, cadmium, copper, and SAR all contributed significantly to the synthesis of PC3 and strongly correlated with one another, accounting for 6.99% of the variation. Sandstone, limestone, and quartzite are all present in the Doon gravels, serving as the region's primary aquifers (Mandal *et al.,* 2021). In investigations conducted in 2019 by Das *et al.,* a higher abundance of carbonates and cations was also found. They concluded that both natural and human interventions may cause the existence of these elements in higher proportions. Additionally, Nagaraju *et al.,* (2014) and Singh *et al.,* (2000) reported that anthropogenic activities, such as the application of chemical fertilizers, industrial discharge, and untreated discharge of sewage water in the agricultural field, have resulted in the presence of cations like calcium and sodium and anions like sulphate and nitrate.

The link between the water quality metrics was determined using Pearson's correlation analysis. A complete linear connection between the variables may be seen if the "r" value is discovered to be between +1 and -1 (Meena *et al.,* 2016; Lakshmi *et al.,* 2021). TH showed a highly positive correlation with Ca^{+2} (r= 0.99), Mg^{+2} (r= 0.98), SO_4^{2-} (r=0.97), HCO^-_3(r=0.90) at sampling site GW1, alkalinity showed a positive correlation with HCO^{-3} (r=0.98), SO_4^{2-} (r=0.95), Cl^- (r=0.91), K^+(r=0.91) showing a significant impact on the groundwater quality compared to other physical parameters. This highly substantial positive correlation shows that rock-water interaction, weathering, and precipitation of minerals affect groundwater. A high correlation between TH and Mg^{2+}, according to similar research by Chakraborty *et al.,* published in 2021, suggests that the carbon dissolving process at the rock-water interface is accelerated. It was discovered that Fe and TU are closely linked. The corrosion of underground pipes and metal surfaces at a few water sources precipitated iron corrosion products and rendered them water-soluble. This resulted in water that was turbid and yellow in color. The strong correlation between Fe and Cl^- suggests that ferrous metal pipes and other equipment rust in the presence of chloride ions.

WQI Interpretation

The WQI findings have proven to be highly helpful in determining the groundwater safety status for drinking (Yadav *et al.,* 2010; Adimalla and

Qian, 2019; Verma *et al.*, 2020). Based on the Contamination Index (C_d), Comprehensive Pollution Index (CPI), and Heavy Metal Pollution Index (HPI), assessments of the water quality of the examined water samples for suitability for drinking purposes have been made. Groundwater quality and pollution levels may be assessed using the HPI, CPI, and C_d methods (Kumar *et al.*, 2019; Shil and Singh, 2019).

Contamination Index (C_d): The contamination index (C_d) of the groundwater for all the sampling locations is shown in Table 4.44. During the study period, at sampling sites GWCS, GW2, and GW3, the groundwater samples had a low degree of contamination, and at sampling sites GW1, GW4, and GW5, the groundwater samples had a medium degree of contamination. In this study, some parameters were found to exceed the standards and have the most influence on groundwater quality. The low indices values of the analyzed samples were mainly due to the lesser concentration of the metals (Rahman *et al.*, 2022). Areas with high levels of potentially harmful geogenic elements or anthropogenic pollutants can be the reason for medium contamination, as measured by the contamination index C_d (Mester *et al.*, 2018).

Comprehensive Pollution Index (CPI): The comprehensive pollution index (CPI) of the groundwater for all the sampling locations is shown in Table 4.45. During the study period, sampling sites GWCS were found in the sub-clean category, sampling sites GW2, GW3, and GW4 were found in the slightly polluted category, and sampling sites GW4 and GW5 were found in the medium category. The current research's findings on CPI show that the groundwater quality is contaminated as a result of metal concentrations in industrial effluents and wastewater that have been rapidly diluted over a short distance. Natural processes, such as the water-soil interface, and human activities are also involved in the contamination of metals (Egbueri *et al.*, 2020; Ram *et al.*, 2021; Karnena *et al.*, 2022).

Heavy Metal Pollution Index (HPI): The heavy metal pollution index (HPI) of the groundwater for all the sampling locations is shown in Table 4.46. During the study period, control sites GWCS and GW3 were found in the low contamination category, and GW1, GW2, GW4, and GW5 were found in the medium contamination category. Heavy metal poisoning in groundwater is a serious hazard because of its toxic nature, capacity for bioaccumulation, and amplifying effects at tropical levels. Heavy metal poisoning can alter a groundwater system's natural characteristics, posing a serious health danger to individuals who rely on it (Bhuyan *et al.*, 2019; Matta *et al.*, 2022). Some parameters are found to exceed the standards and have the most influence on groundwater quality. The HPI values and subsequent ratings and grades provide quite good evidence to either continue to use the water of the studied site or source as a drinking water source or discontinue its use as a source of drinking water for public supply. According to related research by Hasan and Rai (2020), the variance in HPI values in more industrialized regions may be caused by significant pollution from human activities and industrial wastes. The decreasing water quality of these sample locations may be due to the extensive and indiscriminate application of chemical fertilizers and the leaching of solid waste. Many researchers also concluded that leaching processes and agricultural practices like chemical fertilizers and untreated sewage water affect groundwater (Gao *et al.,* 2022; Singh *et al.,* 2020).

Irrigation Suitability of Groundwater Quality for Both the Years

The amount of dissolved ions contained in groundwater determines its appropriateness for irrigation. According to Snousy *et al.,* (2021) and Shan *et al.,* (2021), an excessive ion concentration can impact the soil, plant development, and agricultural output. In the study area, groundwater is the primary water source for irrigation, and recently increased agricultural activities, as well as excessive use of chemical fertiliser, pesticides, and livestock waste, have hurt the quality of groundwater (CGWB, 2013; Jasrotia *et al.,* 2018; Raheja *et al.,* 2022). Electrical conductivity, sodium percentage, sodium adsorption ratio, Gibbs ratio, and magnesium hazard were examined to determine whether groundwater was suitable for irrigation. Typically, the sodium adsorption ratio (SAR) represents the sodium/alkali risk. Using irrigation water with a high salt content may raise the sodium content of the soil; as a result, the soil's permeability and texture may be changed, making it difficult to till and unsuitable for seedling emergence (Kouser *et al.,* 2022). In both years, the SAR values in the study region were less than 10, indicating irrigation suitability. The Na^{+} ions in the water can instantly displace the Mg^{2+} and Ca^{2+} ions existing in the soil when groundwater with a high SAR value is introduced. This will make it difficult to develop a solid soil aggregate and, as a result, severely damage the soil's structure. This can further reduce the infiltration rate or water permeability into the soil, resulting in a significant decrease in the overall crop yield (Wang *et al.,* 2022). The class/type of water used for irrigational purposes can also be determined graphically by plotting the obtained values of EC and SAR on the US Salinity Laboratory (USSL) diagram. A higher concentration of sodium in water affects the soil's characteristics and reduces its permeability (Adimalla *et al.,* 2018; Fang *et al.,* 2020). The salinity hazard (C) is classified into 4 subzones: very low salinity (C1: 250 μS/cm), medium salinity (C2: 250 to 750 μS/cm), high salinity (C3: 750 to 2250 μS/cm), and very high salinity (C4: above 2250 μS/cm). Similarly, the sodium hazard is classified into 4 subzones: low sodium hazard (S1: < 10), medium sodium hazard (S2: 10–18), high sodium hazard (S3:18–26), and very high sodium hazard (S4: > 26). In the study area, most of the samples fall under the C1 S1 category of the USSLS diagram, indicating low conductivity and very low sodium hazards, indicating that the groundwater of the study area is suitable for irrigation. Groundwater with high salinity is unsuitable for plants and poses a salinity hazard. Water with a low EC has a significant impact on crop productivity. In the study area, all samples fall under the in good category, and hence, groundwater is suitable for the growth of crops. Magnesium is regarded as a necessary nutrient for plant growth. Lower magnesium levels can cause plant growth to slow, affecting crop yield in the long run. As a result, the magnesium content of water is likely to influence its suitability for irrigation. A value of MAR > 50 is unsuitable for agriculture as it adversely affects crop yield and quality. In the present study, GWCS, GW1, and GW2 had a MAR value less than 50 and were suitable for irrigation. The remaining two samples, GW4 and GW5, yielded MAR values greater than 50 and were considered unsuitable for irrigation. The Gibbs plot shows that rocks dominate the research region's whole sample. The Gibbs diagram is drawn (Figure 4.41 c,d) to explain hydrochemical faces and aquifer lithology (Gibbs, 1970; Raheja *et al.*, 2022; Zhang *et al.*, 2020).The cation or anion ratios are measured on the horizontal plane, and the TDS is shown on the vertical plane. In these graphs, the water chemistry is primarily formed by precipitation dominance, rock dominance, and evaporation dominance. The majority of the groundwater samples were taken in an area where rocks predominated. This indicates that the main processes dictating groundwater chemistry in the study region are leaching and rock weathering.

Health Risk assessment on human health

The population boom, urbanization, and industrialization seriously threaten human health. According to several studies (Liu *et al.*, 2022; Ramalingam *et al.*, 2022; Gao *et al.*, 2020; Adimalla *et al.*, 2020), the demand for groundwater is rising every day. Constant groundwater drinking negatively affects human health, including carcinogenic and non-carcinogenic dangers. The HHRE is a useful tool for assessing how eating polluted groundwater could affect people's health. In contrast to HQ, which evaluates the potential risk of exposure to heavy metals to human health, CDI and DAD measure individuals' daily exposure and dermal exposure to metal contaminants in mg L^{-1} day^{-1}. Values greater than one (>1) are regarded as detrimental. Similarly, HI assesses the risks caused by specific heavy metals. The HI and HQ values for the two groups are summarised in Tables 4.47 to 4.50. Since the HQ and HI values for heavy metals were within acceptable ranges (1), there is no immediate danger to human health.

The HQ and HI values of children for each metal were generally observed to be higher than the HQ and HI values for adults at each location. The HI_{CDI} (2019-2020) value for adults obtained by this study ranged from 3.34E-02 to 6.89E-02. For children, the HI_{CDI} value ranged from 1.82E-01 to 3.75E-01. The HI_{DAD} (2019-2020) value calculated for adults at all the sampling locations ranged from 1.23E-02 to 8.25E-03. For children, the HI_{DAD} value varied from 1.23E-03 to 9.11E-02. The HI_{CDI} (2020-2021) value for adults obtained by this study ranges from 3.38E-02 to 6.64E-02. For children, the HI_{CDI} value was found to range from 1.84E-01 to 3.61E-01. The HI_{DAD} (2019-2020) value calculated for adults at all the sampling locations varied from 1.23E-02 to 8.26E-03. For children, the HI_{DAD} value was found to be 1.37E-02 to 9.10E-02. The results showed no threat to human health from GW consumption-related heavy metal exposure, and both 'groups' HI_{CDI} and HI_{DAD} values were below one. Children may be more susceptible than adults to specific contaminants. For instance, their developing bodies may be more sensitive to the effects of particular substances or contaminants. The pollutants may originate from natural sources, such as minerals leaching from rocks and soil, or from human activities like agricultural practices, sewage discharge, and industrial wastewater discharge. There are several ways that pollutants contribute to health problems. Fe, Cr, Mn, and Cd had the lowest non-carcinogenic risk contribution to adults and children, whereas the remaining metals had the lowest overall HI impact. Higher cancer rates (Tirkey *et al.*, 2017), neurologic consequences, and behavioural repercussions are some of the non-carcinogenic and carcinogenic effects of zinc (Zn) and iron (Fe) on or within the human body (CDC, 2012; Matta *et al.*, 2022; Zeng *et al.*, 2015).

Health Risk assessment by Questionnaire survey

At the national, regional, and local levels, access to clean drinking water is essential for development and health (WHO, 2022). In the last several decades, drinking water contamination incidents have occurred often in India and many other developing nations (Sridhar *et al.*, 2021; Wang *et al.*, 2018; Osta *et al.*, 2022; Gevera *et al.*, 2022; Hussain *et al.*, 2023). Government, business, academia, and non-profit organisations in various sectors have made significant efforts to address the problems posed by drinking water contamination incidents. Public engagement in environmental conservation been particularly established in India, nevertheless, due to how society is governed. As municipalities and educators have discovered, it is not sufficient to merely offer customers clean drinking water. According to Doria (Doria, 2009; Nasir *et al.*, 2019; Yang *et al.*, 2021), the public at large plays a significant role in water management. The World Health Organisation's drinking water quality recommendations include public acceptance of the water as well (WHO, 2022). Therefore, it is important to consider how the general population feels about drinking water safety and contamination incidents (Mahler *et al.*, 2015; Chen *et al.*, 2022).

Properly understanding environmental concerns is essential for the effectiveness of public environmental engagement (Khalid *et al.*, 2022; Wang *et al.*, 2018). Six sample locations in the Dehradun district were chosen as pilot sites for the inquiry to gauge public perceptions of the safety of drinking water and water pollution incidents in industrial regions. This study conducted a questionnaire survey, composed of 15 questions, in these sampling locations. 960 people (480 people in both years) participated in this survey in 2019-2020 and 2020-2021. A short survey to assess the personal information includes the name, age, gender, and occupation of the person. People of age groups (15-60 years) had participated in this survey. 58% male and 42% female in 2019-2020 and 62% male and 38% female in 2020-2021 participated in this survey. The people who answered these questions were 44 % servicemen, 7% laborers, 7% students, 13% others such as maids, housewives, etc., and 36% businessmen and small vendors who run their restaurants and shops in the studied region in the year 2019-2020. In this survey, 62% and 70% of people found the industrial sector to be one of the biggest sources of pollution in both years in the context of water because of their wastewater discharge. Most of the participants have found excellent and good health conditions, but due to the development of the industrial sector and urbanization, there might be chances of increasing cases of health risk in the studied region. Numerous studies have evaluated the satisfaction of customers concerning drinking water quality (Praveena *et al.*, 2021; Alameddine *et al.*, 2017). The public impression of drinking water quality has been shown to be largely influenced by a variety of factors, including the accessibility and safety of water sources, taste, and attitudes towards chemicals frequently present in water (Wright *et al.*, 2012; Hussain *et al.*, 2023; Gevera *et al.*, 2022). In the monsoon season, 54% of people find water polluted in terms of visualization and smell. They found that water became more turbid-like, yellowish, and cloudy during the monsoon period. 23% of people find foul smells in water. They found it unable to be used for drinking purposes. Water quality and human well-being are closely related, and having access to clean water is crucial for good health. Many of the population had frequent waterborne and skin diseases, especially in the monsoon season followed by the summer, autumn, and winter seasons, which were comparatively high among people living and working near industrial areas or using them for drinking and household purposes. All respondents showed interest in knowing about drinking water quality, the effects of low water quality, and improved preventative strategies after learning about the health consequences linked to their drinking water during our survey. All survey respondents provided accurate responses, which is one of the study's main presumptions. Some responses provided by the participants that may not be accurate but are appealing to them may constitute a significant research constraint. For instance, most participants worried that their water sources may be closed if they were discovered to contain potentially dangerous substances; thus, they might have stated that they were satisfied with the taste and quality of their water. In addition, drinking dirty water was believed to be the only source of the diarrhoea and gastrointestinal discomfort that some participants had experienced. In such situations, bacterial contamination and unsanitary conditions might still occur. Future research might focus on a larger or more varied sample group to properly analyze and eliminate any sampling or statistical flaws.

As a result of these adverse effects, there has been an alarming increase in the number of people suffering from other types of problems. Thus, it is strictly recommended that regular water quality monitoring and awareness of emerging contaminants be provided to the general public. The government is definitely spending a lot, but it should be checked to see how far it benefits sustainable development. According to research (Wang *et al.,* 2018; Nasir *et al.,* 2019), a small amount of information and a few facts are insufficient to alter public opinion significantly. According to several studies (Darko *et al.,* 2022; Eilam and Trop, 2014; Hussain *et al.,* 2023), enhancing community environmental knowledge and comprehension is essential to the efficacy of avoiding pollution. Promoting household water treatment, household decisions on drinking water sources, and avoiding water contamination incidents all depend on the public understanding of drinking water safety (Wright *et al.,* 2012; Praveena *et al.,* 2021). There is little research on drinking water safety and the connection between public knowledge and the prevention of drinking water accidents in industrial areas. To analyze information on public skills and mindsets regarding drinking water and water pollution, this research was created to look into the facts of the public awareness of the safety of drinking water and contamination of water accidents in the Dehradun region of Uttrakhand state. These findings will help decision-makers increase their monitoring of drinking water quality to safeguard the water supply's safety. These findings will help decision-makers increase their monitoring of drinking water quality to safeguard the water supply's safety. Additionally, it provides them with information on how to improve emergency response for incidents involving drinking water contamination, boost education, and expand understanding about drinking water safety. Public satisfaction might be raised by making the significance of the yearly drinking water quality report supplied to the public by the local government more widely known via local print and television media.

16

Case Study – Conclusion

The present investigation findings provide essential information and useful insights about the water quality status of the Dehradun industrial region of Uttarakhand State and the possible pollution sources. A comparable approach of commonly used C_d, CPI, and HPI worked well in elaborating on the seasonal variation in water quality. The WQI estimation revealed that the groundwater quality at sampling sites GW1, GW2, GW4, and GW5 was moderately contaminated in both years during the research period. The WQIs go into detail regarding the way anthropogenic and industrial activity affect groundwater quality and the way that contamination occurred. Groundwater in the research location is suitable for agricultural growth because all samples fulfil the criteria for "good" irrigation appropriateness. The two samples, GW4 and GW5, exhibited MAR values above 50, rendering them inappropriate for irrigation. The applicability of multivariate statistical approaches provided the lesser variables to reduce the project's cost-effectiveness. The results showed no threat to human health from GW consumption-related heavy metal exposure, and both 'groups' HI_{CDI} and HI_{DAD} values were below one. Children may be more susceptible than adults to specific contaminants. For instance, their developing bodies may be more sensitive to the effects of particular substances or contaminants. According to a questionnaire survey, the majority of participants reported excellent and good health; however, because of the growth of the industrial sector and urbanization, there may be more instances of health risk in the area under study. In order to ensure the security of the water supply, decision-makers will be able to better monitor the quality of the water being used for drinking. It also gives them knowledge on how to enhance education, expand knowledge about drinking water safety, and improve emergency response for occurrences involving contaminated water. The current study can be useful in educating the people of Uttarakhand about maintaining their health and hygiene.

Bibliography

Abbasi, T., & Abbasi, S. A. (2012). Water quality indices. Elsevier Science

Adimalla, N. (2020). Spatial Distribution, Exposure, and Potential Health Risk Assessment from Nitrate in Drinking Water from Semi-Arid Region of South India. Hum. Ecol. Risk Assess. 26, 310–334

Adimalla, N., & Qiana, H. (2019). Groundwater quality evaluation using water quality index (WQI) for drinking purposes and human health risk (HHR) assessment in an agricultural region of Nanganur, South India. Ecotoxicol Environ Saf. 176:153–161. https:// doi. org/10.1016/j.ecoenv.2019.03.066

Adimalla, N., & Taloor, A. K. (2020). Hydrogeochemical investigation of groundwater quality in the hard rock terrain of South India using Geographic Information System (GIS) and groundwater quality index (GWQI) techniques. Groundwater for Sustainable Development, 10, 100288

Adimalla, N., Li, P., & Venkatayogi, S. (2018). Hydrogeochemical evaluation of groundwater quality for drinking and irrigation purposes and integrated interpretation with water quality index studies. Environmental Process, 5(2), 363–383. https:// doi. org/ 10. 1007/ s40710- 018- 0297-4

Adimalla, N., Manne, R., Zhang, Y., Xu, P., & Qian, H. (2022). Evaluation of groundwater quality and its suitability for drinking purposes in semi-arid region of Southern India: an application GIS. Geocarto Int. https:// doi. org/ 10.1080/ 10106 049. 2022. 20406

Adimalla, N., Qian, H., & Nandan, M.J. (2020). Groundwater chemistry integrating the pollution index of groundwater and evaluation of potential human health risk: A case study from hard rock terrain of south India. Ecotoxicology and Environmental Safety, 206, 111217, https://doi.org/10.1016/j.ecoenv.2020.111217

Afzal, M.S., Ashraf, A., & Nabeel, M. (2018).Characterization of industrial effluents and groundwater of Hattar industrial estate, Haripur. Adv Agr Environ Sci.;1(2): 70–77. DOI: 10.30881/aaeoa.00013

Agency for Toxic Substances and Disease Registry (Atsdr) (2012). Toxicological profile for cadmium. Department of Health and Human Services, Atlanta, GA, Public Health Service

Ahirwar, R., Malik, M.S., Ahirwar, S., & Shukla, J. (2021). Groundwater potential zone mapping of Hoshangabad and Budhni industrial area, Madhya Pradesh, India. Groundwater for Sustainable Development. 14. 100631. 10.1016/j.gsd.2021.100631

Akhigbe, S., Udom, G.J., & Nwankwoala, H.O. (2018). Impact of Domestic and Industrial Waste on Surface and Ground Water QualityWithin Slaughter Area, Trans-Amadi Industrial Layout, Port Harcourt, Nigeria. Int J Waste Resour 8: 327. doi: 10.4172/2252-5211.1000327

Alameddine, I., Jawhari, G., & El-Fadel, M. (2017). Social Perception of Public Water Supply Network and Groundwater Quality in an Urban Setting Facing Saltwater Intrusion and Water Shortages. Environ. Manag. 2017, 59, 571–583. [CrossRef]

Ali, M.M., Ali, M.L., Islam, M.S., & Rahman, M.Z. (2016). Preliminary assessment of heavy metals in water and sediment of Karnaphuli River, Bangladesh. Environmental Nanotechnology, Monitoring & Management, 5: 27-35.doi:10.1016/j.enmm.2016.01.002

Alsuhaimi, A. O., Almohaimidi, K. M., & Momani, K. A. (2019). Preliminary assessment for physicochemical quality parameters of groundwater in Oqdus Area, Saudi Arabia. Journal of the Saudi Society of Agricultural Sciences, 18(1), 22–31. https://doi.org/10.1016/j.jssas.2016.12.002

Anoop, Y., & Renu, D. (2014). Effect of sugar mill on physico-chemical characteristics of ground water of surrounding area. Int. Res. J. Env. Sci, 3(6), 62-66

Anh, P.K., & Giao, N.T. (2018). Groundwater quality and human health risk assessment related to groundwater consumption in An Giang province, Vietnam J. Viet. Env. 10(2):85-91, DOI: 10.13141/jve.vol10.no2.pp85-91

Anwar, K. M., & Vanita, A. (2014). Analysis of groundwater quality using statistical techniques: A case study of Aligarh city (India). International journal of technical research and applications, 2(5), 100-106

Anyanwu, E.D., & Nwachukwu, E.D. (2020). Heavy metal content and health risk assessment of a South-eastern Nigeria River. Appl. Water Sci .10, 210 (2020). https:// doi. org/ 10. 1007/ s13201- 020- 01296-y

APHA, (2017). Standard Methods for Examination of Water and Wastewater. 23rd Edn., APHA,AWWA,WPCF

Armstrong, M. (2022). Infographic: Where Water Stress Will Be Highest by 2040. Statista Infographics. https://www.statista.com/chart/26140/water-stress-projections-global/

Asadollahfardi, G. (2015). Water Quality Management. 978-3-662-44724-6, https://doi.org/10.1007/978-3-662-44725-3

Asmamaw, M., & Debie, E. (2023). Characterizing groundwater quality for a safe supply of water using WQI and GIS in Bahir Dar city, northwest Ethiopia. Water Practice & Technology, Vol 18 No 4, 859 doi: 10.2166/wpt.2023.046

Awol, M.J., Adugna, A.T., & Kamaraj, M. (2021). Assessing the Industrial Effluent Effect on Irrigation Water Quality and Farm Soil near Kombolcha Town, Ethiopia. Hindawi Journal of Chemistry Article ID 5586213, 9 pages, https://doi.org/10.1155/2021/5586213

Backman, B., Bodis, D., Lahermo, P., & Rapant, T. T. (1998). Application of a groundwater contamination index in Finland and Slovakia. Environmental Geology, 36(1), 55–64

Bahukhandi, D. K., & Bartarya, S. K. (2014). Major ion chemistry of Asan river catchment of Dehradun District. Octa Journal of Environmental Research, ISSN 2321 3655

Banda, T.D., & Kumarasamy, M. (2020). Development of a Universal Water Quality Index (UWQI) for South African River Catchments. Water, 12, 1534. https://doi.org/10.3390/w12061534

Bartarya, S. K. (1995). Hydrogeology and Water Resources of Intermontane Doon Valley Barzegar, R., Moghaddam, A.A., Soltani, S., Fijani, E., Tziritis, E., & Kazemian, N. (2019).

Heavy Metal (loid)s in the Groundwater of Shabestar Area (NW Iran): Source Identification and Health Risk Assessment. Exposure and Health, 11:251– 265https://doi.org/10.1007/s12403-017-0267-5

Berhe, B.A. (2020). Evaluation of groundwater and surface water quality suitability for drinking and agricultural purposes in Kombolcha town area, eastern Amhara region, Ethiopia. Applied Water Science,10:127, https://doi.org/10.1007/s13201-020-01210-6

Bhadula, S., & Joshi, B.D. (2013). Zooplankton diversity of Ganga River water with special reference to pollution status. Journal of Environmental Bio-Science, 27(1): 111-114

Bhardwaj, R.K., & Sharma, S. (2022). Impact of Rainfall on Ground Water Level Fluctuations in the State of Uttarakhand, India. International Research Journal of Engineering and Technology (IRJET), e-ISSN: 2395-0056

Bharti, P. K., Kumar, P., & Singh, V. (2013). Impact of industrial effluents on ground water and soil quality in the vicinity of industrial area of Panipat city, India. Journal of Applied and Natural Science, 5(1), 132-136

Bhatia, D., Sharma, N.R., Kanwar, R., & Singh, J. (2018). Physicochemical assessment of industrial textile efuents of Punjab (India). Applied Water Science 8:83 https://doi.org/10.1007/s13201-018-0728-4

Bhutiani, R. Ram, K., & Ahamad, F. (2021). Quality assessment of groundwater at Laksar Block, Haridwar in Uttarakhand, India using Water Quality Index: A case study. Journal of Applied and Natural Science, 13(1): 197 - 203. https://doi.org/10.31018/jans.v13i1.2435

Bhutiani, R., Kulkarni, D. B., Khanna, D. R., & Gautam, A. (2016). Water quality, pollution source apportionment and health risk assessment of heavy metals in groundwater of an industrial area in North India. Exposure and Health, 8(1), 3-18

Bhutiani, R., Ram, K., & Ahamad, F. (2019). Assessment of suitability of ground water quality in and around Laksar, Haridwar, Uttarakhand on the basis Water Quality Index (WQI). Environment Conservation Journal, 41–46. https://doi.org/10.36953/ECJ.2019.1008.1207

Bhutiani, R., Ram, K., Taygi, V., Ahamad, F. & Kaushik, P. (2018). Assessment of groundwater quality of Laksar block in district Haridwar, Uttarakhand. Environment Conservation Journal, 19(3), 123-128. https://doi.org/10.3 6953/ECJ.2018.19316

Bhuyan, S., Abu, M., & Rashed, B. (2019). Monitoring and assessment of heavy metal contamination in surface water and sediment of the Old Brahmaputra River, Bangladesh. Appl Water Sci 9, 1–13, https:// doi. org/ 10. 1007/ s13201- 019- 1004-y

BIS, (2012). Drinking water specifications, Bureau of Indian Standards, 1991, IS:10500 (revised 2012)

Bodrud-Doza, M., Didar-Ul Islam, S.M., Rume, T., Quraishi, S.B., Rahman, M.S., & Bhuiyan, M.A.H. (2020). Groundwater quality and human health risk assessment for safe and sustainable water supply of Dhaka City dwellers in Bangladesh. Groundw. Sustain. Dev 10:100374. https://doi.org/10.1016/j.gsd.2020.100374

Bowes, M.J., Read, D.S., Joshi, H., Sinha, R., Ansari, A., Hazra, M., Simon, M., Vishwakarma, R., Armstrong, L.K., Nicholls, D.J.E., Wickham, H.D., Ward, J., Carvalho, L.R., & Rees, H.G. (2020). Nutrient and microbial water quality of the upper Ganga River, India: identification of pollution sources. Environ Monit Assess 192(8):533. https://doi.org/10.1007/ s10661-020-08456-2

Briffa, J., Sinagra, E., Blundell, R. (2020). Heavy metal pollution in the environment and their toxicological effects on humans. Heliyon, e04691,.https://doi.org/10.1016/j.heliyon.2020.e04691

Bunkar, N., & Kumar, V. (2019). Water Quality Index for Assessment of Groundwater Quality Parameters in Udham Singh Nagar District of Uttrakhand. International

Journal of Current Microbiology and Applied Sciences, ISSN: 2319-7706 Special Issue-8 pp. 68-72

Centers for Disease Control and Prevention, (2012). What do parents need to know to protect their children? http://www.cdc.gov/nceh/lead/ ACCLPP/blood_lead_levels.htm. Accessed date 10 Nov 2018

CGWB, (2013). Ground Water Information Booklet of Kurukshetra District, Haryana. Central Ground Water Board, Ministry of Water Resources Government of India, North Western

Region, Chandigarh. Available from: http://cgwb.gov.in/District_ Profile/Haryana/ Kurukshetra.pdf

CGWB, (2019). Central Ground Water Board. Retrieved May 5, 2021, from https:// www. National herald india. com/ natio nal/ water- crisis- groundwater- levels- dip- acute ly- in- north- india

Chakraborti, D., Singh, S.K., Rahman, M.M., Dutta, R.N., Mukherjee, S.C., Pati, S., & Kar,

P.B. (2017). Groundwater Arsenic Contamination in the Ganga River Basin: A Future Health Danger. International Journal of Environmental Research and Public Health, doi: 10.3390/ijerph15020180

Chakraborti, R. K., Kaur, J., & Kaur, H. (2019). Water Shortage Challenges and a Way Forward in India. Journal AWWA, 111(5), 42–49. https://doi.org/10.1002/awwa.1289

Chandan, R. (2017). Groundwater quality deterioration due to unplanned industrialization: a case study of district Samba, J&K, India. J Emerg Technol Innov Res 4(12):266–271

Chauhan, M. (2010). A perspective on watershed development in the Central Himalayan State of Uttarakhand, India. Int J Ecol Environ Sci 36:253–269

Chen, F., Yao, L., Mei, G., Shang, Y., Xiong, F., & Ding, Z. (2021). Groundwater Quality and Potential Human Health Risk Assessment for Drinking and Irrigation Purposes: A Case Study in the Semiarid Region of North China. Water, 13, 783. https://doi.org/10.3390/w13060783

Chen, K., Yu, S., Ma, T., Ding, J., He, P., Li, Y., Dai, Y., & Zeng, G. (2022). Modeling the Water and Nitrogen Management Practices in Paddy Fields with HYDRUS-1D. Agriculture 2022, 12, 924. [CrossRef]

Chowdhury, P., Mukhopadhyay, B.P., Nayak, S., & Bera, A. (2021). Hydrochemical characterization of groundwater and evaluation of health risk assessment for fluoride contamination areas in the eastern blocks of Purulia district, India. Environ Dev Sustain. https:// doi. org/ 10. 1007/ s10668-021-01911-1

Corcoran, E., Nellemann, C., Baker, E., Bos, R.,Osborn, D., & Savelli, H. (2010). Sick water? The central role of wastewater management in sustainable development - A rapid response assessment. UNEP (United Nations Environment Programme), UN-HABITAT, Nairobi, Kenia

Darko, G., Obiri-Yeboah, S., Takyi, S.A., Amponsah, O., Borquaye, L.S., Amponsah, L.O., & Fosu-Mensah, B.Y. (2021). Urbanizing with or without nature: Pollution effects of human activities on water quality of major rivers that drain the Kumasi Metropolis of Ghana. Environ Monit Assess.; 194(1):38. doi: 10.1007/s10661-021-09686-8. PMID: 34935079

Das, R., Laishram, B., & Jawed, M. (2019). Perception of groundwater quality and health effects on willingness to procure: The case of upcoming water supply scheme in Guwahati, India. Journal of Cleaner Production., 226, 615–627. https:// doi. org/ 10. 1016/j. jclep ro. 2019. 04. 097

Deepali, & Joshi, N. (2012). Study of Groundwater quality in and around SIDCUL Industrial Area,Haridwar,Uttarakhand,India. Journal of Applied Technology in Environmental Sanitation, 2(2), 129-134

Deoli, B. K., Bartarya, S. K., & Siddiqui, N. A. (2017). Assessment of surface and ground water Quality of Haridwar district of Uttarakhand. International Journal of Chem. Tech. Research, 10, 95-118

Dheeraj, V.P., Singh, C.S., Kishore, N., & Sonkar, A.K. (2023). Groundwater Quality Assessment in Korba Coalfield Region, India: An Integrated Approach of GIS and Heavy Metal Pollution Index (HPI) Model. Nat. Env. & Poll. Tech. e-ISSN: 2395- 3454

Dohare, D., Deshpande, S., & Kotiya, A. (2014). Analysis of ground water quality parameters: a Review. Research Journal of Engineering Sciences ISSN, 2278, 9472.

Doria, M.d.F., Pidgeon, N., & Hunter, P.R. (2009). Perceptions of drinking water quality and risk and its effect on behaviour: A cross-national study. Sci. Total Environ. 407, 5455–5464. [CrossRef] [PubMed]

Dubey, R. K., Hussain, J., Malhotra, N., & Mehta, A. (2014). Ground water quality and water quality index of Dwarka district of national capital of India. International journal of Research in Engineering and Technology, 3(04), 2321-7308

Dudeja, D. S., Bartarya, K., & Biyani, A. K. (2011). Hydrochemical and water quality assessment of groundwater in Doon valley of outer Himalaya Uttarakhand, India. Environmental Monitoring and Assessment, 181(1–4), 183–204

Duggal, V., & Rani, A. (2018). Carcinogenic and Non-carcinogenic Risk Assessment of Metals in Groundwater via Ingestion and Dermal Absorption Pathways for Children and Adults in Malwa Region of Punjab. J Geol Soc India, https://doi.org/10.1007/s12594-018-0980-0

Edet, A.E., & Offiong, O.E. (2002). Evaluation of water quality pollutionindices for heavy metal contamination monitoring. A study casefrom Akpabuyo-Odukpani area, Lower Cross River Basin (southeastern Nigeria) Geoj, 57:295–304. https://doi.org/10.1023/B:GEJO.0000007250.92458

Egbueri, J.C., Ezugwu, C.K., Ameh, P.D., Unigwe, C.O., & Ayejoto, D.A. (2020). Appraising drinking water quality in Ikem rural area (Nigeria) based on chemometrics and multiple indexical methods. Environ Monit Assess. 192, 308, https:// doi. org/ 10.1007/ s10661-020- 08277-3

Ehya, F., & Marbouti, Z. (2016). Hydrochemistry and contamination of groundwater resources in the Behbahan plain, SW Iran. Environmental Earth Sciences, 75, 455–467

Eilam, E., & Trop, T. (2014). Factors influencing adults' environmental attitudes and behaviors and the role of environmental schools in influencing their communities. Educ. Urban Soc. 2014, 46, 234–263. [CrossRef]

Elumalai, V., Vhonani, G., Nethononda, V., Manivannan, N., Rajmohan, Peiyue, Li., & Elango, L. (2020). Groundwater quality assessment and application of multivariate statistical analysis in Luvuvhu catchment, Limpopo, South Africa, Journal of African EarthSciences,Volume171,ISSN1464343Xhttps://doi.org/10.1016/j.jafrearsci.2020.103967

Fang, Y., Zheng, T., Zheng, X., Peng, H., Wang, H., Xin, J. & Zhang, B. 2020 Assessment of the hydrodynamics role for groundwater quality using an integration of GIS, water quality index and multivariate statistical techniques. Journal of Environmental Management, 273, 111185. https://doi.org/10.1016/j.jenvman.2020.111185

Ferrer, N., Folch, A., Lane, M., Olagod, D., Katuva, J., Thomson, P., Jou, S., Hope, R., & Custodio, E. (2019). How does water-reliant industry affect groundwater systems in coastal Kenya? Science of the Total Environment, 694, 133634, https://doi.org/10.1016/j.scitotenv.2019.133634

Field, A. (2009). Discovering statistics using IBMSPSS statistics, 3th edn. SAGE, London Fito, J., Said, H., Feleke, S., & Worku, A. (2019). Fluoride removal from aqueous solution

onto activated carbon of Catha edulis through the adsorption treatment technology.

Environ Syst Res 8, 25, https://doi.org/10.1186/s40068-019-0153-1Gad, M., Gaagai, A., Eid, M. H., Szűcs, P., Hussein, H., Elsherbiny, O., Elsayed, S., Khalifa,

M. M., Moghanm, F. S., Moustapha, M. E., Tolan, D. A., & Ibrahim, H. (2023). Groundwater Quality and Health Risk Assessment Using Indexing Approaches, Multivariate Statistical

Analysis, Artificial Neural Networks, and GIS Techniques in El Kharga Oasis, Egypt. Water, 15(6), 1216. https://doi.org/10.3390/w15061216

Gao, S., Li, C., Jia, C., Zhang, H., Guan, Q., Wu, X., Wang, J., & Lv, M. (2020). Health Risk Assessment of Groundwater Nitrate Contamination: A Case Study of a Typical Karst Hydrogeological Unit in East China. Environ. Sci. Pollut. Res. 2020, 27, 9274–9287.

Gao, Y., Qiana, H., Zhouc, Y., Chen, J., Wanga, H., Rena, W., & Qu, W. (2022). Cumulative health risk assessment of multiple chemicals in groundwater based on deterministic and Monte Carlo models in a large semiarid basin. Journal of Cleaner Production, 352, 131567, https://doi.org/10.1016/j.jclepro.2022.131567

Garcia, C.A.B., Silva, I.S., Mendonça, M.C.S., & Garcia, H.L. (2018). Evaluation of Water Quality Indices: Use, Evolution and Future Perspectives. Environmental Monitoring and Assessment, DOI: 10.5772/intechopen.79408

Gevera, P.K., Dowling, K., Gikuma-Njuru, P., & Mouri, H. (2022). Public Knowledge and Perception of Drinking Water Quality and Its Health Implications: An Example from the Makueni County, South-Eastern Kenya. Int. J. Environ. Res. Public Health, 4530. https://doi.org/10.3390/ ijerph19084530

Ghalib, H. B. (2017). Groundwater chemistry evaluation for drinking and irrigation utilities in east Wasit province, Central Iraq. Applied Water Science, 7(7), 3447–3467. https://doi.org/10.1007/s13201-017-0575-8

Gharbi, A., Ibn, A.Z., & Zairi, M. (2019). Groundwater suitability for drinking and agriculture purposes using irrigation water quality index and multivariate analysis: case of Sidi Bouzid aquifer, central Tunisia. Environ Earth Sci, 78:692–711. https://doi.org/10.1007/s12665-019-8733-y

Gibbs, R. J. (1977). Mechanism controlling world water chemistry. Science, 170, 1081–1090

Giri, N., & Singh, O. P. (2013). Urban Growth and Water Quality in

Thimpu, Bhutan Journal of Urban and Environmental Engineering, 7(1), 82–95. http://www.jstor.org/stable/26203392

Gltau, M.W., Chen, J., & Zhao, M. (2016). Water Quality Indices as Tools for DecisionMaking and Management, Water Resour Manage, 30:2591–2610DOI 10.1007/s11269-016-1311-0

Gogoi, L. (2015).Industrial Effluents and Their Impact on Ground Water Quality and Human Health in Guwahati City, India: A Geographical Analysis. INNOVATIVE ENERGY TECHNOLOGY SYSTEMS AND ENVIRONMENTAL CONCERNS: A SUSTAINABLE APPROACH. ISBN: 978-93-84144-81-4

Goswami, R., Neog, N., & Thakur, R. (2022). Hydrogeochemical assessment of groundwater quality for drinking and irrigation in Biswanath and Sonitpur district of the Central Brahmaputra Plain, India. Front. Water 4:889128.doi: 10.3389/frwa.2022.889128

Govil, P.K., Sorlie, J.E., Murthy, N.N., Sujatha, D., Reddy, G.L.N., RudolphLund, K., Krishna, A.K., & Rama, M.K. (2008). Soil contamination of heavy metals in the Katedan industrial development area, Hyderabad, India. Environ Monit Assess, 140:313e323

Gugulothu, S., Subbarao, N., Das, R., & Dhakate, R. (2022). Geochemical evaluation of groundwater and suitability of groundwater quality for irrigation purpose in an agricultural region of South India. Applied Water Science 12:142 https://doi.org/10.1007/s13201-022-01583-w

Gupta, A., Singh, R., Singh, P., & Dobhal, R. (2017a). Heavy Metals in Drinking Water Sources of Dehradun, Using Water Quality Indices, Analytical Chemistry Letters, 7:4, 509-519, DOI: 10.1080/22297928.2017.1372209

Gupta, N., Pandey, P. and Hussain, J. (2017b). Effect of physicochemical and biological parameters on the quality of river water of Narmada, Madhya Pradesh, India. Water Science, 31: 11–23. doi:10.1016/j.wsj.2017.03.002

Gyeltshen, S., Tran, T. V., Gunda, G. K. T., Kannaujiya, S., Chatterjee, R. S., & Champatiray, P. K. (2019). Groundwater potential zones using combination of geospatial technology and geophysical approach: Case study in Dehradun India. Hydrological Sciences Journal, 65(2), 169–182. https://doi. org/ 10.1080/ 02626 667. 2019. 16883 34

Haque, S., Kannaujiya, S., Taloor, A.K., Keshri, D., Bhunia, R.K., Ray, P.K.C., & Chauhan, P. (2020). Identification of groundwater resource zone in the active tectonic region of Himalaya through earth observatory techniques. Groundw Sustain Dev 10:100337. https:// doi. org/ 10.1016/j. gsd. 2020. 100337

Hasan, M.S., & Rai, A. (2020). Groundwater quality assessment in the Lower Ganga Basin using entropy information theory and GIS. Journal of Cleaner Production, DOI: 10.1016/j. jclepro.2020.123077

Hassan, G.Z., Hassan, F.R., Jehangir, F., Shabir, G., & Khan, M.A.H. (2022).Groundwater Contamination from Drainage Effluents in Punjab Pakistan. Journal of Agricultural Research Pesticides and Biofertilizers, 4(1); DOI:http;//doi.org/06.2022/1.1075

Herojeet, R., Rishi, M.S., Lata, R., & Sharma, R. (2016). Application of environmetrics statistical models and water quality index for groundwater quality characterization of alluvial aquifer of Nalagarh Valley, Himachal Pradesh, India. Sustain. Water Resour. Manag. 2:39–53, DOI 10.1007/s40899-015-0039-y

Hossain, M., & Patra, P.K. (2020).Water pollution index – A new integrated approach to rank water quality, Ecological Indicators, Volume 117,ISSN 1470-160X, https://doi. org/10.1016/j.ecolind.2020.106668

Hossain, S.M.S., Haque, M.E., Pramanik, M.A.H., Uddin, M.J., & Harun, A.Y. (2020) Assessing the groundwater quality and health risk: A case study on Setabganj sugar mills limited, Dinajpur, Bangladesh, Water Science, 34:1, 110-123, DOI: 10.1080/11104929.2020.1790184

Hu, J., Dong, H., Xu, Q., Ling, W., Qu, J., & Qiang, Z. (2018). Impacts of water quality on the corrosion of cast iron pipes for water distribution and proposed source water switch strategy. Water Research, 129, 428–435. https:// doi. org/ 10. 1016/j. watres. 2017. 10. 065

Hussain, I., Hayat, W., Gong, S., Yang, X., & Lai, W.F. (2023). A Comparative Analysis of Public Awareness Level about Drinking Water Quality in Guangzhou (China) and Karachi (Pakistan).Sustainability, 2023, 15, 8408. https://doi.org/10.3390/su15108408

Hutton, G., & Varughese, M. (2016). The Costs of Meeting the 2030 Sustainable Development Goal Targets on Drinking Water, Sanitation, and Hygiene. © World Bank, Washington, DC. http://hdl.handle.net/10986/23681 License: CC BY 3.0 IGO

Ibrahim, A.A., Ibrahim, M.A., & Yusuf, A.G. (2021). Implications of industrial effluents on surface water and ground water. World Journal of Advanced Research and Reviews, 09(03), 330–336

International Association of Hydrogeologists, (2020). Groundwater—More about the Hidden Resource.https://iah.org/education/general-public/groundwater-hidden-resource

IPCC Report, (2021). Climate change 2021: the physical science basis. https:// www. ipcc. ch/ report/ sixth-asses sment-report-working-group-i/

JEESR-158. DOI: https://doi.org/10.47363/JEESR/2021(3)143

Jain, C. K., Bandyopadhyay, A., & Bhadra, A. (2012). Assessment of ground water quality for irrigation Purpose, district Nainital, Uttarakhand, India. Journal of Indian Water Resources Society, 32(3-4), 8-14

Jain, C.K., Singh, R.D., Mehta, R., Sharma, S.K., Singh, Y., & Sharma, B. (2015). Ground Water Quality Mapping and Surveillance for Safe Water Supply in Bahadrabad Block, District Hardwar, Uttarakhand. International Journal of Engineering Research & Technology, (IJERT), ISSN: 2278-0181

Jan, H.A., Ullah, N., & Rehan, R. (2020). To Assess the Effect of Industrial Waste on The Quality of Ground Water. A Case Study of Hayatabad Industrial Estate, Peshawar. Pakistan. International Conference on Sustainability in Process Industry

Jareda, G., Mahapatra, S.P., & Dhekne, P.Y. (2018). Water quality index, heavy metal pollution index and seasonal variation correlation of groundwater of Bailadila iron ore mine area and its peripherals: Dantewada district, Chhattisgarh, India. Desalination and Water Treatment. doi: 10.5004/dwt.2018.21632

Jasrotia, A. S., Taloor, A. K., Andotra, U., & Bhagat, B. D. (2018). Geoinformatics based groundwater quality assessment for domestic and irrigation uses of the Western Doon valley, Uttarakhand, India.Groundwater for Sustainable Development, 6, 200–212. https:// doi. org/ 10. 1016/j. gsd. 2018. 01. 003

Jasrotia, A.S., Taloor, A.K., Andotra, U., & Kumar, R. (2019). Monitoring and assessment of groundwater quality and its suitability for domestic and agricultural use in the Cenozoic rocks of Jammu Himalaya, India: a geospatial technology based approach. Groundw Sustain Dev, 8:554–566

Jayarajan, P., & Kuriachan, L. Exposure and health risk assessment of nitrate contamination in groundwater in Coimbatore and Tirupur districts in Tamil Nadu, South India. Environ Sci Pollut Res 28, 10248–10261 (2021). https://doi.org/10.1007/s11356-020-11552-y

JJM, (2019). Jal Jeewan Mission, Department of Drinking Water and Sanitation, Ministry of Jal Shakti, Government of India. Retrieved November 24, 2021, from https:// jalsh akti-ddws. gov. in/ sites/ default/ files/ JJM_ note. pdf

Joshi, D.M., Kumar, A., & Agrawal, N. (2009). Assessment of the irrigation water quality of river Ganga in Haridwar District. Rasayan J Chem 2(2):285–292, Journal of Himalayan Geology Vol. 6(2)

Kabir, M.M., Hossain, N., Islam, A.R.T., Akter, S., Fatema, K.J., Hilary, L.N., Hasanuzzaman, M., Didar-ul-Alam & Choudhury, T.R. (2021). Characterization of groundwater hydrogeochemistry, quality, and associated health hazards to the residents of southwestern Bangladesh. Environmental Science and Pollution Research, https://doi.org/10.1007/s11356-021-15152-2

Kachroud, M., Trolard, F., Kefi, M., Jebari, S., & Bourrié, G. (2019). Water Quality Indices: Challenges and Application Limits in the Literature. Water, 11, 361. https://doi.org/10.3390/w11020361

Kadam, A., Wagh, V., Patil, S., Umrikar, B., & Sankhua, R. (2021). Seasonal assessment of groundwater contamination, health risk and chemometric investigation for a hard rock terrain of western India. Environmental Earth Sciences, 80:172, https://doi.org/10.1007/s12665-021-09414-y

Kainthola, S., Rawat, G., Bhandari, K & Pant, H.B., (1988). Water Requirement of Doon valley. Project report of the Center for Science and Environment under NEM-87 project, 46.

Kala, C.P. (2014). Deluge, disaster and development in Uttarakhand Himalayan region of India: challenges and lessons for disaster management. Int J Disaster Risk Reduct 8:143–152. https://doi.org/10.1016/j.ijdrr.2014.03.002

Kamble, R.K. (2020). Health Risk Assessment of Groundwater Iron and Manganese in Chandrapur District, Central India. Sustainability, Agri, Food and Environmental Research, (ISSN: 0719-3726), 8(X), http://dx.doi.org/10.7770/safer-V0N0-art2072

Kanagaraj, G., & Elango, L. (2016). Hydrogeochemical processes and impact of tanning industries on groundwater quality in Ambur, Vellore district, Tamil Nadu, India. Environmental Science and Pollution Research, 23(23), 24364-24383.

Kanagaraj, G., & Elango, L. (2016). Hydrogeochemical processes and impact of tanning industries on groundwater quality in Ambur, Vellore district, Tamil Nadu, India. Environmental Science and Pollution Research, 23(23), 24364-24383

Karanth, K.R. (1987). Groundwater assessment, development and management. Tata McGraw-Hill Publishing, New Delhi

Karnena, M. K., Konni, M., Dwarapureddi, B. K., Satyanarayana, Y. & Saritha, V. 2022 GIS-based approach qualitative features of sub-surface water from coastal district in Andhra Pradesh. Applied Water Science, 12(3), 1–15. https://doi.org/10.1007/ s13201- 021-01506-1

Karpagam, V., & Ramesh, K. (2015). Assessment of Groundwater Quality of Chrompet Industrial Area by Water Quality Index Method. International Journal of Engineering Technology, Management and Applied Sciences, 3(7), 123-132

Karunanidhi, D., Aravinthasamy, P., Subramani, T., & Muthusankar, G. (2021). Revealing drinking water quality issues and possible health risks based on water quality index (WQI) method in the Shanmuganadhi River basin of South India. Environ Geochem Health 43(2):931–948

Kaur, T., Bhardwaj, R., & Arora, S. (2017). Assessment of groundwater quality for drinking and irrigation purposes using hydrochemical studies in Malwa region, southwestern part of Punjab, India. Applied water science, 7(6), 3301-3316

Khalid, S., Murtaza, B., Shaheen, I., Imran, M., & Shahid, M. (2022). Public Perception of DrinkingWater Quality and Health Risks in the District Vehari, Pakistan. VertigO. Available online: http://journals.openedition.org/vertigo/21171

Khan, A., Govil, H., Taloor, A.K., & Kumar, G. (2020). Identification of artificial groundwater recharge sites in parts of Yamuna river basin India based on remote sensing and geographical information system. Groundw Sustain Dev, 10:100415. https:// doi. org/ 10. 1016/j.gsd. 2020. 100415

Kothari, V., Vij, S., Sharma, S. K., & Gupta, N. (2021). Correlation of various water quality parameters and water quality index of districts of Uttarakhand. Environmental and Sustainability Indicators, 9, 100093. https:// doi. org/ 10. 1016/j. indic. 2020. 100093

Kouser, B., Bala, A., Verma, O., Prashanth, M., Khosla, A., & Pir, R.A. (2022). Hydrochemistry for the assessment of groundwater quality in the Kathua region, Jammu and Kashmir, India. Applied Water Science, https://doi.org/10.1007/s13201- 022-01673-9

Krishan, G., Sudarsan, N., Ghosh, N.C., Kumar, S., Singh, S., Sharma, A., Mittal, S., Sidhu, B.S., & Vashisth, R. (2021). Concentration of heavy metals in groundwater and heavy metal pollution index in Punjab. e-Journal of Geohydrology, 2(1)

Krishna, A.K., Mohan, K.R., & Dasaram, B. (2019). Assessment of groundwater quality, toxicity and health risk in an industrial area using multivariate statistical methods. Environ Syst Res, 8, 26, https://doi.org/10.1186/s40068-019-0154-0

Krishna, G., Singh, S., Gurjar, S., & NC, G. (2016c). Assessment of Water Quality Index (WQI) of Groundwater in Rajkot District. Gujarat, India. journal of earth science and climate change, 7(3), 123-134

Kumar, A., Bhambri, R., Tiwari, S.K., Verma, A., Gupta, A.K., & Kawishwar, P. (2019). Evolution of debris flow andmoraine failure in the Gangotri Glacier region, Garhwal Himalaya:hydrogeomorphologicalaspects.Geomorphology,https://doi.org/10.1016/j.geomorph.2019.02.015

Kumar, D., Kumar, S., Sharma, A., Rastogi, D., & Kandpal, V. (2020). Physico-chemical study of Ground Water of SIDCULRudrapur of Udham Singh Nagar district, Uttarakhand. Res. J. Chem. Env. Sci. Vol 8, 19-24

Kumar, D., Malik, S., Madan, M., Pandey, A., & Asthana, A. K. (2013). Bacteriological analysis of drinking water by MPN method in a tertiary care hospital and adjoining area Western UP, India. J Environ Sci Toxicol Food Technol, 4(3), 17-22

Kumar, G., Ismudeen, A.R., & Natarajan, V. (2017). Physico-Chemical Analysis of Groundwater Pollution in Cuddalore District, Tamil Nadu, India. Int J Chem Sci. 15(3):162

Kumar, J., Biswas, B., & Verghese, S. (2021). Assessment of groundwater quality for drinking and irrigation purpose using geospatial and statistical techniques in a semi- arid region of Rajasthan, India. J Geol Soc India 97:416–427

Kumar, S., Kumar, V., Saini, R.K., Kumar, C.P., Raju, M., Singh, S., Singh, O., & Chakravorty, B. (2019). Detection of Arsenic In Groundwatero of Laksar Area, Haridwar District, Uttarakhand. J. Indian Water Resour. Soc, Vol. 39

Kumar, V., Kumar, S., Srivastava, S., Singh, J. & Kumar, P. (2018). Water quality of River Ganga with reference to physico-chemical and microbiological characteristics during Kanwar Mela 2017 at Haridwar, India: A case study. Archives of Agriculture and Environmental Science, 3(1): 58-63. doi:10.26832/24566632.2018.030108

Kumari, S., & Rani, J. (2014). Assessment of water quality index of ground water in Smalkhan, Haryana. International Journal of Latest Research in Science and Technology, 3(6), 169-172

Lakshmi, R.V., Raja, V., Sekar, C.P., & Neelakantan, M.A. (2021). Evaluation of groundwater quality in Virudhunagar Taluk, Tamil Nadu, India by using statistical methods and GIS technique. J Geol Soc India 97(5):527–538

Li, F., Qiu, Z., Zhang, J., Liu, W., Liu, C., & Zeng, G. (2017). Investigation, pollution mapping and simulative leakage health risk assessment for heavy metals and metalloids in groundwater from a typical brownfeld, middle China. Int J Environ Res Public Health 14:768

Li, P., Karunanidhi, D., Subramani, T., & Srinivasamoorthy, K. (2021). Sources and consequences of groundwater contamination. Archives of Environmental Contamination and Toxicology, 80(3),1–10. https:// doi. org/ 10. 1007/ s00244- 020- 00805-z

Liu, C.W., Lin, K.H., & Kuo, Y.M. (2003). Application of factor analysis in the assessment of groundwater quality in a blackfoot disease area in Taiwan. Sci Total Environ 313:77–89. https://doi.org/10.1016/S0048-9697(02)00683-6

Liu, F., Zhen, P., & Wang, S. (2022). Groundwater Quality Assessment and Health Risks from Nitrate Contamination in the Heilongdong Spring Basin, a Typical Headwater Basin of the North China Plain. Environ. Sci. Pollut. Res., 29, 17655–17670.

Lorestani, B., Merrikhpour, H., & Cheraghi, M. (2020) Assessment of heavy metals concentration in groundwater and their associated health risks near an industrial area. Environmental Health Engineering and Management Journal; 7(2): 67–77. doi: 10.34172/EHEM.2020.09

Mahler, R.L., Barber, M.E., & Shafii, B. (2015). Urban public satisfaction with drinking water since 2002 in the Pacific northwest, USA. Int. J. Sustain. Dev. Plan, 10, 620–634.

Maji, K.J., & Chaudhary, R. (2019). Principal component analysis for water quality assessment of the Ganga river in Uttar Pradesh, India. Water Res 46:789– 8 0 6. http s : / / doi. org/10.1134/S0097807819050129

Małecki, J.J., Kadzikiewicz-Schoeneich, M., Eckstein, Y., Szostakiewicz-Hołownia, M., & Gruszczynski, T. (2017). Mobility of copper and zinc in near-surface groundwateras a function of the hypergenic zone lithology at the KampinosNational Park (Central Poland). Environ Earth Sci, 76:276DOI 10.1007/s12665-017-6527-7

Mallick, J., Singh, C., Almesfer, M., Kumar, A., Khan, R., Islam, S., & Rahman, A. (2018). Hydro-geochemical assessment of groundwater quality in aseer region. Saudi Arabia. Water, 10(12), 1847.https://doi.org/10.3390/w10121847

Mandal, D., Chandrakala, M., Alam, N. M., Roy, T., & Mandal, U. (2021). Assessment of soil quality and productivity in different phases of soil erosion with the focus on land degradation neutrality in tropical humid region of India. CATENA, 204, 105440. https:// doi. org/ 10. 1016/j. catena. 2021105440

Mao, X., Zhang, S., Wang, S., Li, T., Hu, S., & Zhou, X.(2023). Evaluation of Human Health Risks Associated with Groundwater Contamination and Groundwater Pollution Prediction in a Landfill and Surrounding Area in Kaifeng City, China. Water, 723. https://doi.org/10.3390/w15040723

Matta, G., Kumar, A., Nayak, A., Kumar, P., & Pant, G. (2022). Pollution complexity quantification using NPI and HPI of River Ganga system in Himalayan Region. Proceedings of the Indian National Science Academy, https://doi.org/10.1007/s43538-022-00111-3

Maurya, P. K., Ali, S. A., Alharbi, R. S., Yadav, K. K., Alfaisal, F. M., Ahmad, A., Ditthakit, P., Prasad, S., Jung, Y. K., & Jeon, B. H. (2021). Impacts of land use change on water quality index in the upper Ganges River near Haridwar, Uttarakhand: A GIS-Based Analysis. Water,13(24), 3572.https:// doi. org/ 10. 3390/ w1324 3572

Meena, P.L., Jain, P.K., & Meena, K.S. (2016). Assessment of groundwater quality and its suitability for drinking and domestic uses by using WQI and statistical analysis in river basin area in Jahzpur Tehsil, Bhilwara District (Rajasthan, India). Int J Curr Microbiol App Sci, 5:415–423

Meijerink, A.M.J. (1974). Photohydrological Reconnaissance Surveys. Ph.D. Thesis, University of Amasterdam (Unpublished)

Meshabaz, R.A., & Umer, M.I. (2022). Assessment of industrial effluent impacts on soil physiochemical properties in Kwashe Industrial Area, Iraq Kurdistan Region. Water Resources in Iraq: Perspectives and Prognosis, doi:10.1088/1755- 1315/1120/1/012037

Mishra, A. (2010). Assessment of water quality using principal component analysis: a case study of the river Ganges. J Water Chem Technol 32:227–234. https://doi.org/10.3103/S1063455X10040077

Mishra, S., Sharma, M.P., & Kumar, A. (2015). Assessment of Surface Water Quality in Surha Lake using Pollution Index,India, J. Mater. Environ. Sci. 7 (Y) 713-719

Mitra, S., Ghosh, S., Satpathy, K.K., Bhattacharya, B.D., Sarkar, S.K., Mishra, P. and Raja, P. (2017). Water quality assessment of the ecologically stressed Hooghly river estuary, India: A multivariate approach. Marine Pollution Bulletin, 126: 592-599

Mohamed, A.K., Liu, D., Song, K., Mohamed, M.A., Aldaw, E., & Elubid, B.A. (2019). Hydrochemical analysis and fuzzy logic method for evaluation of groundwater quality in the North Chengdu Plain, China. Int J Environ Res Public Health 16(302):1–21. https:// doi. org/ 10.3390/ ijerp h1603 0302

Mohan, K. C., Suresh, J., & Venkateswarlu, P. (2014). Physio-chemical analysis of bore-well water of Kurnool environs, Andhra Pradesh. Journal of Chemical and Pharmaceutical Research, 6(9), 77-80

Mohan, S.V., Nithila, P., & Reddy, S.J. (1996). Estimation of heavy metal indrinking water and development of heavy metal pollution index. J Environ Sci Health 31:283–289

Mukherjee, A., Duttagupta, S., Chattopadhyay, S., Bhanja, S.N., Bhattacharya, A., Chakraborty, S., Sarkar, S., Ghosh, T., Bhattacharya, J., & Sahu, S. (2019). Impact of sanitation and socio-economy on groundwater fecal pollution and human health towards achieving sustainable development goals across India from ground- observations and satellite-derived nightlight. Sci Rep, 9(1):1–11. https:// doi. org/10. 1038/ s41598-019-50875-w

Muller, K., & Cornel, P. (2017). Setting water quality criteria for agricultural water reuse purposes. Journal of Water Reuse and Desalination, DOI: 10.2166/wrd.2016.194

Nagaraju, A. K., Kumar, S., & Ejaswi, A. (2014). Assessment of groundwater quality for irrigation: A case study from Bandalamottu lead mining area Guntur District, Andhra Pradesh, South India. Applied Water Science, 4(4), 385–396

Naidu, S., Gupta, G., Singh, R., Tahama, K., & Erram, V.C. (2021). Hydrogeochemical processes regulating the groundwater quality and its suitability for drinking and irrigation purpose in parts of coastal Sindhudurg District, Maharashtra. J Geol Soc, 97:173–185

Nasir, M., F.A., Praveena, S.M., & Aris, A.Z. (2019). Public awareness level and occurrence of pharmaceutical residues in drinking water with potential health risk: A study from Kajang (Malaysia). Ecotoxicol. Environ. Saf. 185, 109681.

Natarajan, P. M., Kallolikar, S., & Ganesh, S. (2016). Transforming Ganges to be a Living River through Waste Water Management. World Academy of Science, Engineering and Technology International Journal of Environmental and Ecological Engineering, 10(2), 251–260. https://doi.org/10.5281/zenodo.1124311

Nayak, A., Matta, G, Uniyal, D.P., Kumar, A., Kumar, P., & Pant, G. (2023). Assessment of potentially toxic elements in groundwater through interpolation, pollution indices, and chemometric techniques in Dehradun in Uttarakhand State. Environmental Science and Pollution Research, https://doi.org/10.1007/s11356-023-27419-x

Nayak, A., Matta, G., & Uniyal, D.P. (2022). Hydrochemical characterization of groundwater qualityusing chemometric analysis and water quality indices in the foothills of Himalayas. Environment, Development and Sustainability https://doi.org/10.1007/s10668-022-02661-4

Noori, R., Khakpour, A., & Omidvar, B. (2010). Comparison of ANN and principal component analysis-multivariate linear regression models for predicting the river flow based on developed discrepancy ratio statistic. Expert Syst Appl, 37:5856–5862

Ogungbe, A.S., Babatunde, O.S., Alabi, A.A., & Coker, J.O. (2020). Impact of sand mining on soil and water quality along addo-baddore road, Ajah, Lagos State, Nigeria. African Journal of Environment and Natural Science Research, 3(3): 73- 91.

Osta, EL., M., Masoud, M., Alqarawy, A., Elsayed, S., & Gad, M. (2022). Groundwater Suitability for Drinking and Irrigation UsingWater Quality Indices and Multivariate Modeling in Makkah Al-Mukarramah Province, Saudi Arabia. Water, 14, 483. [CrossRef]

Oyem, H.H., Oyem, I.M. & Usese, A.I. (2015). Iron, manganese, cadmium, chromium, zinc and arsenic groundwater contents of Agbor and Owa communities of Nigeria. Springer Plus 4, 104. https://doi.org/10.1186/s40064-015-0867-0

Packialakshmi, S., Deb, M., & Chakraborty, H. (2015). Assessment of groundwater quality index in and around Sholinganallur area, Tamil Nadu. Indian Journal of Science and Technology, 8(36), 1-7

Panghal, V., & Bhateria, R. (2020). A multivariate statistical approach for monitoring of groundwater quality: A case study of Beri block, Haryana India. Environmental Geochemistry and Health, 43(2615), 2629

Panneerselvam, B., Ravichandran, N., Kaliyappan, S.P., Karuppannan, S., & Bidorn, B. (2023). Quality and Health Risk Assessment of Groundwater for Drinking and Irrigation Purpose in Semi-Arid Region of India Using EntropyWater Quality and Statistical Techniques. Water,601. https://doi.org/10.3390/w15030601

Papazotos, P. (2021). Potentially toxic elements in groundwater: a hotspot research topic in environmental science and pollution research. Environ Sci Pollut Res Int 28(35):47825–47837

Parihar, S. S., Ajit, K., Ajay, K., Gupta, R. N., Manoj, P., Archana, S., & Pandey, A. C. (2012). Physico-chemical and Microbiological analysis of underground water in and around Gwalior city, MP, India. Research Journal of Recent Sciences ISSN, 2277, 2502

Patel, P., Raju, J., Reddy, S.R., Suresh, U., Sankar, D.B., Reddy, T.V.K. (2018). Heavy metal contamination in river water and sediments of the Swarnamukhi River Basin, India: risk assessment and environmental implications. Environ. Geochem. Health. 40, 609– 623. https:// doi. org/ 10. 1007/ s10653- 017- 0006-7

Pathak, H., & Limaye, S.N. (2011). Study of Seasonal Variation in GroundwaterQuality of Sagar City (India) by Principal Component Analysis. ISSN: 0973-4945; CODEN ECJHAOE-Journal of Chemistry http://www.e-journals.net 2011, 8(4), 2000-2009

Patil, V.B.B., Pinto, S.M., Govindaraju, T., Hebbalu, V.S., Bhat, V., & Kannanur, L.N. (2020). Multivariate statistics and water quality index (WQI) approach for geochemical assessment of groundwater quality a case study of Kanavi Halla Sub- Basin, Belagavi, India. Environ Geochem Health, 42(9):2667–2684

Paul, J. M., Arya, V. S., Jesteena, G., Reji, K. J., & Sumitha, K. S. (2014). Assessment of Ground Water Quality in Nellikkuzhy Panchayat of Kerala State, India. International Journal of Engineering Science Invention, 3(4), 21-28

Perez′Avila, V. J. J. A., IsradeAlc, A. I., & Buenrostro, D. O. (2019). Nitrate as a parameter for diferentiating groundwater fow systems in urban and agricultural areas: The case of Morelia-Capula area Mexico. Hydrogeology Journal, 27, 1767–1778

Pirlea, A. F., U. Serajuddin, D. Wadhwa, M. Welch, eds. (2023). Atlas of Sustainable Development Goals 2023. Washington, DC: World Bank. License: Creative Commons Attribution CC BY 3.0 IGO. https://datatopics.worldbank.org/sdgatlas/

Popoola, L.T., Yusuff, A.S., & Aderibigbe, T.A. (2019). Industrial and Residential Ground Water Physico-Chemical Properties Assessment in Lagos Metropolis, Drink. Water Eng. Sci. Discuss. [preprint], https://doi.org/10.5194/dwes-2019-2

Prasad, B., & Bose, J.M. (2001). Evaluation of heavy metal pollution indexfor surface and spring water near a limestone mining area of the lower Himalayas. Environ Geol 41:183–188

Praveena, S.M., Rashid, M.Z.M., Nasir, F.A.M., Wee, S.Y., & Aris, A.Z. (2021). Occurrence, Human Health Risks, and Public Awareness Level of Pharmaceuticals in Tap Water from Putrajaya (Malaysia). Expos. Health, 13, 93–104. [CrossRef]

Pundir, S., Kandari, V., Singh, P., Singh, R., & Siddique, M.S. (2022). Assessment of groundwater quality of Dehradun Valley, Uttarakhand, India, for drinking purposes using multivariate statistical techniques. Environment, Development and Sustainability. https://doi.org/10.1007/s10668-022-02784-8

Pundir, S., Singh, R., Singh, P., & Kandari, V. (2021). Risk assessment and water safety planning for rural water supply in Uttarakhand, India. Environmental Monitoring and Assessment, 193, 795.https:// doi. org/ 10. 1007/ s10661- 021- 09609-7

Raghunath, H.M. (1987). Ground Water, Vilely Easteren Ltd., New Delhi, India, 2nd edition

Rahaman, S.K.M. (2023). Groundwater Quality Assessment of Sirsa River Basin for Baddi-Barotiwala-Nalagarh Industrial Region. J Ind Pollut Control. 39:005

Raheja, H., Goel, A., & Pal, M. (2021).Prediction of groundwater quality indices using machine learning algorithms. Water Practice & Technology Vol 00 No 0, 1 doi: 10.2166/ wpt.2021.120

Rahman, M., Khan, M.S.I., Hossain, M.S., Hossain, I.S., Hasan, M., Hamli, H., & Mustafa, M.G. (2022). Groundwater Contamination and Health Risk Evaluation of Naturally Occurring Potential Toxic Metals of Hatiya Island, Bangladesh. Journal of Ecological Engineering, 23(6), 223–236 https://doi.org/10.12911/22998993/148192

Ram, A., Tiwari, S.K., Pandey, H.K., Chaurasia, A.K., Singh, S., & Singh, Y.V. (2021). Groundwater quality assessment using water quality index (WQI) under GIS framework. Appl Water Sci. https:// doi. org/ 10.1007/ s13201-021-01376-7

Ramadan, J., & Haruna, A. (2019). Health Risk Assessment from Exposure to Heavy Metals in Surface and Groundwater Resources within Barkin Ladı, North Central Nigeria. Journal of Geoscience and Environment Protection, 7, 1-21. doi: 10.4236/gep.2019.72001

Ramalingam, S., Panneerselvam, B., & Kaliappan, S.P. (2022). Effect of High Nitrate Contamination of Groundwater on Human Health and Water Quality Index in Semi-Arid Region, South India. Arab. J. Geosci. 15, 242. [CrossRef]

Rameeza, S., Srikant, V. N. V., Rao, D. M., & Ramakrishna, C. (2012). Study of ground water quality in industrial zone of Visakhapatnam. Adv. Appl. Sci. Res, 3(4), 2463-2467

Ravikumar, P., &Somashekar, R.K. (2017). Principal component analysis and hydrochemical facies characterization to evaluate groundwater quality in Varahi river basin, Karnataka state, India. Appl Water Sci 7(2):745–755

Ravindra, K., & Mor, S. (2019). Distribution and health risk assessment of arsenic and selected heavy metals in Groundwater of Chandigarh, India. Environmental Pollution, 250, 820–830

Rawat, M., Gupta, A., Gupta, S., Khare, A.B., & Jain, S. (2017). ANALYSIS OF WATER QUALITY OF DIVERSE SOURCES OF WATER IN DEHRADUN CITY.

International Journal of Recent Scientific Research Research, Vol. 8, Issue, 4, pp. 16618-16623

Richard, L.A. (1954). Diagnosis and improvement of saline alkali soils. Agriculture handbook, 60. US Department of Agriculture, Washington DC, USA

Ritchie, H. & Roser, M. (2017).Water Use and Stress. Published online at Our World In Data. org. Retrieved from: 'https://ourworldindata.org/water-use-stress' [Online Resource]

Rzętała, M. (2016). Cadmium contamination of sediments in the water reservoirs in Silesian Upland (southern Poland). J Soils Sediments, 16 (10). doi: 10.1007/s11368-016-1477- 3

Saha, R., Dey, N. C., Rahman, S., Galagedara, L., & Bhattacharya, P. (2018). Exploring suitable sites for installing safe drinking water wells in coastal Bangladesh. Groundwater for Sustainable Development, 7, 91–10. https :// doi.org/ 10.1016/ j.gsd. 2018.03. 002

Saleem, M., Ahmed, M., Mahmood, G., & Rizvi, S. A. M. (2012). Analysis of Groundwater Quality Improvement Using Rainwater Harvesting: A Case Study of Jamia Millia Islamia. International Journal of Modern Engneering Research, 3912-3916

Sangewar, C.V., Shukla, S.P., & Singh, R.K. (2009). Geological Survey of India, Inventory of the Himalayan glaciers: a contribution to the international hydrological programme. Director General, Geological Survey of India, Calcutta

Sathyaseelan, R., Mundepi, A. K., & Kumar, N. (2017). Quantifying seismic vulnerability, dynamical shear strain and liquefaction of the quaternary deposits in the Doon valley near the main boundary thrust in the Northwest Himalaya, India. Quaternary International, 462, 162–175. https://doi.org/10. 1016/j. quaint. 2017. 05. 018

Saxena, P.B. (1974). Water resources and potential utilization programme in Dehradun valley (Garhwal Himalaya). Indian Sci. Cong. Nagpur, 74

Selvaraj, A., Saravanan, S., & Jennifer, J.J. (2020). Mamdani fuzzy based decision support system for prediction of groundwater quality: an application of soft computing in water resources. Environmental Science and Pollution Research. https://doi. org/10.1007/s11356-020-08803-3

Şener, S., Şener, E., & Davraz, A. (2017). Assessment of groundwater quality and health risk in drinking water basin using GIS. J Water Health, 15 (1): 112–132. doi: https://doi.org/10.2166/wh.2016.148

Shan, V., Singh, S. K. & Haritash, A. K. (2021). Evaluation of water quality and potential metal contamination in ecologically important Bhindawas bird sanctuary, India. Applied Water Science 11(1), 1–9. https://doi.org/10.1007/s13201-020-01334-9

Sharma, B. (2016a). Sustainable Drinking Water Resources in Difficult Topography of Hilly State Uttarakhand, India. American Journal of Water Resources, vol. 4, 16-21. doi: 10.12691/ajwr-4-1-2

Sharma, B., Savera, K.K., Kausik, S., Saini, P., Bhadula, S., Sharma, V., & Singh, P. (2016b). Assessment of Ground Water Quality of Bhagwanpur Industrial Area of Haridwar in Uttarakhand, India. Applied Ecology and Environmental Sciences, 4(4), 96-101

Sharma, M., Kansal, A., Jain, S., & Sharma, P. (2015). Application of multivariate statistical techniques in determining the spatial temporal water quality variation of ganga and Yamuna rivers present in Uttarakhand State, India. Water Qual Expo Health, 7:567–581. https://doi.org/10.1007/s12403-015-0173-7

Sharma, P.K., & Naithani, P. (2023). Effect of Industrial Effluent on the Groundwater Quality in Haridwar. Journal of Graphic Era University, 6(1), 47–54. Retrieved from https://www.journal.riverpublishers.com/index.php/JGEU/article/view/89

Sharma, S., & Chhipa, R. C. (2016). Seasonal variations of ground water quality and its agglomerates by water quality index. GJESM, 2, 79–86. https://doi.org/10.7508/gjesm.2016.01.009

Sharma, S., Nagpal, A. K., & Kaur, I. (2019). Appraisal of heavy metal contents in groundwater and associated health hazards posed to human population of Ropar wetland, Punjab, India and its environs. Chemosphere, 227, 179–190. https://doi.org/10.1016/j.chemosphere.2019.04.009

Shawai, S. A. A., Nahannu, M. S., Mukhtar, H. I., & Ismai'l, I. M. (2019). Assessment of groundwater samples from Sa'adatu Rimi college of education, Kumbotso, Kano. Journal of Medicinal and Chemical Sciences, 2, 96–100. https:// doi. org/ 10. 26655/ JMCHE MSCI. 2019.4.5

Shil, S., & Singh, U.K. (2019). Health risk assessment and spatial variations of dissolved heavy metals and metalloids in a tropical river basin system. Ecol Indicat. https://doi.org/10.1016/j.ecolind. 2019.105455

Shrestha, A.M., Neupane, S., & Bisht, G. (2017). An Assessment of Physicochemical Parameters of Selected Industrial Effluents in Nepal. Hindawi, Journal of Chemistry Article ID 3659561, 9 pages, https://doi.org/10.1155/2017/3659561

Singh, A.K., Jasrotia, A.S., Taloor, A.K., Kotlia, B.S., Kumar, V., Roy, S., Ray, P.K.C., Singh, K.K., Singh, A.K., & Sharma, A.K. (2017). Estimation of quantitative measures of total water storage variation from GRACE and GLDAS-NOAH satellites using geospatial technology. Quat Int, 444:191–200

Singh, D., Panwar, A., Kainthola, A., & Bartwal, S. (2014). Study of Water Quality Index with the help of Remote Sensing and GIS for Ground Water Sources between Ganga and Yamuna River Siwalik region in Doon Valley in Outer Himalaya. Int. Res. J. Environment Sci. ISSN 2319–1414

Singh, G., & Kamal, R.K. (2016). Heavy metal contamination and its indexing approach for groundwater of Goa mining region, India. Appl Water Sci. DOI 10.1007/s13201-016-0430-3

Singh, G.T., Kumar, S., Busa, R., Chaturvedi, A., Rathore, S.S., Singh, R.K., & Gangwar, A. (2023). Understanding urban groundwater pollution in the Upper Gangetic Alluvial Plains of northern India with multiple industries and their impact on drinking water quality and associated health risks. Groundwater for Sustainable Development,ISSN 2352-801X, https://doi.org/10.1016/j.gsd.2023.100902

Singh, K.K., Tewari, G., & Kumar, S. (2020). Evaluation of Groundwater Quality for Suitability of Irrigation Purposes: A Case Study in the Udham Singh Nagar, Uttarakhand. Hindawi Journal of Chemistry, Article ID 6924026, 15 pages https://doi.org/10.1155/2020/6924026

Singh, K.P., Malik, A., & Sinha, S. (2005). Water quality assessment and apportionment of pollution sources of Gomti River (India) using multivariate statistical techniques: a case study. Anal Chim Acta, 538:355–374

Singh, K.P., Malik, A., Mohan, D., & Sinha, S. (2004). Multivariate statistical techniques for the evaluation of spatial and temporal variations in water quality of Gomti River (India)-a case study. Water Res 38:3980–3992. https://doi.org/10.1016/j.watres.2004.06.011

Singh, R., & Pandey, J. (2019). Non-point source-driven carbon and nutrient loading to Ganga River (India). Chem Ecol 35(4):344–360. https:// doi.org/10.1080/02757540.2018.155 4061

Singh, S., Gautam, P., Sarkar, T., & Taloor, A.K. (2021). Characterization of the Groundwater Quality in Udham Singh Nagar of Kumaun Himalaya, Uttarakhand. Pre-print, DOI: https://doi.org/10.21203/rs.3.rs-407828/v1

Sinha, P.R., Kumar, E.K., & Uniyal, V.P. (2020). Groundwater hydrochemistry of Naulas and Dharas (Springs) of Almora Kumaon Himalayas in Uttrakhand their evaluation using graphical tools. Eco. Env. & Cons. 27 (May Suppl. Issue), pp. (S80-S83), ISSN 0971– 765X

Sirajudeen, J., Manikandan, S. A., & Manivel, V. (2013). Water quality index of ground water around Ampikapuram area near Uyyakondan channel Tiruchirappalli District, Tamil Nadu, India. Archives of Applied Science Research, 5(3), 21-26.

Snousy, M.G., Wu, J., Su, F., Abdelhalim, A., & Ismail, E. (2021). Groundwater quality and its regulating geochemical processes in Assiut Province, Egypt. Expo Health. https:// doi. org/ 10. 1007/s12403-021-00445-1

Sobhanardakani, S., Taghavi, L., Shahmoradi, B., & Jahangard, A. (2017). Groundwater quality assessment using the water quality pollution indices in Toyserkan Plain. Environmental Health Engineering and Management Journal, 4(1): 21–27. doi: 10.15171/EHEM.2017.04

Song, Y., Jiang, G., Chen, Y., Zhao, P., & Tian, Y. (2017). Effects of chloride ions on corrosion of ductile iron and carbon steel in soil environments. Scientific Reports. https:// doi. org/ 10. 1038/s41598- 017- 07245-1

Soo, C.L., Ling, T.Y., & Lee, N.A.K. (2016). Assessment of the characteristic of nutrients, total metals, and fecal coliform in Sibu Laut River, Sarawak, Malaysia. Applied Water Science, 6: 77-96.

Souid, F., Hamdi, M. & Moussa, M. (2022). Groundwater pollution risk assessment using a calculated contamination index and geostatistical analysis: Jerba Island case study (southeast of Tunisia). Arab J Geosci, 15, 01576, https://doi.org/10.1007/s12517-022-10862-4

Souza, S. O., Lima, G. R. R., Alencar, F. K. M., Batista, A. C. O. N., & Silva, F. J. A. (2020). Assessment of water quality for industrial and irrigation purposes based on ionic indices in the Sítios Novos Reservoir, Ceará Brazil. Sustainable Water Resources Management, 6, 95. https://doi.org/10.1007/s40899-020-00452-1

Sridhar, N., Ezhisaivallabi, K., & Poongothai, S. (2021). A Questionnaire-based Survey on Groundwater Characterization. Eco. Env. & Cons. 28: 2022; pp. (S279-S283) ISSN 0971–765X

Sureshjani, M.K., Amanipoor, H., & Battaleb-Looie, S. (2020). The Effects of Industrial Wastewater on Groundwater Quality of the Boroujen Aquifer, Southwest Iran. Natural Resources Research, Vol. 29, No. 6, https://doi.org/10.1007/s11053-020-09665-9

Syed, Md., Hossain, Md., Haque, Md., Pramanik, A.H., Uddin, J., & Harun, M.Y. (2020) Assessing the groundwater quality and health risk: A case study on Setabganj sugar mills limited, Dinajpur, Bangladesh, Water Science, 34:1, 110-123, DOI: 10.1080/11104929.2020.1790184

Szabolcs, I., & Darab, C. (1964).The influence of irrigation water of high sodium carbonate content on soils. In: Szabolics I (eds) Proceedings of the 8th international congress soil science sodics soils. Res Inst Soil Sci Agric Chem Hungarian Acad Sci, ISSS Trans II, pp 802–812

Tabassum, R.A., Shahid, M., Dumat, C., Niazi, N.K., Khalid, S., Shah, N.S., Imran, M., & Khalid, S. (2019). Health risk assessment of drinking arsenic-containing groundwater in Hasilpur, Pakistan: effect of sampling area, depth, and source. Environ. Sci. Pollut. Res. 26, 20018–20029. doi: 10.1007/s11356-018-1276-z .

Taloor, A.K., Kotlia, B.S., & Kumar, K. (2021). Water, cryosphere, and climate change in the Himalayas: a geospatial approach. Springer, Switzerland

Taloor, A.K., Pir, R.A., Adimalla, N., Ali, S., Manhas, D.S., Roy, S., & Singh, A.K. (2020). Spring water quality and discharge assessment in the Basantar watershed of Jammu Himalaya using geographic information system (GIS) and water quality index (WQI). Groundw Sustain Dev, 10:100364. https:// doi. org/ 10. 1016/j. gsd. 2020. 100364

Tesseme, A.T., Vinti, G., & Vaccari, M. (2022). Pollution potential of dumping sites on surface water quality in Ethiopia using leachate and comprehensive pollution indices. Environ Monit Assess, 194(8):545. doi: 10.1007/s10661-022-10217-2

Thatte, C. D. (2018). Water resources development in India. International Journal of Water Resources Development, 16–27. https://doi.org/10.1080/07900627.2017.1364987

Tirkey, P., Bhattacharya, T., Chakrabarty, S., & Baraik, S. (2017). Assessment of groundwater quality and associated health risks: a case study of Ranchi City, Jharkhand, India. Groundwater Sustain Dev 5:85–100. https://doi.org/10.1016/j.gsd.2017.05.002

Tiwari, A.K., Singh, P.K., Singh, A.K., & De, M.M. (2016). Estimation ofheavy metal contamination in groundwater and development of a Heavy Metal Pollution Index by using GIS technique. Bull Envi-ron Contam Toxicol. https://doi.org/10.1007/s00128-016-1750-6

Tripathi, M., & Singal, S.K. (2019). Allocation of weights using factor analysis for development of a novel water quality index. Ecotoxicol Environ Saf. 2019 Nov 15;183:109510. doi: 10.1016/j.ecoenv.2019.109510. Epub 2019 Aug 8. PMID: 31401332

Trivedy, R.K. & Goel, P.K. (1986). In: Chemical and biological methods for water pollution studies. Environmental Publication, Carad.

Ubuoh, E.A., Ezenwa, L.I., Ndukwu, M.C., & Emeka-Chris, C.C. (2019). Assessment of cation chemistry of groundwater near hospital wastes dumpsites in Umuahia Nigeria using multivariate and analytical index approach, Environmental Technology & Innovation, ISSN 2352-1864,https://doi.org/10.1016/j.eti.2019.100371

Uddin, G., Olbert, AI., Nash, S., & Rahman, A. (2022) .Development of a water quality index model - a comparative analysis of various weighting methods. In: Mediterranean geosciences union annual meeting, Istanbul, vol 19. Springer, pp 1–6

United Nations, (2015). Good Practices for Regulating Wastewater Treatment: Legislation, Policies and Standards Copyright

United Nations, (2018). Sustainable Development Goal 6 synthesis report on water and sanitation (p.10017). United Nations

United Nations, (2022). The Sustainable Development Goals Report. Published by the United Nations New York, United States of America. Retrieved August 15, 2022, from https:// unsta ts. un. org/ sdgs/ report/2022/ The- Susta inable- Devel opment- Goals- Report-2022. pdf

UN-Water Analytical Brief, (2016). 3A UN-Water Analytical Brief outlines the target-level links between water and other SDGs. Water and sanitation inter linkages across the 2030 Agenda for Sustainable Development

USEPA, (2004). Risk assessment guidance for superfund, vol 1, Human Health Evaluation Manual. Part E (Supplemental Guid-ance for Dermal Risk Assessment). Office of Superfund Remediation and Technology Innovation, Washington, DC, USAEPA/540/R/99/005

USEPA, (2005). Risk assessment guidance for superfund, vol 1. Human Health Evalu-ation Manual (Part A). Report EPA/540/1-89/002, Washington, DC

USEPA, (2018). 2018 Edition of the Drinking Water Standards and Health Advisories. Office of Water, U.S. Environmental Protection Agency, Washington, DC EPA 822-F-18-001 USGS, (2019). U.S. Geological Survey Water Mission Area: Washington, D.C., The National

Academies Press, 96 p. [Also available at https://doi.org/10.17226/25134

Varol, S. (2021). Potential health risk assessment related to arsenic pollution and hydrogeochemistry of groundwaters in Akşehir and surroundings (Konya/Turkey). J Water Health, 19 (1): 97–107. doi: https://doi.org/10.2166/wh.2020.107

Varol, S., & Şekerci, M. (2018). Hydrogeochemistry, water quality and health risk assessment of water resources contaminated by agricultural activities in Korkuteli

(Antalya, Turkey) district center. J Water Health, 16 (4): 574–599. doi: https://doi.org/10.2166/wh.2018.003

Vasanthavigar, M., Srinivasamoorthy, K., Vijayaragavan, K., Ganthi, R. R., Chidambaram, S., Anandhan, P., & Vasudevan, S. (2010). Application of water quality index for groundwater quality assessment: Thirumanimuttar sub-basin, Tamilnadu, India. Environmental monitoring and assessment, 171(1), 595-609

Verma, A.K. & Saksena, D.N. (2010). Impact of pollution on sewage collecting River Kalpi (Morar) Gwalior (M.P.) with special reference to water quality and Macrozoobenthic fauna. Asian Journal of Experimental Biological Sciences, 1: 155-161.

Verma, A.K., & Saksena, D.N. (2010). Impact of pollution on sewage collecting River Kalpi (Morar) Gwalior (M.P.) with special reference to water quality and Macro-zoobenthic fauna. Asian Journal of Experimental Biological Sciences, 1: 155-161

Verma, O. (2021). Climate change and its impacts with special reference to India. In: Taloor AK, Kotlia BS, Kumar K (eds) Water, Cryosphere, and Climate Change in the Himalayas: a geospatial approach. Springer, Cham, pp 39–55. https://doi. org/ 10. 1007/978-3-030-67932-3_3

Verma, P., Singh, P. K., Sinha, R. R., & Tiwari, A. K. (2020). Assessment of groundwater quality status by using water quality index (WQI) and geographic information system (GIS) approaches: a case study of the Bokaro district, India. Applied Water Science, 10(1), 1-16.

Vyas, V. G., Hassan, M. M., Vindhani, S. I., Parmar, H. J., & Bhalani, V. M. (2015). Physicochemical and microbiological assessment of drinking water from different sources in Junagadh City, India. American Journal of Microbiological Research, 3(4), 148-154.

Wagh, V.M., Panaskar, DB., Mukate, S.V., Gaikwad, S.K., Muley, A.A., & Varade, A.M. (2018). Health risk assessment of heavy metal contamination in groundwater of Kadava river basin Nashik, India. Model Earth Syst Environ. https://doi.org/10.1007/s40808-018-0496-z

Wang, Li., Zhang, L., Lv, J., Zhang, Y., & Ye, B. (2018). Public Awareness of Drinking Water Safety and Contamination Accidents: A Case Study in Hainan Province, China. Water, 2018, 10, 446; doi:10.3390/w10040446

Wang, X., Xu, X. Q., Gao, C. H., Li, L. H., Liu, Y., Zhang, N., Xia, Y., Fang, X. & Zhang, X.
G. 2022 Assessing the drinking water quality in the Inner Mongolia autonomous
region from 2014 to 2018. Journal of Water and Health, 00(0), 1–10. https://doi. org/10.2166/wh.2022.217

Wang, Y., & Li, P. (2021). Appraisal of shallow groundwater quality with human health risk assessment in different seasons in rural areas of the Guanzhong Plain (China). Environ Res. https:// doi. org/ 10.1016/j. envres. 2021. 112210

Wekesa, A.M., & Otieno, C. (2022). Assessment of Groundwater Quality Using Water Quality Index from Selected Springs in Manga Subcounty, Nyamira County, Kenya. Hindawi, Scientific World Journal, Article ID 3498394, 7 pages https://doi.org/10.1155/2022/3498394

WHO, (2011) Fourth edition. Water Sanitation and Hygiene Links to Health, WHO Press WHO, (2022). Drinking-water. https://www.who.int/news-room/fact-sheets/detail/drinking- water

WHO/UNICEF, (2020). Joint Monitoring Programme for Water Supply, Sanitation and Hygiene (JMP) https://ourworldindata.org/water-sanitation-2020-update

Wilcox, L.V. (1955). Classification and use of irrigation water Washington (Vol. 19). US Department of Agriculture Circular

WorldBank, (2022). Water Overview [Text/HTML]. World Bank. https://www.worldbank.org/en/topic/water

Wright, J.A., Yang, H., Rivett, U., & Gundry, S.W. (2012). Public perception of drinking water safety in South Africa 2002–2009: A repeated cross-sectional study. BMC Public Health, 12, 556. [CrossRef] [PubMed]

Wu, H., Liao, Q., Chillrud, S.N., Yang, Q., Huang, L., Bi, J., & Yan, B. (2016). Environmental exposure to cadmium: health risk assessment and its associations with hypertension and impaired kidney function. Sci Rep 6:29989. https://doi.org/10.1038/srep29989

Wu, J., & Sun, Z. (2016). Evaluation of Shallow Groundwater Contamination and Associated Human Health Risk in an Alluvial Plain Impacted by Agricultural and Industrial Activities, Mid-west China. Expo Health 8, 311–329, https://doi.org/10.1007/s12403-015-0170-x

Wugan, L., & Tao, L. (2012). The use of chromium minerals in the 4th3rd century BC China. A preliminary study of bronze. J Raman Spectrosc 43:303e306

Xiao, Y., Yin, S., Hao, Q., Gu, X., Pei, Q., & Zhang, Y. (2020). Hydrogeochemical appraisal of groundwater quality and health risk in a near-suburb area of North China. Journal of

Water Supply: Research and Technology-Aqua, 69 (1): 55–69. doi: https://doi.org/10.2166/aqua.2019.101

Xu, X., Liu, S., Smith, K., Cui, Y., & Wang, Z. (2020). An overview on corrosion of iron and steel components in reclaimed water supply systems and the mechanisms involved. Journal of Cleaner Production, 276, 124079. https:// doi. org/ 10. 1016/j. jclep ro. 2020.124079

Yadav, A.K., Khan, P., & Sharma, S.K. (2010). Water quality index assessment of groundwater in Todaraisingh Tehsil of Rajasthan state, India-a greener approach. E J Chem 7(s1):S428–S432. https:// doi. org/10. 1155/ 2010/ 419432

Yang, K.H., Liu, H., Ma, L.H., Wang, S., Tian, Y.L., Zhang, F.F., Li, Z.Y., Song, Y.Y., &

Jiang, X.L. (2021). Knowledge, attitude and practice of residents in the prevention and control of COVID-19: An online questionnaire survey. J. Adv. Nurs.77, 1839–1855

Yao, T., Thompson, L.G., Mosbrugger, V., Zhang, F., Ma, Y., Luo, T., Xu, B., Yang, X., Joswiak, D.R., & Wang, W. (2012). Third pole environment (TPE). Environ Develop 3:52–64

Yasir, & Srivastava, S. (2016). Monitoring of ground water quality in the province of district Dehradun, (Uttarakhand), India. Archives of Agriculture and Environmental Science, 1(1): 43-48

Zacchaeus, O.O., Adeyemi, M.B., Adedeji, A., Adegoke, K.A., Anumah, A.O., Taiwo, A.M., & Ganiyu, S.A. (2020). Effects of industrialization on groundwater quality in Shagamu and Ota industrial areas of Ogun state, Nigeria. Heliyon. 6(7):e04353. doi: 10.1016/j.heliyon.2020.e04353

Zakir, H.M., Sharmin, S., Akter, A., & Rahman, M.S. (2020). Assessment of health risk of heavy metals and water quality indices for irrigation and drinking suitability of waters: a case study of Jamalpur Sadar area, Bangladesh. Environmental Advances, 100005, https://doi.org/10.1016/j.envadv.2020.100005

Zeng, X., Liu, Y., You, S., Zeng, G., Tan, X., Hu, X., Hu, X., Huang, L., & Li, F. (2015). Spatial distribution, health risk assessment and statistical source identification of the trace elements in surface water from the Xiangjiang River China. Environ. Sci. Pollut. Res. 22(12), 9400–9412, https:// doi. org/ 10. 1007/ s11356- 014- 4064-4

Zhang, Q., Xu, P., & Qian, H. (2020). Groundwater quality assessment using improved water quality index (WQI) and human health risk (HHR) evaluation in a semi-arid region of northwest China. Expo Health 12:487–500

Zhang, Q., Zhou, B., Dong, F., Liu, Z., & Ostovari, Y. (2021). Spatial variability of groundwater quality for freshwater production in a semi-arid area. Water, 13(21), 3024. https:// doi. org/ 10. 3390/ w1321 3024

Zhang, Y., Wu, J., & Xu, B. (2018). Human health risk assessment of groundwater nitrogen pollution in Jinghui canal irrigation area of the loess region, northwest China. Environmental Earth Sciences, 77:273, https://doi.org/10.1007/s12665-018-7456-9

Zhao, H., Song, F., Su, F., Shen, Y., & Li, P. (2021). Removal of cadmium from contaminated groundwater using a novel silicon/aluminum nanomaterial: an experimental study. Arch Environ Contam Toxicol, 80 (1):234–247

Zhao, Y., Xia, X. H., Yang, Z. F., & Wang, F. (2012). Assessment of water quality in Baiyangdian Lake using multivariate statistical techniques. Procedia Environmental Sciences, 13(1213), 1226.